*

Understanding
Food

*

* * *

* * *

*

Understanding Food

*

Lendal H. Kotschevar
Food Consultant

Margaret McWilliams
Professor and Chairman
Home Economics Department
California State College, Los Angeles

JOHN WILEY AND SONS, INC.
NEW YORK   LONDON   SYDNEY   TORONTO

Library of Congress Catalog Card
Number: 69–19238   SBN 471 50530 7
Printed in the United States of America

3 4 5 6 7 8 9 10

*

Preface

Good food can provide great pleasure in life. A gourmet is a person with a keenly developed appreciation of good food, just as other people are connoisseurs of good art, music, and literature. Today in this country we possess more resources than ever before to eat well. We have an amazingly wide variety of food and food products, the income to purchase them, and the technical competence to make them of highest quality and excellence. We also have more leisure time to learn about food and the gracious art of the gourmet.

To eat well, we need to know:

1. How to shop wisely for food.
2. How to prepare and serve appetizing and tempting meals.
3. The important roles of the various nutrients in the body.
4. Which foods are good sources of these nutrients.

As you can see, this subject is a big area for study!

This book has been written to help you gain knowledge of foods and confidence in your ability to prepare them. It is designed to lead you through the broad range of information regarding institutional and home food preparation in a course you can consume and digest at your own rate. Through careful study, you will gain the knowledge needed to control flavor, color, texture, shape, form, and nutritional value of food. You will learn to plan, prepare, and serve meals that are both nutritionally adequate and aesthetically satisfying.

For your convenience, the chapters are designed to be complete discussions of a topic; the programmed questions and the review questions at the end of each chapter will help you test your understanding of each subject. If you find that your knowledge of some portions of a chapter is weak, you may wish to study that chapter in more detail before proceeding. By the time you have completed this book, you will have a sound basis for confidently preparing all types of food. Such confidence is essential for good results in the kitchen.

As you study, you will no doubt find certain items that are of particular interest to you. You may then wish to seek additional information by reading supplemental references. If you have an interest in chemistry, you might wish to delve into the science of food preparation from the chemist's point of view. For maximum understanding and appreciation of food, you should explore both the scientific and artistic contributions to the subject of food preparation.

Now we wish you well as you launch into this introduction to the broad subject of *Understanding Food!*

L. KOTSCHEVAR
M. MCWILLIAMS

February 1969

*

Contents

1

Food and Your Body

THE IMPORTANCE OF FOOD

You are what you eat. This is a common expression, but have you ever stopped to think just what it really means? The food you have eaten since birth, and indeed the nutrients received from your mother during the prenatal period, have all contributed to make you what you see in the mirror today. Good nutrition throughout life is essential if your image is to be that of a well-formed, healthy person with shining hair, bright eyes, sound teeth, and a clear, glowing complexion.

True or false. Foods not only affect your health but also your appearance.

True.

*

The food you have eaten throughout your life has influenced your size, general health, activity, and perhaps even your personality. Heredity imposes natural limitations on skeletal growth, but only through good nutrition will children be supplied the building materials necessary to achieve their potential maximum growth.

What two factors work together to determine how tall you will be when growth stops?

Heredity and diet together determine your final height.

*

Evidence of the importance of a good diet during growth is not difficult to find. The Japanese people are now consuming a better diet than was usually eaten earlier in this century and they are becoming taller. The good nutrition

1

children have in this country is producing taller generations. At one time we made beds only five feet long for women in dormitories; now they are six feet. And we no longer put six-foot beds in men's dorms, but have seven-foot ones.

Are you (or will you be) taller than your parents?

If you have been eating a good diet throughout the growth period, it is likely that you will be taller than your parents.

*

Various physical ailments such as scurvy and beriberi have been shown to result when the diet is inadequate. Coaches stress the importance of good diet in promoting vitality and endurance in athletes.

Would a good diet help an athlete win a sports contest or make a girl or boy more attractive?

Yes, good diets help make strong, well-formed bodies and promote the vitality that is so important to an attractive appearance.

*

The influence of diet on personality is subtle, but increased irritability commonly occurs when a nutritional deficiency develops. Overweight people may show changes in personality because of being self-conscious about their appearance.

Is the idea that fat people are always jolly really true?

No. Often over-weight people are inwardly concerned and unhappy about their appearance; they may develop a jolly exterior as a cover for their real feelings.

*

Food plays an important role at social events and in economic affairs. What would a party be without food? Even informal, spur-of-the-moment gatherings usually center around food. Probably some of your warmest childhood memories include family gatherings at the table to celebrate a birthday or special event. In old Ireland, a guest was greeted with an offering of a pinch of salt and a cup of hot wine to indicate the pleasure of his company. An Arab merchant considers it an insult to a prospective buyer if he does not offer him tea or coffee before discussing a sale. In today's business world, it is common practice to take clients to a fine restaurant for lunch or dinner.

True or false. We eat food only because it has nutritional value.

False. In our society, food also is used to promote sociability and business ventures.

Food is important to most healthy persons because of the psychological satisfaction it gives. We feel more secure and have a stronger feeling of belonging to our particular culture when we are fed our favorite foods in adequate amounts. Many adults enjoy the feelings of success that come from being able to set a table with an abundance of food for company. Others delight in the cosmopolitan knowledge of the gourmet.

True or false.   The psychological value of food should be considered when planning meals.

True. Appetite and food acceptance are best when the menu includes the foods that are psychologically satisfying to the diner.

*

When you have eaten a well-prepared, appetizing meal and are just comfortably sated, you experience a satisfied feeling. Some meals leave you with this sense of well-being for only a short time while other meals make you feel satisfied for three or four hours. This characteristic of food to satisfy is known as the satiety value of food. The amount and the kind of food eaten determine the satiety value of a meal.

Which of the following is the best definition of the satiety value of food?
A.   *A full stomach.*
B.   *Eager anticipation of eating.*
C.   *Hunger pangs.*
D.   *A satisfied feeling after eating.*
E.   *Discomfort after eating a large quantity.*

D. A satisfied feeling after eating.

*

Sometimes the psychological value of food becomes so important to a person that he eats more than his body needs, and he gradually becomes too heavy. Overeating is probably the greatest nutritional problem in the United States today. This is just the opposite of the nutritional difficulties in many areas of the world where there is simply not enough food to go around. Underfed people often have an anxious, worried look and they seldom laugh or smile. This "hunger look" is all too common in over-populated areas, but it does disappear when the people have enough food to eat.

Can the "hunger look" be made to disappear from the earth if enough food is available for all people?

Yes, but one major concern today is the complex problem of supplying ever-increasing quantities of food for the exploding world population.

3

The psychological and social values of food are important, but still more vital to life is its physiological role. Food is the source of your body's fuel; it supplies energy to keep you warm and to move your muscles. It furnishes the building materials for growth and maintenance, and it provides the nutrients necessary for your body to function normally.

_____ *is the source of the body's fuel and energy, the building materials we need to grow, and the substances that regulate our bodies.*

Food.

*

The best sources of food and energy for the body are carbohydrates and fats. Protein, water, and various minerals are all important structural or skeletal materials. Normal operation of the body can occur only when the body is supplied with adequate amounts of vitamins, minerals, proteins, carbohydrates, fats, and water. These important nutrients are examined in the following sections of this chapter.

If a person lacks an essential nutrient, what may be the result?

Growth, health, and body vitality will be impaired and death could result.

*

BASIC FOUR FOOD PLAN

The term "well-balanced diet" is frequently used and most people readily agree that this is the desirable way to eat. But just what foods should be eaten if you are to have a well-balanced diet? Use of the Basic Four Food Plan is a good way to check to see how adequate your diet is. In this plan the foods eaten each day are divided into four categories: (1) milk and dairy products, (2) meat and other protein foods, (3) vegetables and fruits, and (4) breads and cereals. If you eat the recommended number of servings in each group during each day, you will be eating the variety of foods necessary for good nutrition.

What are the four food groups in the Basic Four Food Plan?

Milk and dairy products, meat and other protein foods, vegetables and fruits, and breads and cereals.

4

Milk and Dairy Products

The milk and dairy products group is very important for children and adolescents because it helps to provide the protein and minerals needed for growth. However, adults also need milk to keep their bones and teeth strong. The following quantities of milk are recommended for different ages: babies—up to one quart each day; young children—two to three glasses; teenagers—four or more glasses; adults —two or more glasses.

A teenager should drink approximately:
A.   *Two glasses of milk daily.*
B.   *Three glasses of milk daily.*
C.   *Four or more glasses of milk daily.*
D.   *Two or more glasses of milk at each meal.*
E.   *Four or more glasses of milk at each meal.*

C. Four or more glasses of milk daily.

*

One glass equals eight fluid ounces or one cup. It is possible to meet part of your need in this food group by eating cheese, ice cream, and other foods high in milk. The calcium content of one ounce of cheese (such as American or Swiss) is approximately equal to six ounces of milk; one-half cup of creamed cottage cheese is about the equivalent of one-third cup of milk; and one-half cup of ice cream may be substituted for one-fourth glass of milk.

If you eat a pint of ice cream (two cups) in a day, what part of an adult's daily requirement for milk is met?

A cup of milk or half the daily requirement.

*

Meat and Other Protein Foods

Several foods may be used to provide the two or more servings recommended daily in the meat group. Beef, pork, and lamb are meats commonly served, but any type of game such as venison or rabbit is also a suitable source of protein. All kinds of fish and poultry, eggs, and cheese are other good sources of animal protein.

True or false. To obtain all the nutrients you need in the meat group, you must eat meat; no other kind of food will suffice.

False. Eggs, cheese, and other dairy products are valuable protein sources. In addition, vegetable foods high in protein may be used to meet part of the body's need for protein.

5

Some vegetables such as legumes (split peas, lentils, navy beans, pinto beans, kidney beans, and lima beans) are high in vegetable protein. Nuts of any type are also good sources of protein. Three-eighths cup of roasted peanuts or three-fourths cup baked beans contain about the same amount of protein as is found in one and one-half ounces (half a serving) of cooked lean beef.

True or false. It is possible to include some less expensive protein foods such as legumes in the diet.

True, but it is wise to include some animal protein food in the diet each day.

*

Fruits and Vegetables

Four or more servings of fruits and vegetables are recommended each day. It is important that one of these servings be a citrus fruit or citrus juice (or berries).

For one of the four or more servings of fruits and vegetables each day, you should eat a _____ fruit or _____.

Citrus, berries.

*

At least every other day, the diet should include a dark-green, leafy vegetable such as spinach or a yellow vegetable such as carrots or sweet potatoes. The other two (or three) servings each day in the fruit and vegetable group may be any type of fruit or vegetable.

Some dark-green, leafy vegetable or yellow vegetable should be eaten:
A. Every day.
B. At least every other day.
C. Only when no citrus fruit or its equivalent is eaten.
D. Only during winter months.

B. At least every other day.

*

A serving or portion in the fruit and vegetable group is considered to be one-half cup if the food is cooked or one cup if it is served raw.

6

If you ate a cup of freshly sliced peaches, a small potato with creamed peas, a half cup of mashed squash, and a cup of shredded lettuce in a day, would you meet your daily requirement in the fruit and vegetable group?

Not quite; you should have had a citrus fruit or some berries. A cup of cole slaw instead of the lettuce would have provided about two-thirds of the vitamin C (ascorbic acid) that was lacking because of the absence of the citrus fruit.

*

Breads and Cereals

Breads and cereals are of value in the diet when they are either enriched or whole-grain cereal products. Most breads and cereal products made from refined cereals are enriched today. Enrichment is the addition of vitamins and minerals. We sometimes call this fortification. The label on the food will tell what and how much has been added.

True or false. Only whole-grain cereals and cereal products should be eaten, because foods made with white flour or other refined cereals have no nutritive value.

False. Enriched bread and cereal products have had enough of the B vitamins and minerals added back to them to provide useful amounts of these nutrients.

*

The Basic Four Food Plan recommends four or more servings in this group daily. A serving is defined as one slice of bread, three-fourths cup dry cereal, or one-half cup cooked cereal.

If you had one-half cup of oatmeal for breakfast, a sandwich (two slices of bread) for lunch, and a roll at dinner, would you have eaten the recommended minimum servings of breads and cereals?

Yes, this totals to four servings.

*

Butter, margarine, and other fats are not mentioned in the Basic Four Food Plan because it was desirable to keep the plan as simple as possible. It seems safe to assume that most persons who eat four or more servings of bread daily plus perhaps a cooked vegetable would eat an adequate quantity

7

of butter or margarine to supply the body's need for fat. Additional support for this logic is given by the fact that many Americans apparently consume approximately twice as much fat as is deemed necessary nutritionally.

Do you eat more fat than is necessary for good nutrition?

If you eat French fries, potato chips, doughnuts, and fatty meats frequently, and if you also use generous amounts of butter or margarine on bread and potatoes, you are probably eating more fat than is recommended by nutritionists.

Many people, particularly adolescents, will still feel hungry if they eat only the minimum number of servings recommended for each of these four categories during a day because this basic diet may provide as little as 1200 Calories. Larger servings or more variety may be added to provide the necessary quantity of food.

By avoiding rich foods and just eating the foods in the portions recommended in the Basic Four Food Plan, you will consume about _____ Calories per day.

1200.

*

The Basic Four Food Plan should be viewed as an outline to guide you in the selection of a diet that will supply all the protein, vitamins, and minerals you need. Then it is up to you to select the additional food needed to provide the energy necessary for your physical activity. Keep in mind that your intake should be regulated to keep your weight within a desirable normal range; it should be neither so little that you are too thin nor so much that you begin to become overweight.

Write down your entire food intake for one day and compare it with the recommendations of the Basic Four Food Plan. How do your dietary habits compare with this outline?

If you are not eating the minimum number of servings in each of the four food groups, you will be lacking in one or more nutrients essential for optimum growth and health.

*

If you are a normal, healthy person who eats a diet carefully based on the Basic Four Food Plan, you should not

8

need any vitamin pills, food supplements, or health foods. Each year in the United States millions of dollars are wasted on diet fads and food supplements that purportedly will do something special that regular foods cannot do. Foods carefully selected in a grocery store and tastefully prepared are very adequate sources of the nutrients needed by the normal individual. The key to good nutrition is planned, organized meals every day. A vitamin pill gulped down hastily or an organically grown tomato will not provide a quick cure-all to poor nutritional practices.

Could consuming vitamin pills and other dietary supplements be harmful as well as economically wasteful?

Which of the following is not true?
A. Foods make our bodies grow.
B. Most individuals should take dietary supplements.
C. Foods are important for social, economic, and psychological reasons as well as for physiological needs.
D. Foods are necessary to maintain our health and vigor.
E. Nourishing foods that are excellent sources of nutrients are available in regular grocery stores.

Yes, it is possible to take toxic doses of vitamins A and D. Sometimes persons who really need medical help will delay important medical care while treating themselves with food supplements.

B is not true. When healthy individuals eat a good diet, they have no need for dietary supplements.

*

CALORIES AND YOUR BODY

Many Americans conscientiously count Calories and worry about how many Calories they can eat without getting too heavy, but they often have little knowledge of what a Calorie actually is. A Calorie is a unit of heat. In nutrition, the word Calorie is capitalized because the large Calorie is the unit used to express the amount of heat (or the amount of energy) available from a food. The large Calorie is defined as the amount of heat required to raise one kilogram of water (2.2 pounds) 1° Centigrade. The small calorie is one-thousandth of the large Calorie; i.e., a small calorie is the amount of energy required to raise one gram of water (about 30 drops) 1° C.

Is a Calorie a nutrient?

No, a Calorie is a measure of energy provided by food and is not a nutrient performing a specific function in the body.

*

The number of Calories you need each day is determined by adding together the Calories required for your basic

9

maintenance needs (basal metabolism), physical activity, and the specific dynamic effect of food.

If you want to gain weight, should you just eat enough for your basic maintenance requirements, your physical activity, and the specific dynamic effect of the food you eat?

No, you should eat something over and above this amount which could go into added weight.

*

Basal Metabolism

The body requires a basic amount of energy to exist, because some processes such as breathing and the beating of the heart must go on continuously. This is the energy that is said to be needed for basal metabolism (or basic maintenance) of the body.

The amount of energy you need just to maintain your body is based on the speed at which the basic functions of the body occur; this basic functioning is called _____ _____.

Basal metabolism.

*

Determination of the Calories needed for your basal metabolism can be done by having you rest quietly, but not sleep, in a comfortably warm room. Since an accurate value can only be obtained when an individual has completed all digestion, it is customary to perform the test early in the morning after a 12-to-15-hour fasting period. The quantity of oxygen inhaled is measured; this indicates how much energy you are using just for breathing, maintaining body temperature, and operating the heart and other vital organs.

To ascertain your basal metabolic needs, we have you **rest/sleep** *in a comfortably* **cool/warm** *room after* **a full meal/fasting,** *and then we measure the* **oxygen/ carbon dioxide** *inhaled.*

Rest, warm, fasting, oxygen.

*

The number of Calories required for basal metabolism will be determined by a person's surface area or total body size, sex, age, and secretions from the endocrine glands. Not all bodies function at the same speed; even though two people are exactly the same height and body size, their energy needs for basal metabolism can vary.

If the basal metabolic needs of a person are known, can we tell how many Calories he should consume in a day?

No, the body requires energy for any physical activity as well as for the basal metabolic needs and the digestion of food.

10

True or false. If two persons are the same height and body size but vary in age and sex, their basal metabolic needs will be much the same.

False.

*

The rate at which the basic body processes proceed varies from person to person, although the large majority of the population has a basal metabolic rate considered to be within a normal range. In rare instances, the thyroid gland malfunctions and the basal metabolic rate is altered. This gland, located at the front of the throat, may be overactive and secrete too much of its hormone, thyroxine. This speeds up the metabolic rate, a circumstance known as hyperthyroidism.

Would you think a person with an overactive thyroid gland would need more Calories than one who had a normally functioning gland?

Yes, the Calorie requirement for basal metabolism is higher for persons with hyperthyroidism.

*

The opposite condition can also occur in which metabolism is retarded. This condition, termed hypothyroidism, results in fewer Calories being used for basal metabolic needs. Persons with hypothyroidism may eat as many Calories daily as normal people do, but the result is a gradual increase in body weight. Many rotund persons blame their weight difficulties on poorly functioning glands, but there are not many people who are actually overweight because of hypothyroidism. Most cases are simply the result of excessive food intake and little exercise.

Which of the following persons may reasonably be expected to require the fewest Calories to provide the energy necessary for basal metabolic needs?
A. Person with hypothyroidism.
B. Person with normally functioning thyroid gland.
C. Person with hyperthyroidism.

A. Person with hypothyroidism.

*

Specific Dynamic Effect

The body is not 100 per cent efficient in converting potential food Calories into energy available to the body. The energy which is required to digest and utilize the food you eat is referred to as the specific dynamic effect of food; these Calories needed to utilize food must be calculated in

11

the day's total needs. Although this is a rather small fraction of the day's needs, specific dynamic effect still should be considered when calculating total caloric needs.

True or false. Specific dynamic effect is the term used to explain the energy available from food.

False. Specific dynamic effect of food is the factor that accounts for the energy in food which is not available to the body to use for energy.

*

Activity Needs

In addition to the need for Calories to meet the energy needs of basal metabolism, the body also requires Calories for any movement. When you are exercising vigorously, you require more Calories than you do if you are sitting or studying. Table 1.1 lists the number of Calories expended per hour per pound of body weight for some typical activities.

Table 1.1 Activity and Energy Expenditure per Hour per Pound of Body Weight*

Activity	Cal/Hr/Lb	Activity	Cal/Hr/Lb
Sleeping	0.4	Walking, 3 miles per hour	1.5
Awake, lying still	0.5	Painting furniture	1.3
Sitting at rest	0.6	Bicycling, moderate speed	1.7
Reading aloud	0.7	Dancing, slowly	2.0
Standing relaxed	0.8	Walking, 4 miles per hour	2.2
Hand sewing	0.7	Dancing, fast	2.4
Machine sewing	0.7	Horseback riding, trot	2.6
Dressing or undressing	0.9	Swimming	4.5
Skating	2.2	Running	4.0
Driving car	1.0	Eating	0.7
Typewriting fast	1.0	Playing ping-pong	2.7
Dishwashing	1.0	Playing piano	1.2
Ironing	1.0	Tailoring	1.0
Laundry, light	1.1	Writing	0.7
Sweeping, vacuum cleaner	1.9		

*Adapted from Taylor, MacLeod, and Rose, *Foundations of Nutrition*, (5th ed.), Macmillan, New York, 1956. (Calculated by adding to original figures 1 cal/kg/hour for basal metabolic rate and 10 per cent for influence of food.)

At which activity would you be using the most energy: bicycling at moderate speed, swimming, or typewriting fast?

Swimming.

*

If you wish to determine the number of Calories you use in any of these activities, multiply your body weight times the Calories per hour per pound for the particular activity times the hours (or fractions of hours) spent in the activity. Thus, if you weigh 135 pounds and bicycle for one-half hour at moderate speed, the calculation should be as follows:

weight × Calories × time (hours) = Calories expended
$$135 \times 1.7 \times \tfrac{1}{2} = 115 \text{ Calories}$$

A 120-pound girl who swam for 45 minutes would need:
A. 107 Calories.
B. 45 Calories.
C. 405 Calories.
D. 29.2 Calories.
E. 2922 Calories.

C. 4.5 × ¾ ×
120 = 405
Calories.

*

If you wish to calculate a day's caloric needs, list your activities; then put down the Calories consumed per hour. Total this and multiply by your body weight. Thus, if you weighed 130 pounds and had a 24-hour activity pattern as follows, you would show:

Activity	Hours	Cal/ Hr/Lb	Total Cal/Lb in Activity
Sleep	8	0.4	3.2
Light activity (walking, shopping, etc.)	3	1.3	4.5
Dancing, fast	1½	2.4	3.6
Typewriting	5	1.0	5.0
Sitting, reading	3½	0.6	2.1
Dressing, bathing, etc.	1	0.9	0.9
Driving car	2	1.0	2.0
Total	24		21.3 × 130 = 2769 Calories

Note that in these calculations, as well as in Table 1.1, the energy values for different activities also include the Calories necessary for basal metabolic needs. Therefore, the final

13

Table 1.2 Recommended Daily Dietary Allowances,* Food and Nutrition Board, National Academy of Sciences—National Research Council (Revised 1968)

	Age,† Years From	Up to	Weight Kg	Lbs	Height cm	in	Kcalories	Protein, gm	Fat Soluble Vitamins Vitamin A Activity, I.U.	Vitamin D, I.U.	Vitamin E Activity, I.U.
Infants	0 –	⅙	4	9	55	22	kg × 120	kg × 2.2‡	1500	400	5
	⅙ –	½	7	15	63	25	kg × 110	kg × 2.0‡	1500	400	5
	½ –	1	9	20	72	28	kg × 100	kg × 1.8‡	1500	400	5
Children	1 –	2	12	26	81	32	1100	25	2000	400	10
	2 –	3	14	31	91	36	1250	25	2000	400	10
	3 –	4	16	35	100	39	1400	30	2500	400	10
	4 –	6	19	42	110	43	1600	30	2500	400	10
	6 –	8	23	51	121	48	2000	35	3500	400	15
	8 –	10	28	62	131	52	2200	40	3500	400	15
Males	10 –	12	35	77	140	55	2500	45	4500	400	20
	12 –	14	43	95	151	59	2700	50	5000	400	20
	14 –	18	59	130	170	67	3000	60	5000	400	25
	18 –	22	67	147	175	69	2800	60	5000	400	30
	22 –	35	70	154	175	69	2800	65	5000	...	30
	35 –	55	70	154	173	68	2600	65	5000	...	30
	55 –	75+	70	154	171	67	2400	65	5000	...	30
Females	10 –	12	35	77	142	56	2250	50	4500	400	20
	12 –	14	44	97	154	61	2300	50	5000	400	20
	14 –	16	52	114	157	62	2400	55	5000	400	25
	16 –	18	54	119	160	63	2300	55	5000	400	25
	18 –	22	58	128	163	64	2000	55	5000	400	25
	22 –	35	58	128	163	64	2000	55	5000	...	25
	35 –	55	58	128	160	63	1850	55	5000	...	25
	55 –	75+	58	128	157	62	1700	55	5000	...	25
Pregnancy							+ 200	65	6000	400	30
Lactation							+1000	75	8000	400	30

*The allowance levels are intended to cover individual variations among most normal persons as they live in the United States under usual environmental stresses. The recommended allowances can be attained with a variety of common foods, providing other nutrients for which human requirements have been less well defined.

†Entries on lines for age range 22-35 years represent the reference man and woman at age 22. All other entries represent allowances for the midpoint of the specified age range.

Table 1.2 (continued)

Water Soluble Vitamins							Minerals				
Ascorbic Acid, mg	Folacin,§ mg	Niacin, mg equiv¶	Ribo-flavin, mg	Thia-mine, mg	Vitamin B$_6$, mg	Vitamin B$_{12}$, µg	Cal-cium, gm	Phos-phorus, gm	Iodine, µg	Iron, mg	Mag-nesium, mg
35	0.05	5	0.4	0.2	0.2	1.0	0.4	0.2	25	6	40
35	0.05	7	0.5	0.4	0.3	1.5	0.5	0.4	40	10	60
35	0.1	8	0.6	0.5	0.4	2.0	0.6	0.5	45	15	70
40	0.1	8	0.6	0.6	0.5	2.0	0.7	0.7	55	15	100
40	0.2	8	0.7	0.6	0.6	2.5	0.8	0.8	60	15	150
40	0.2	9	0.8	0.7	0.7	3	0.8	0.8	70	10	200
40	0.2	11	0.9	0.8	0.9	4	0.8	0.8	80	10	200
40	0.2	13	1.1	1.0	1.0	4	0.9	0.9	100	10	250
40	0.3	15	1.2	1.1	1.2	5	1.0	1.0	110	10	250
40	0.4	17	1.3	1.3	1.4	5	1.2	1.2	125	10	300
45	0.4	18	1.4	1.4	1.6	5	1.4	1.4	135	18	350
55	0.4	20	1.5	1.5	1.8	5	1.4	1.4	150	18	400
60	0.4	18	1.6	1.4	2.0	5	0.8	0.8	140	10	400
60	0.4	18	1.7	1.4	2.0	5	0.8	0.8	140	10	350
60	0.4	17	1.7	1.3	2.0	5	0.8	0.8	125	10	350
60	0.4	14	1.7	1.2	2.0	6	0.8	0.8	110	10	350
40	0.4	15	1.3	1.1	1.4	5	1.2	1.2	110	18	300
45	0.4	15	1.4	1.2	1.6	5	1.3	1.3	115	18	350
50	0.4	16	1.4	1.2	1.8	5	1.3	1.3	120	18	350
50	0.4	15	1.5	1.2	2.0	5	1.3	1.3	115	18	350
55	0.4	13	1.5	1.0	2.0	5	0.8	0.8	100	18	350
55	0.4	13	1.5	1.0	2.0	5	0.8	0.8	100	18	300
55	0.4	12	1.5	0.9	2.0	5	0.8	0.8	90	18	300
55	0.4	10	1.5	0.9	2.0	6	0.8	0.8	80	10	300
60	0.8	15	1.8	+0.1	2.5	8	+0.4	+0.4	125	18	450
60	0.5	20	2.0	+0.5	2.5	6	+0.5	+0.5	150	18	450

‡Assumes protein equivalent to human milk. For proteins not 100 per cent utilized, factors should be increased proportionately.

§The folacin allowances refer to dietary sources as determined by *Lactobacillus casei* assay. Pure forms of folacin may be effective in doses less than one-fourth of the RDA.

¶Niacin equivalents include dietary sources of the vitamin itself plus 1 mg equivalent for each 60 mg of dietary tryptophan.

figure is approximately the total number of Calories needed for the person in the example.

If the person in this example had walked two hours instead of driving the car for two hours, would his need for Calories have been greater or less than that shown in the calculations?

The need would have been greater. Walking at 4 miles per hour requires 2.2 Calories per hour per pound, but driving a car requires only 1.0 Calorie per hour per pound.

*

Weight Control

As children grow, their energy needs gradually rise to meet the demands of a larger body and increasing activity. As can be seen in Table 1.2, males reach their maximum need for Calories between the ages of 15 and 18.

The largest requirement for Calories will be in males between the ages of _____ and _____.

Fifteen, 18.

*

Due to the fact that girls mature physically somewhat earlier than boys, their maximum need for Calories occurs between the ages of 12 and 15.

At age 17, does a boy usually need more Calories daily than does a 17-year-old girl?

Yes.

*

Notice in Table 1.2 that the energy needs of adults gradually diminish with increasing age. This trend is caused by two factors: (1) the basal metabolic rate slowly decreases and (2) most older persons gradually become less active and consequently need fewer Calories for physical work.

*As an adult ages, his caloric requirements **increase/decrease.***

Decrease.

*

In addition to age, sex, and activity, the total size of the body and its composition also influence the number of Calories needed. A person weighing 150 pounds usually needs more Calories than someone weighing 110 pounds.

16

A person who is lean will lose heat more rapidly than will a fat person. Remember, fat is effective as an insulator against heat loss and will help keep a fat person warm.

Indicate in each comparison which person will require more Calories:
A. *Football player versus chess player.*
B. *Tall, thin person versus short, plump person (both the same weight).*
C. *A 160-pound man versus a 140-pound man (both the same age).*
D. *Woman cleaning house versus woman reading a book.*

The first person in each comparison probably requires more Calories than the second one.

*

Weight control is an important nutritional problem in the United States at the present time. Low weight in relation to height is of concern for a rather small fraction of our population; the problem much more commonly is that of excess weight. Overweight is the result of eating more food than is required to meet the body's energy needs over an extended period of time.

True or false. There are more overweight people in this country than underweight.

True.

*

A single day with too much food will not cause obesity, but regular excessive intake gradually will cause fat to accumulate. The best way to manage your weight is to avoid letting your weight creep above the recommended weight for your height. If you notice that you are beginning to be too heavy, it is time to cut back on the quantity of food that you are eating. Table 1.3 lists recommended weight ranges for adults of various heights. Figures 1.1 and 1.2 show height and weight ranges for boys and girls up to the age of 18.

How does your weight compare with the recommended weight for your height?

If you are more than five pounds too light or too heavy, it would be wise to see your doctor and have him help you modify your diet to achieve and maintain an optimum weight for your height.

*

The wise dieter will begin to watch food intake when the gain first begins rather than after 5 or 10 pounds have been acquired. The best weight-reduction plan will be a slight modification in diet but still following the Basic Four Food Plan. This way you can retrain your appetite while slowly losing the excess poundage. Then, you should be able to maintain your desired weight once you slim down to it.

Table 1.3 Desirable Weights for Men and Women 25 and Over*

Height (In Shoes)	Weight in Pounds (In Indoor Clothing)		
	Small Frame	Medium Frame	Large Frame
Men			
5' 2"	112-120	118-129	126-141
3"	115-123	121-133	129-144
4"	118-126	124-136	132-148
5"	121-129	127-139	135-152
6"	124-133	130-143	138-156
7"	128-137	134-147	142-161
8"	132-141	138-152	147-166
9"	136-145	142-156	151-170
10"	140-150	146-160	155-174
11"	144-154	150-165	159-179
6' 0"	148-158	154-170	164-184
1"	152-162	158-175	168-189
2"	156-167	162-180	173-194
3"	160-171	167-185	178-199
4"	164-175	172-190	182-204
Women			
4' 10"	92- 98	96-107	104-119
11"	94-101	98-110	106-122
5' 0"	96-104	101-113	109-125
1"	99-107	104-116	112-128
2"	102-110	107-119	115-131
3"	105-113	110-122	118-134
4"	108-116	113-126	121-138
5"	111-119	116-130	125-142
6"	114-123	120-135	129-146
7"	118-127	124-139	133-150
8"	122-131	128-143	137-154
9"	126-135	132-147	141-158
10"	130-140	136-151	145-163
11"	134-144	140-155	149-168
6' 0"	138-148	144-159	153-173

*Reproduced by permission of the Metropolitan Life Insurance Company. Data are based on weights associated with lowest mortality. To obtain weight for adults younger than 25, subtract one pound for each year under 25.

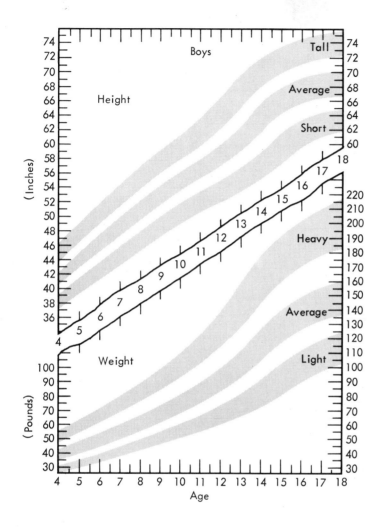

Figure 1.1 *Height and weight tables for boys age 4 to 18 in Iowa City, Iowa, schools (data collected from 1961 to 1963). Prepared for the Joint Committee on Health Problems in Education of the NEA and AMA by Howard V. Meredith and Virginia B. Knott, State University of Iowa.*

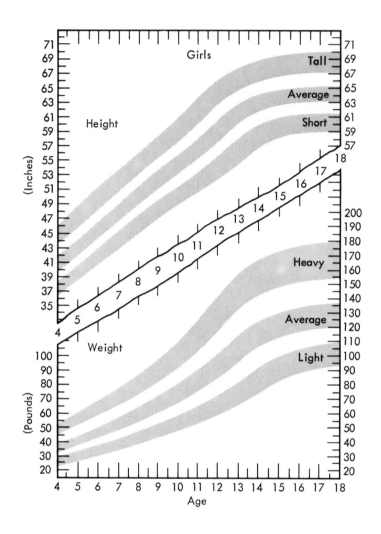

Figure 1.2 *Height and weight tables for girls age 4 to 18 in Iowa City, Iowa, schools (data collected in 1961). Prepared for the Joint Committee on Health Problems in Education of the NEA and AMA by Howard V. Meredith and Virginia B. Knott, State University of Iowa.*

The best way to lose excess weight is to:
A. *Skip breakfast.*
B. *Skip lunch.*
C. *Go on a crash diet with the goal of losing at least five pounds the first week.*
D. *Trim down your intake but be sure to follow the Basic Four Food Plan to ensure that your diet provides all the necessary nutrients.*
E. *Eat only bananas.*   *D.*

*

Three different types of food materials contribute Calories: carbohydrates, fats, and proteins. These organic substances occur naturally in foods in varying proportions along with non-caloric substances (water, minerals, and vitamins).

Which of the following do not contribute Calories in the diet?
A. *Water.*
B. *Vitamins.*
C. *Fats.*
D. *Proteins.*
E. *Minerals.*
F. *All contribute.*   *A, B, and E.*

*

In Appendix B you will find a table showing the composition of various foods. The fats in any food will provide approximately nine Calories per gram; proteins and carbohydrates contribute approximately four Calories per gram.

Per gram, fats contribute _____ Calories, proteins _____ Calories, and carbohydrates _____ Calories.   *Nine, four, four.*

*

These values are for the pure substances, which actually account for only a fraction of the weight of many foods. For instance, milk is 87.5 per cent water by weight and contains less than 5 per cent carbohydrate, 3.5 per cent protein, and 3.5 per cent fat. According to the approximate values given above, 100 grams of milk will then have about 19.5 Calories from carbohydrate, 14 Calories from protein, and 31.5 Calories from fat for a total of 65 Calories. It can be seen from these calculations that water contributes to the volume of the food we eat without adding Calories.

21

Which would provide the most Calories: 10 grams of pure carbohydrate, 10 grams of protein, or 10 grams of fat?

Ten grams of fat would provide 90 Calories, but 10 grams of carbohydrate or protein would each yield only 40 Calories.

If you eat a slice of bread that has in it a gram of fat, 10 grams of carbohydrate, and 2 grams of protein, buttered with 10 grams of butter which is 80 per cent fat, how many Calories do you consume?

$(1 \times 9) + (10 \times 4) + (2 \times 4) + (10 \times 0.80 \times 9) = 9 + 40 + 8 + 72 = 129.$

*

CARBOHYDRATES

Carbohydrates are an important group of organic substances that contain carbon, hydrogen, and oxygen. Some of the more common carbohydrates in foods are various sugars, dextrins, starch, pectic substances, and cellulose.

Which substance is not a carbohydrate: sugar, starch, pectin, or oil?

Oil.

*

Although all carbohydrates are related in a chemical sense, their behavior in foods is quite varied. Sugars taste sweet and are soluble in water, but different sugars vary in sweetness and solubility. You are already familiar with the sweet taste and easy solubility of granulated sugar. The chemical name for this sugar is sucrose.

The chemical name of the granulated sugar we buy in the store is _____.

Sucrose.

*

Lactose, the sugar found in milk, is not very sweet and dissolves less readily than sucrose. The type of sugar we carry in our blood is glucose; the solubility and sweetness of glucose are unlike either sucrose or lactose. The other sugars in foods also have their own unique characteristics.

Sugars **differ/are alike** in their sweetness and **differ/are alike** in their solubility.

Differ, differ.

22

Starch and cellulose, in comparison with sugars, are quite insoluble carbohydrates that lack a sweet taste. However, regardless of the difference in characteristics of carbohydrates, if we can digest them, they will give four Calories per gram.

A gram of sucrose, glucose, lactose, or starch will each give _____ Calories per gram in the diet; cellulose will provide _____ Calories.

Four, zero. You receive no energy from cellulose because humans cannot digest it; you will read more about cellulose later.

*

Carbohydrates are very important in your regular diet for they are excellent sources of energy. To use fats in the body efficiently, we need to burn some carbohydrate with them. Without enough carbohydrate, fat is not completely oxidized (burned) in the body and a toxic condition called ketosis develops.

If a fat is burned in your body without using some carbohydrate in the process also, the oxidation (burning) is/ is not complete and you may develop a condition called _____.

Is not, ketosis.

*

Carbohydrates also spare protein; with adequate amounts of carbohydrate available for energy, the body will not use protein as a source of energy. This spares the protein for its unique functions.

_____ have a protein-sparing action in the body; if we have enough of them, we do not burn protein for energy and heat.

What are three functions of carbohydrates in the diet?

Carbohydrates.

Carbohydrates: (1) provide energy and heat for the body, (2) aid in the complete oxidation (metabolism or burning) of fats, and (3) spare protein by providing sufficient energy (along with fat) to meet the body's demands.

One type of carbohydrate, cellulose, performs a unique function in the body. Cellulose is an important structural carbohydrate that helps to make plant structures more rigid. This type of plant material can be digested by cows and some other animals, but humans are unable to digest cellulose. Cellulose, known as roughage, is actually an important dietary material because it aids in maintaining motility in the digestive tract and helps to prevent constipation.

True or false. Because cellulose cannot be digested by human beings, it is of no benefit in the diet.

False. Although cellulose is a waste product, it aids in moving food through the digestive tract and facilitates elimination of residue from the intestine.

*

The percentage of the total number of Calories available from carbohydrates in a diet varies considerably from one culture to another, but it is common for at least 50 per cent of the day's Calories to come from carbohydrate-rich foods. In some areas of the world, particularly in the rice-eating countries, 80 per cent or more of the Calories consumed is from carbohydrate. The Irish and Polish populations have relatively high carbohydrate intakes because they eat an abundance of potatoes; Italians are characterized by a diet high in carbohydrate because of their preference for pastas.

Are these people living only on carbohydrate then?

No. Potatoes, rice, and pastas contain other nutrients in addition to carbohydrate. Also these diets contain other foods to help meet the body's total nutritional needs.

*

Foods high in sugar and low in moisture content are important sources of carbohydrate, although they are usually poor sources of other nutrients. Granulated sugar is nearly 100 per cent carbohydrate. Jams, jellies, and candies are high in carbohydrate.

Sweet foods that are low in moisture are **high/low** *in carbohydrate.*

High.

*

Cereals on a dry basis may be 80 per cent carbohydrate, but when macaroni, spaghetti, rice, oatmeal, cornmeal, and other cereals are moistened and cooked, they contain approximately 10 to 20 per cent carbohydrate. However, cereals are important in the diet for their vitamin and protein content as well as for their carbohydrate contribution.

Cooked cereals are **80/20** *per cent carbohydrate.*

Usually less than 20 per cent.

*

Some fresh fruits are high in sugar and some fresh vegetables are high in starch. Figs, grapes, bananas, and dates contain around 15 to 20 per cent sugar. Corn, peas, potatoes, dried beans, and legumes contain 15 to 20 per cent carbohydrate, largely in the form of starch.

Which is not high in carbohydrate: candy, dried figs, macaroni, lettuce, or sweet potatoes?

Lettuce is very high in water content and much of its carbohydrate content is in the form of cellulose which cannot be used for energy by humans.

*

There is a wide range in the size of carbohydrate molecules. Molecules of starch, pectic substances, cellulose, and some other carbohydrates are very large compared with sugars. However, even among sugars there is a difference in size. Sucrose (or granulated sugar), maltose, and lactose molecules are almost twice as large as the molecules of simple sugars such as glucose, fructose, and galactose. This size of the carbohydrate molecule is important during digestion as we shall see.

True or false. Molecules of starch and cellulose are very large, molecules of sugar such as granulated sugar are much smaller, and molecules of simple sugars are still smaller.

True.

*

A carbohydrate may sometimes be referred to as a "saccharide." A monosaccharide (one saccharide unit) is a sugar such as glucose, the sugar we burn in our bodies. A disaccharide (two units) is a more complex sugar such as sucrose. A polysaccharide (many units) is a carbohydrate even more complex than disaccharides. Starch, pectin, and cellulose are some of the more familiar polysaccharides. A polysaccharide can be broken down into a number of monosaccharides by chemical reaction.

Select the simplest of the following carbohydrates: pectin, sucrose, glucose, and starch.

Glucose.

25

How many monosaccharides can be obtained when a disaccharide is broken down? How many monosaccharides from a polysaccharide?

Two. Many.

*

Many chemical changes must take place in complex carbohydrates before you can use them in your body. Some foods high in sugar may contain a large quantity of the simple sugar, glucose. This sugar does not require any change to be absorbed by the body and some even can be absorbed through the stomach lining. This gives us quick energy.

Why would an athlete eat a candy bar before running a race?

To get extra energy quickly.

*

Limited digestion of starch takes place in the mouth and continues until the food reaches the stomach. Saliva contains an enzyme that starts to change this carbohydrate into a simpler form. Actually, only a small breakdown of starch occurs as the food travels through the mouth, esophagus, and stomach. Most of the chemical breakdown (digestion) from complex carbohydrates to simple sugars takes place in the small intestine.

Most of the digestion of a carbohydrate takes place in the **mouth/stomach/small intestine.**

Small intestine.

*

To digest a carbohydrate, it is necessary for the right enzyme to be present at each stage of the carbohydrate's breakdown. In the mouth an enzyme called salivary amylase begins to break down the starch molecule. Several other enzymes function in the small intestine to complete the digestive process.

True or false.   Salivary amylase completes the breakdown of starch in the small intestine.

False. Action of salivary amylase is stopped by hydrochloric acid in the stomach. Pancreatic and intestinal enzymes take over to complete the digestive process in the small intestine.

*

Enzymes are organic catalysts or specific protein materials that enable a particular chemical reaction to take place without being changed themselves. The body makes many different enzymes to catalyze the numerous and varied chemical reactions that are constantly taking place in the body. The suffix "-ase" is used to indicate that a substance is an enzyme.

If a word ends in "-ase" it is an _____; this substance is an organic catalyst or specific protein material that makes a chemical reaction take place **with/without** *being changed itself.*

Enzyme, without.

*

Carbohydrates are absorbed through the intestinal wall in the form of simple sugars and then are carried in the blood stream to the liver where they are converted to glucose. Glucose is the type of simple sugar used by the body for energy.

After _____ sugars are absorbed through the intestinal wall, the blood carries them to the _____ to be converted into glucose.

Simple, liver.

*

Glucose must be present in the blood stream for you to feel energetic. When the level of blood sugar drops, as it does about three hours after you have eaten, you will notice that you begin to feel rather tired and hungry.

After eating a meal, you should have a good supply of blood sugar (glucose) in your blood stream for approximately **three to five/five to eight** *hours.*

Three to five.

*

For you to obtain energy or heat from glucose, it must be oxidized. This metabolism of glucose goes through several chemical reactions and eventually results in the conversion of glucose to carbon dioxide, water, and energy.

_____, _____, and _____ are produced in the body when glucose is metabolized.

Carbon dioxide,
water, energy.

*

We can exhale the carbon dioxide from the lung; the water, known as metabolic water, can be used in the body or excreted in perspiration, tears, urine, and feces. Some water even escapes from the body in the air we exhale.

27

When we exhale air, we excrete **one/two** *products obtained in the metabolism of glucose:* _____ *and* _____.

Two: carbon dioxide, water.

*

FATS

Fats are another important group of foods made of carbon, oxygen, and hydrogen. As you will note, these are the same elements found in carbohydrates, yet the physical characteristics of fats are quite different from carbohydrates.

The chemical substances or elements in fats are **the same as/different from** *carbohydrates; fats* **are/are not** *similar to carbohydrates in their behavior.*

The same as, are not.

*

The key to this difference is in the ratio of the elements as well as in the way they are linked together. There is much less oxygen in proportion to carbon in fats than in carbohydrates. As a result of this low oxygen-to-carbon ratio, fats provide more than twice as many Calories per gram as do carbohydrates. You will recall that pure carbohydrates yield four Calories per gram and pure fats nine Calories.

Fats are composed of three elements: _____, _____, *and* _____, *but they have much less* _____ *than do carbohydrates.*

Carbon, oxygen, hydrogen, oxygen.

*

One very important function of fat in the diet is to provide energy; its effectiveness in this regard has been mentioned above. Gram for gram, fat gives more than twice as much energy as any other type of food material.

A gram of fat gives nine Calories, while a gram of carbohydrate gives _____ *Calories.*

Four.

*

Fats occur in rather high proportions in some foods common in the diet; for instance, lard, hydrogenated fats, and oils are 100 per cent fat and butter is 81 per cent fat. It is readily apparent that visible fats (fats that are eaten as an observable part of meat, butter, or margarine) are a highly concentrated source of energy.

28

Which is lowest in Calories: olive oil, lard, or butter?

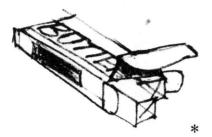

Because butter is 81 per cent fat, it contributes just over seven Calories per gram while olive oil and lard each have nine Calories per gram.

*

Fats also serve other essential functions in the diet. Some of the vitamins needed by humans are fat-soluble substances; therefore it is necessary to eat some fat in order to have dietary carriers of the fat-soluble vitamins. Vitamins A, D, E, and K are the fat-soluble vitamins and will be discussed subsequently.

Some/all vitamins are soluble in fats or oils.

Some.

*

Researchers have found that there is one fatty acid that is essential both for growth and for health of the skin. With inadequate amounts of linoleic acid (an essential fatty acid) in the diet, human infants will show poor growth and develop skin lesions.

Linoleic acid is an essential/a non-essential fatty acid.

An essential.

*

Since linoleic acid occurs abundantly in vegetable oils (hence also in margarines) and in lesser quantities in animal fats, older children and adults will, in all probability, eat more than adequate amounts of this fatty acid. Nevertheless, it is important to remember that fats are necessary in the diet to ensure that at least the minimum requirement of linoleic acid is available.

True or false. Most persons eat enough linoleic acid in their diets to meet the body's need for this fatty acid.

True.

*

In summary, the functions of dietary fats are: (1) to provide energy, (2) to carry the fat-soluble vitamins, and (3) to provide a source of the essential fatty acid, linoleic acid.

Fats not only provide energy but also are sources of fat-soluble _____ and an essential _____ acid.

Vitamins, fatty.

*

It is recommended that approximately 20 per cent of the total caloric intake should be from fats, but this level is far exceeded in most American diets. Foods that are particularly good sources of fats include butter, margarine, salad dressings, fats from meats, fried foods, rich pastries, and nuts. Most fruits and vegetables are low in fat; avocado and coconut are two exceptions and are reasonably high in fat content.

Which of the following would contain the most fat?
A. *Boiled potatoes.*
B. *Baked potatoes.*
C. *French-fried potatoes.*
D. *Scalloped potatoes.*

C. French-fried potatoes would be highest in fat, but if gravy, butter, or sour cream is used generously on boiled or baked potatoes the difference rapidly disappears.

*

If you were to examine the chemical formula of a simple fat, you would discover that there are two basic parts of each molecule. The simple fats, which occur commonly in foods, all contain an alcohol called glycerol (also occasionally called glycerine). This is one part. The other part is fatty acid. Several different fatty acids occur commonly in various simple fats. The variety in flavor and physical characteristics that we find in butter, margarines, and various animal fats is due to the fatty acids in the simple fat molecules.

What are the two basic chemical substances found in a simple fat molecule?

Glycerol and fatty acids.

*

When fat is eaten, it undergoes little actual chemical change until it reaches the small intestine. As fat moves through the mouth and esophagus and into the stomach, it becomes gradually more fluid due to the warm environment within the body. During the time the food is in the stomach, the fat becomes still more fluid because of the continued warming taking place.

*As fat moves through the body, it becomes **firmer/softer.***

Softer.

30

The stomach is useful in digesting fats and other materials because its churning action aids in working the food into a softer, more mobile mass. This action continues as long as food remains in the stomach (usually from three to four and one-half hours after a meal).

The normal length of time food is in the stomach is _____ to _____ hours.

Three, four and one-half.

*

Carbohydrates will leave the stomach first, followed somewhat later by proteins, with fat being the last material to proceed from the stomach into the small intestine. During this prolonged period in the stomach, there is ample time for gastric lipase (the fat-digesting enzyme) to begin the digestion of fat. However, this enzyme is limited in its action, and most of the fat will remain intact until it reaches the small intestine.

Briefly describe fat digestion from the mouth until it leaves the stomach.

Fat becomes gradually more fluid as it passes through the digestive tract, but only a very limited amount of digestion is accomplished by gastric lipase.

*

In the small intestine, bile salts join the fat and cause it to be dispersed in very, very small droplets. This emulsion provides a large surface area for the pancreatic and intestinal lipases to digest the fat molecules.

*Bile salts **emulsify/digest** fats.*

Emulsify.

*

The fat-digesting enzymes cause some of the fatty acids to split from the glycerol portion of fat; the fat components are then ready for absorption. Absorption of the products of fat digestion, as well as the final products released from carbohydrates and proteins, is greatly aided by structures in the small intestine known as villi. Villi are the convolutions in the wall of the small intestine which greatly increase the surface area available for absorption of nutrient materials.

*Villi are located in the **stomach/small intestine** and give a **greater/smaller** surface area to its wall.*

Small intestine, greater.

*

As soon as the fatty materials (glycerol and fatty acids) are absorbed, they are recombined into fats and trans-

31

ported throughout the body where they are burned for energy. Excess fat, beyond the body's need for immediate energy, can be stored as fatty deposits in the body.

True or false. We burn up excess fats.

False. We store them.

*

Fatty deposits in the body are important for appearance and health as long as they do not become excessive. These deposits provide a protective layer that helps to insulate our bodies. They also are found internally around the various body organs where they help protect them from shock.

*Some fatty deposits in the body are **helpful/harmful.***

Helpful.

*

Although glycogen is the first source of stored energy used to meet your energy requirements, you also can draw upon your stores of fat to meet additional energy needs. As you can see, fatty deposits are useful as long as they are not allowed to interfere with good health or to become large and unattractive.

Which of the following is not a function of fat?
A.   Furnishes heat or energy.
B.   Protects the body and cushions internal organs.
C.   Provides a source of essential fatty acid and certain vitamins.
D.   Acts as an insulator against heat loss from the body.
E.   Furnishes essential amino acids.

E. Amino acids are components of protein, not of fat.

*

PROTEIN

Protein, like carbohydrates and fats, is an organic material which means that carbon is an important part of the structure. Analysis also shows that hydrogen and oxygen are always present.

What other food substances have we learned contain carbon, hydrogen, and oxygen?

Carbohydrates and fats.

32

The unique element that is a part of any protein is nitrogen, and it is this element that is of particular significance in body tissues. Proteins are complex organic compounds which occasionally may contain some other elements besides the four that are always present. Sulfur is frequently found in protein, and iron, iodine, phosphorus, and cobalt occur in a few proteins.

What elements are found in all molecules of protein?

Carbon, hydrogen, oxygen, and nitrogen.

*

The basic building blocks of proteins are small units called amino acids. Although there are just over 20 amino acids that commonly occur in foods, there is actually tremendous variation in the arrangement and amounts of the various amino acids in different proteins.

True or false.   All proteins contain the same amino acids and in equal quantities.

False.

*

The relative amounts and pattern of occurrence of amino acids in each type of protein are different for each protein. Muscle proteins in cattle are unique to that species; human proteins are found only in humans. This characteristic of proteins is known as the *specificity* of proteins.

A protein found in sheep flesh will be **different from/the same as** *a protein in pigs.*

Different from.

*

Protein molecules are very large molecules which are often, in their natural state, curled up into a generally spherical or globular shape. In this shape, some of the component amino acids are trapped in the center of the molecule and do not greatly influence the chemical properties of the natural or native protein. However, during the cooking of food or during its digestion in the body, there is an unwinding of this globular shape and the protein molecule may show some different characteristics. Such a change is readily apparent when egg white is cooked; you can easily see the loss of transparency and the thickening that takes place as a result of this change in the protein molecule.

A protein that has not been heated or changed in any way from its natural form is called a _____ protein.

Native (or natural).

33

An extremely important function of protein is to provide the amino acids needed by the body to build new tissue and to maintain existing tissue. If total protein in the diet is limited, growth will be poor and the general condition of the body will deteriorate. Unfortunately, this dietary deficiency is all too common in some areas of the world. Numerous small children are suffering from the protein-deficiency condition known as *kwashiorkor*. In severe cases, young children die from lack of protein in the diet.

Why are young children more likely to develop a protein deficiency than are adults?

Children need protein for growth as well as for maintenance of tissue. In some cultures this need is not recognized and children may not be given the protein foods because they cannot chew them easily or because the family simply does not have the money to feed these more expensive foods to everyone in the family.

*

Some of the amino acids needed for tissue development can be formed from other amino acids if there is an adequate source of protein in the diet. These amino acids, which can be made in the body by converting one amino acid into another amino acid, are called non-essential amino acids. This term does not mean that these amino acids are unimportant nutritionally, but simply tells us that we can make these necessary amino acids if we have eaten a sufficient quantity of protein to supply the basic building materials.

*We **can/cannot** build some amino acids from other amino acids.*

Can.

*

In contrast to the amino acids we can make, there are the essential amino acids which we cannot make. These amino acids also are needed to make our specific body proteins but, unlike the non-essential amino acids, the essential amino acids must be eaten as a part of proteins in the diet itself. We are unable to make the essential amino acids in our bodies, even though we have an abundance of other amino acids in the diet.

34

We can/cannot make in our bodies the essential amino acids we must have.

Cannot.

*

Human adults require 8 essential amino acids, human infants need 9 essential amino acids, and white rats have been shown to require 10 essential amino acids in their diets. If an essential amino acid is not available in the diet in adequate amounts, growth will be very limited.

What is the difference between essential and non-essential amino acids?

Essential amino acids must be eaten in your food because the body cannot manufacture them; non-essential amino acids can be produced in the body from other amino acids that have been eaten. Both are important for natural growth and maintenance of tissue in the body.

*

Protein food sources are commonly classed as complete protein and incomplete protein. A complete protein contains all the essential amino acids, whereas an incomplete protein doubtless contains several essential amino acids but is inadequate in at least one of them.

A complete protein has all the essential _____, while an incomplete protein lacks _____ or more.

Amino acids, one.

*

With the exception of gelatin, animal foods are sources of complete protein. This means that milk, cheese, eggs, fish, poultry, and any kind of red meat are complete protein foods.

All animal proteins, except _____, are complete proteins.

Gelatin.

*

Vegetable proteins are found in good supply in the legumes such as navy beans and other types of beans, but vegetable proteins are incomplete proteins. The exception to this classification is soybeans, the one legume that contains a complete protein.

35

The only vegetable protein that has all the essential amino acids is found in the _____.

Soybean.

*

Vegetable protein foods are less expensive than most animal protein sources and can be used to good advantage in the diet. However, it is recommended that a complete protein food (meat, fish, eggs, poultry, or cheese) be used a minimum of one meal each day to ensure a source of all the essential amino acids.

We should have an animal protein in our diet at least: (a) once a day, (b) twice a day, (c) every other day, (d) once a week, or (e) once every other week.

Once a day.

*

It appears that 10 to 15 per cent of the total caloric intake should be from protein, with at least part of this intake being of animal origin (milk and eggs are suitable if you do not wish to eat meat). Contrary to what you might expect, heavy physical activity does not appear to alter significantly the body's protein needs.

Match the items in column B with those in column A.

A	B	
1. Incomplete protein	A. Supplies amino acids	
2. Complete protein	B. Lacking in amino acids needed to maintain life	
3. Essential amino acids	C. Contains all the amino acids needed to maintain life	1. B.
4. Protein	D. Substances in proteins that the body needs but is unable to make.	2. C.
		3. D.
		4. A.

*

You have probably heard of some problems involved in attempting to help achieve good nutrition for all the people of the world. Protein supplies are painfully short and so is money in some areas where nutrition is a particularly difficult problem.

True or false. Getting enough protein in the diet is one of the world's nutritional problems.

True.

Considerable research effort is being directed toward developing complete protein foods that are accepted by people and that can be produced for very little expense. The base for much of this research has been an attempt to find a palatable mixture of protein from various vegetable sources that will provide a balanced mixture of all the essential amino acids.

True or false.   Expense and palatability are two problems in finding a good protein source to relieve world hunger.

True.

*

While it is true that no single vegetable protein except soybeans contains all the essential amino acids, cereals and the other legumes contain significant amounts of various essential amino acids. By combining various cereals and legumes, we can build a complete protein. What one lacks in essential amino acids, another has. Ancient people who lacked good animal sources of proteins soon found that they could exist on combinations of vegetables. Thus, even today we see people living on a diet of legumes and cereals; the combination of proteins from these foods will not support life as well as will animal proteins, but it will keep them alive and continue the race. Other people have found that, by eating a large quantity of cereal protein and supplementing it with some animal protein, they can live.

*The Egyptians who live on whole wheat and beans and the Chinese who live on rice and soybeans **have/have not** a diet providing all the essential amino acids.*

While the Egyptians' diet is slightly less adequate than that of the Chinese, both can live and propagate on it.

*

Through research we are striving for a synthetic protein mixture from vegetable proteins that will give a palatable product at low cost. Experimentation is also continuing using ground fish flour and other low-cost animal proteins. Today we can make products from soybeans that are similar to meat in taste and appearance. Of course, this protein is complete. Such products, designed to simulate ham, bacon, fish, chicken, and hamburger, are being introduced on the market.

*We **can/cannot** make soybean protein products that taste and look like meat.*

Can; recently a convention of meat purveyors was served a luncheon featuring a Swiss steak. After the meal, they were told that the delicious Swiss steak they had eaten was not meat at all but was made from soybeans.

37

Proteins are important for other unique functions besides building and maintaining various tissues in the body. All the enzymes in the body are protein material. Some hormones including insulin, thyroxine, adrenaline, and parathyroid hormone are derived from protein materials.

Our enzymes and many hormones of the body are _____ material.

Protein.

*

Gamma globulin, the important antibody-producing substance that builds up a resistance in our body to diseases such as small pox and measles, is a protein. Because of its effect in regulating osmotic pressure, protein also helps to control the passage of fluid in and out of body cells. By concentrating on the inside of a cell, it can pull in moisture; by concentrating on the outside, it can pull moisture out.

True or false. Proteins in our bodies fight disease and help control fluid pressure in the cells.

True.

*

Another regulatory mechanism of protein is in maintaining the very slight alkalinity of the blood. Protein can also function as a source of energy, although it is not desirable to use protein for this purpose. This is an expensive process because of the energy required to oxidize protein in the body as well as the cost of protein foods in comparison with the relatively inexpensive carbohydrates.

Proteins are important in the body because they may be used to make _____ and _____; they are also essential because of their _____ capabilities, but they are an expensive source of _____.

Hormones, enzymes, regulatory, energy.

*

Digestion of protein begins in the stomach with the action of a gastric protease (protein-digesting enzyme) called pepsin. You may wonder why this enzyme does not destroy the stomach itself since the stomach wall contains protein. The answer is that pepsin is stored in an inactive form (pepsinogen) and only becomes active when it is combined with hydrochloric acid in the lower part of the stomach.

Pepsin is an enzyme in gastric juice that digests **fats/proteins/carbohydrates.**

Proteins.

*

Pepsin works to separate portions of the large protein molecule into shorter chains called polypeptides. Proteases secreted in the pancreatic juice take over the digestive process when the proteins and polypeptides reach the small intestine. The pancreas secretes protein-splitting enzymes into the small intestine.

During digestion, large proteins are split into _____.

Polypeptides.

*

The final breakdown of protein into individual amino acids is accomplished by three proteases from the intestinal juice. These enzymes also function in the small intestine, along with the pancreatic proteases.

Enzymes that split proteins are called _____.

Proteases.

*

Occasionally you may find a person who is unable to digest specific proteins in this manner, but instead will absorb a whole protein molecule before it is split into amino acids. This intact protein is irritating to the body and will cause an allergic reaction.

If a person absorbs a protein molecule into the body before it is split into its individual amino acids, he may have an irritation called an _____.

Allergy.

*

Amino acids are absorbed through the intestinal wall and are then carried in the blood stream to the liver before being circulated to the various other tissues of the body. Within the body the amino acids may be used to build proteins for growth and maintenance of tissues or they may be incorporated in the production of hormones, enzymes, or antibodies.

True or false. The body uses amino acids to build new tissue and make hormones, enzymes, and antibodies.

True.

*

Finally, when amino acids are to be excreted from the body, the nitrogen will be split off, transformed into urea, and eventually excreted in the urine. The non-nitrogenous portion of the amino acid will be oxidized just like carbohydrate.

In order for the nitrogen-containing portion of an amino acid to be excreted from the body, it must be:
A.   Removed from the rest of the molecule and eventually transformed into urea.
B.   Transformed into fat.
C.   Transformed into carbohydrate.
D.   Transformed into a hormone.
E.   Made into a new protein.   *A.*

*

VITAMINS

The word vitamin comes from "vita," meaning life, and "amine," a word used to describe a specific nitrogen-containing chemical group. These two words were combined to form the word "vitamine"; the final "e" was later dropped to give the present word "vitamin." This is actually not a very appropriate name because many vitamins do not contain amino groups, but the name is too firmly established to be changed.

Vitamin comes from two words: "vita" meaning _____ and "amine" meaning nitrogen containing.   *Life.*

*

A vitamin may be defined as an organic substance that is essential for life, is needed in only very small amounts, and must be supplied in the diet.

True or false.   Our bodies manufacture the vitamins we need in large amounts.   *False.*

*

The need for certain types of food was known long before vitamins became the subject of formal scientific research. Dietary problems plagued sailors who ate very limited diets while on long voyages of exploration. When Henry Cabot landed in Newfoundland, his men were ill with scurvy, a condition common among sailors. Friendly Indians fed them a drink brewed from spruce needles and they soon became well. The spruce needles contained vitamin C (ascorbic acid) which cures scurvy. Eventually, it was discovered that citrus fruits were also capable of preventing and curing scurvy, a discovery that led to the practice of feeding limes to the British navy and the ensuing tagging of British sailors as "limeys."

_____ cures scurvy.

Ascorbic acid or vitamin C.

*

The early Japanese navy helped to show another food lack. Many of these sailors were very ill when they ate their normal polished-rice diet. Eventually it was discovered that adding some meat to the diet corrected the problem. Later, it was found that the health-promoting substance in meats was thiamine or vitamin B_1.

True or false. Before the twentieth century, people were aware that certain foods prevented some illnesses.

True, but discovery of the vitamins in the food was to wait until the early part of the twentieth century.

*

Vitamin A

Vitamin A is a fat-soluble vitamin that is needed for growth of skeletal as well as soft tissue. In cases of deficiency during rapid growth, the bones may be affected before the soft tissue and, consequently, the bones in the head may crowd the brain and central nervous system. This condition has been known to result in blindness when the optic nerve is pinched.

We need vitamin _____ to aid in developing the bone structure in our bodies.

A.

*

The skin is also altered to a dry and horny condition in a vitamin A deficiency. The particular dryness of the mucous membranes may be at least partially responsible for a decreased resistance to bacteria. A lack of vitamin A opens the way for infection of the body's mucous tissues. Adequate vitamin A helps fight off colds and other possible infections.

True or false. Vitamin A deficiency can cause skin and mucous membranes to become dry.

True.

41

Xerophthalmia, a dryness of the eye that can result in blindness, is sometimes caused when vitamin A is inadequate.

True or false.   It is sensible to assume that anyone who is blind has a vitamin A deficiency.

False. Vitamin A deficiency is but one of many possible causes of blindness.

*

Probably the function of vitamin A that you are best acquainted with is in the prevention of night blindness. Limited ability to see in dim light is characteristic of a mild vitamin A deficiency.

*Which of the following is **not** a function of Vitamin A?*
A. *Regulates the heart rate.*
B. *Assists in growth.*
C. *Helps us to see in dim light.*
D. *Reduces chance of bacterial infections.*
E. *Helps in maintaining healthy skin.*

A. Vitamin A does not appear to influence the heart beat.

*

Vitamin A itself is found only in animal sources and is especially plentiful in fish-liver oils, liver of all animals, egg yolk, and butter.

Liver, fish-liver oil, egg yolk, and butter are good animal sources of vitamin _____.

A.

*

There are many other potential food sources of vitamin A. Several foods that do not contain vitamin A do contain a substance (precursor) which the body can change into vitamin A. The carotenes or yellow pigments found in many fruits and vegetables are precursors of vitamin A. Thus, yellow vegetables such as carrots and sweet potatoes are high in carotene. It is also true that dark-green, leafy vegetables contain carotene and are potential sources of vitamin A. Their green color is misleading; the chlorophyll pigment in these vegetables masks the lighter color of the carotene.

Classify the following either as sources of vitamin A or as precursors of vitamin A: carrots, egg yolk, butter, spinach, sweet potato, cod-liver oil, cream, and lettuce.

Vitamin A: egg yolk, butter, cod-liver oil, and cream.
Precursor of vitamin A: carrots, spinach, sweet potato, and lettuce.

It is unwise to take doses of vitamin A greatly in excess of the 5000 I.U. recommended for the average adult, because large excesses have been known to cause skin lesions, pain in the joints, thinning of the hair, and hemorrhages just under the skin. The amount of vitamin A that will cause these toxic effects is highly variable among individuals but is usually in excess of 100,000 I.U. daily. However, regular consumption of 50,000 I.U. daily may gradually build up a toxic level and cause hypervitaminosis A. This is definitely one case where the old adage, "if a little bit is good, more is better," should not be followed! The ordinary diet will not contain quantities of vitamin A that even approach the toxic level. The problem arises only when persons take additional vitamin A in large quantities in some therapeutic vitamin supplements.

True or false. It is wise to take vitamin A in a therapeutic supplement just to be sure that you are consuming enough to prevent night blindness.

False. A diet that contains a serving of a yellow or dark-green leafy vegetable at least every other day will supply all the vitamin A (as carotene) needed by the normal person. Therapeutic doses of vitamin A may cause a person to develop hypervitaminosis A, a toxic condition caused by too much vitamin A.

*

The B Vitamins

When vitamin B was first discovered, it was not sufficiently refined to show that what was thought to be one vitamin was in reality a group of substances. Later, several substances were identified in the group now known collectively as the B vitamins.

*Vitamin B was first thought to be **one/more than one** vitamin; today we know it is **one/several** substances, each of which is a separate vitamin.*

One, several.

*

Included in this B-complex group are thiamine, riboflavin, niacin, pyridoxine, folic acid, pantothenic acid, biotin, and vitamin B_{12}. Of these B vitamins, the first three listed are perhaps the best known; the recommended intake for them is listed in Table 1.2. All the B vitamins are water-soluble materials, although they are quite varied in their chemical structures.

The B vitamins include _____, _____, _____, pyridoxine, folic acid, pantothenic acid, biotin, and vitamin B_{12}.

Thiamine, riboflavin, niacin.

43

Thiamine is essential to the body because it functions in the enzyme system involved in the metabolism of carbohydrates; it also seems to be necessary for normal functioning of nerves. An inadequate intake of thiamine is the cause of *beriberi,* a condition typified by constipation and tenderness of the calf muscles. In severe cases the heartbeat is disturbed and death may result.

What is the name of the condition resulting from a thiamine deficiency?

Beriberi.

*

The need for thiamine is directly related to the amount of food eaten, because this vitamin is necessary for energy metabolism. Good sources of this vitamin include whole-grain cereals and enriched cereal products, yeast, meats (especially liver and pork), legumes, and milk.

If you eat whole-grain or enriched cereals, you are obtaining _____, a B vitamin.

Thiamine.

*

Since thiamine is a water-soluble vitamin, it will leach out into cooking water. Consequently, it is recommended that a small quantity of water be used in cooking thiamine-rich foods to ensure maximum retention. When feasible, it is wise to use the cooking water in preparing gravies and soups. Thiamine is also destroyed by an alkaline reaction and heat.

True or false. Broiled meats are better sources of thiamine than are braised meats.

True, but the liquid (containing thiamine) from braised meat can be used to make gravy.

*

Riboflavin, like thiamine, is an important coenzyme involved in energy release from food.

Riboflavin and thiamine are important coenzymes in reactions that release _____ from food.

Energy.

*

A deficiency of riboflavin is termed *ariboflavinosis* and is characterized by skin lesions, particularly at the corners of the mouth, and by conjunctivitis (inflammation causing irritation and a mucous in the eyelids). This deficiency condition can be avoided or corrected by eating adequate amounts of milk (as recommended in the Basic Four Food

44

Plan), cheese, eggs, meats, and whole-grain or enriched cereal products.

If you eat milk, cheese, eggs, meats, or whole-grain or enriched cereals, you are getting _____ in your diet.

Riboflavin (also thiamine).

*

The biggest problem in retaining the vitamin activity of riboflavin in food is its fast destruction when exposed to sunlight. This property has led to the use of tinted glass bottles or cartons for milk since these materials do not transmit the rays that reduce the activity of this vitamin. Of course, it should also be remembered that this is a water-soluble vitamin.

Why is milk marketed in tinted glass rather than clear glass?

To preserve the riboflavin content.

*

Niacin deficiency has been of particular interest in the United States because of its incidence in the southern states. *Pellagra,* the condition due to a niacin deficiency, is diagnosed by the occurrence of a symmetrical[1] dermatitis, diarrhea, and, in advanced cases, dementia and eventually death.

Pellagra is a disease caused by the lack of _____.

Niacin.

*

Adequate amounts of niacin in the diet will prevent pellagra. Particularly good sources of this vitamin are peanut butter and brewer's yeast, but all meats are good sources and enriched cereals also add to the intake of this B vitamin. It is of interest that niacin can be formed by the body from the amino acid tryptophan, and milk is a good source of tryptophan.

Peanuts, yeast, meats, and enriched cereals are good sources of _____; milk is a potential source of this vitamin because milk contains _____, which can be used in the body to make niacin.

Niacin, tryptophan.

*

Niacin, like thiamine and riboflavin, functions in energy metabolism in the body. Although all three are involved

---

[1]Symmetrical, as used here, means the skin irritation occurs in the same place on both sides of the body. For examples, both elbows or both hands may be affected at the same time.

in energy production, each of these vitamins performs a different function in this complex process.

True or false.   Niacin, riboflavin, and thiamine perform identical functions in energy metabolism.

False.

*

Niacin is more stable than the other B vitamins, but it is still readily soluble in water; food should be cooked in the minimum amount of water necessary for palatability.

Do you think that there are any B vitamins in the drip that collects as a frozen pork roast thaws?

Yes, because pork contains B vitamins and they are water soluble.

*

The remaining B vitamins are needed in small amounts in comparison with thiamine, riboflavin, and niacin. Since they are commonly found in the plants and animals used for food, it is unlikely that humans will develop deficiency conditions. However, experimentally-induced deficiencies of these substances have been studied. A brief description of these vitamins will serve our purposes in this chapter.

Why is it unlikely that you will have an opportunity to observe a biotin deficiency?

This vitamin is rather widely distributed and is needed in such small amounts that it is unlikely that a human will have such a deficiency.

*

Pyridoxine (vitamin B$_6$) has been receiving considerable attention in the last five years because of its importance in various metabolic reactions involving protein. It is the factor necessary to convert tryptophan into niacin, and it also appears to have some role in the metabolism of linoleic acid. Convulsions have been observed in persons with a deficiency of pyridoxine.

The conversion of tryptophan to niacin requires the presence of _____.

Pyridoxine (or vitamin B$_6$).

*

Folic acid, as its name implies, is found in abundance in leafy (foliage) vegetables such as spinach and leaf lettuce as well as in liver, legumes, asparagus, and broccoli. This vitamin is necessary for normal blood formation. A deficiency of folic acid prevents normal maturation of red blood cells and results in a reduced oxygen-carrying capacity of the blood.

True or false.   Folic acid is important in the body because of its role in energy metabolism.

False. Folic acid is most important for its role in normal blood formation (maturation of the red blood cells).

*

Since pantothenic acid is found in practically all foods, it is never likely to be inadequate in the diets of persons eating very much food. To study a deficiency of this vitamin, it is necessary to administer an antagonist (a substance that interferes with vitamin action) of the vitamin. Pantothenic acid is important as part of a coenzyme that functions in several phases of metabolism within body cells. At one time it was thought to be related to the greying of hair, but this has proven to be incorrect in man.

True or false.   Grey hair cannot be prevented in man by increasing the intake of pantothenic acid.

True.

*

Thanks to the administration of vitamin B_{12}, pernicious anemia patients can now be treated and the condition is no longer fatal. Like folic acid, vitamin B_{12} is necessary for the maturation of red blood cells. In addition, vitamin B_{12} appears to serve some function in the nerves. Folic acid is not a satisfactory substance for treating the nerve malfunctions associated with pernicious anemia, but vitamin B_{12} corrects the nerve disorder as well as the blood problem. The normal individual is able to make some vitamin B_{12} in his intestine and also is able to absorb vitamin B_{12} from dietary sources. Persons with pernicious anemia are unable to absorb this vitamin, even when it is present in the diet in adequate amounts. Therefore, it is necessary to administer this vitamin by injection to persons with a vitamin B_{12} deficiency.

Would you recommend that a person with a vitamin B_{12} deficiency shop carefully to find a vitamin capsule containing large amounts of this vitamin?

No. A person with this deficiency has to have this vitamin injected because there is some problem that prevents the absorption of vitamin B_{12} from the intestine.

47

Ascorbic Acid (Vitamin C)

The next vitamin in alphabetical sequence is vitamin C, now commonly called ascorbic acid. As mentioned earlier, this is the vitamin that cures and prevents scurvy. Scurvy is a dietary-deficiency condition in which the patient first experiences weakness, shooting pains in arms and legs, and weight loss, to be followed by swelling and reddening of the gums and possibly by loosening of the teeth. Cuts fail to heal normally, and hemorrhaging is common. These symptoms have led to the discovery that ascorbic acid is essential for formation of the connective tissue known as collagen, for normal functioning of the adrenal gland, and for metabolism of the amino acid tyrosine.

If you cut yourself and find that you are not healing as you should and also note that your gums are swollen, you may reasonably suspect that you have a deficiency of _____.

Ascorbic acid.

*

Ascorbic acid is very easily oxidized and is also water soluble. To make matters still worse, heat and alkali destroy ascorbic acid. Of all our vitamins, this is the one most easily destroyed. Therefore, special attention should be given to the preparation of foods that we expect to give us our daily ascorbic acid supply.

Ascorbic acid, also called vitamin_____, is a **stable/unstable** *vitamin.*

C, unstable.

*

Only a small quantity of ascorbic acid (about a half drop) is needed to meet our daily needs. Since it cannot be stored by the body, it is necessary to have a good source of this vitamin each day.

True or false. We should be careful to eat adequate amounts of ascorbic acid each day because the body cannot store it.

True.

*

This is the vitamin that is emphasized in the Basic Four Food Plan by the specific stipulation that a citrus fruit should be eaten daily. Six ounces of grapefruit or orange juice contain a day's supply of ascorbic acid. Tomatoes and fresh berries are also excellent sources. Cabbage and potatoes, when eaten with regularity in large quantities, may be significant sources of this vitamin.

If you ate one cup of raw cole slaw (shredded cabbage salad) and a freshly sliced tomato, would you have a day's supply of ascorbic acid?

Yes.

*

Vitamin D

Vitamin D is a fat-soluble vitamin that is fairly stable. It is important in the formation of bones and teeth because it assists in the absorption and utilization of calcium and phosphorus, minerals that are essential for proper development of bones and teeth. With a deficient quantity of vitamin D, there will be poor growth and the bones will be soft because of lack of calcium and phosphorus. This softness will cause the leg bones to bow under a child's weight. Bowed legs are a characteristic sign of rickets.

A vitamin D deficiency coupled with a calcium and phosphorus deficiency is indicated when you see someone who cannot make his knees touch.

Have you ever seen someone who has had rickets?

*

Vitamin D can be consumed in foods or it can be formed in the skin, from a substance (7-dehydrocholesterol) naturally present in the body, if the skin is exposed to the ultraviolet rays of the sun.

Ultraviolet rays can change 7-dehydrocholesterol into vitamin _____.

D.

*

Growing children and adolescents need to eat a dietary source of vitamin D to be certain that they are optimizing the utilization of calcium and phosphorus. However, adults (except pregnant and lactating women) do not need to have a dietary source of this vitamin.

*Adolescents **do/do not** need to eat foods containing vitamin D.*

Do.

*

Persons with dark skins will not form vitamin D in the sunlight as readily as will light-skinned people, so these children should be particularly careful to have a dietary source of this vitamin. The recommended allowance up to the age of maturity and also during pregnancy and lacta-

49

tion is 400 I.U. daily, an amount that is contained in each quart of irradiated or vitamin D-fortified milk.

Why do pregnant and lactating women have a need for vitamin D when other adults do not have a specified requirement for this vitamin?

These women need vitamin D to ensure optimum absorption of calcium and phosphorus during these periods of heavy nutritional demands on their bodies.

*

Vitamins E and K

Two other fat-soluble vitamins are known, but their importance in human nutrition has not been fully determined. Vitamin E has been shown experimentally to be important in the reproductive performance of white rats, but its role in human reproduction does not seem to be so apparent. At present it appears that the chief value of vitamin E in human nutrition is as an antioxidant. A deficiency of this vitamin seems unlikely if one consumes enough Calories because it is widely distributed in foods.

*The distribution of vitamin E in foods is **limited/wide.***

Wide.

*

Vitamin K is known as the antihemorrhagic vitamin because of its role in blood coagulation. Vitamin K probably can be made in the human intestinal tract. Absorption may be seriously impaired if the gall bladder fails to produce the bile salts needed to favor absorption of this fat-soluble substance. Otherwise, it is plentiful in most foods in the diet and there is usually no deficiency in the normally functioning individual.

The antisterility vitamin is vitamin _____ and the antihemorrhagic vitamin is vitamin _____.

E, K.

*

MINERALS

In tables of food composition, iron and calcium are two essential minerals mentioned. However, foods contain a wide range of inorganic materials with important functions in the body. The minerals found in various foods include not only calcium and iron but also phosphorus, copper, potassium, sodium, chlorine, fluorine, magnesium, sulfur, iodine, cobalt, zinc, manganese, and molybdenum.

The two minerals listed in tables of food composition are _____ and _____.

Calcium, iron.

*

Some minerals help to give the body rigidity when they are deposited in bones and teeth. Without these minerals you would look entirely different from the way you look today for you would not have hard bones to hold your body in its usual posture. You would also be unable to eat some of your favorite foods if you did not have the minerals that make hard teeth. Calcium and phosphorus are particularly important minerals in bones and teeth.

_____ and _____ are important as structural materials in bones and teeth.

Calcium, phosphorus.

*

Other minerals are found as parts of various essential compounds in the body. Two hormones contain a mineral: insulin contains zinc and thyroxine has iodine in its molecule. Cobalt is a part of vitamin B_{12}, sulfur is in the thiamine molecule, and hemoglobin contains iron. Soft tissues in the body also contain minerals. Sulfur and phosphorus occur in muscle tissue, and phosphorus is also a part of nerve tissues.

What mineral is a part of insulin?

Zinc.

*

Minerals are used to aid in regulating the body as well as in making some materials in the body. Other important functions of various minerals include: (1) maintenance of the right balance between acids and bases in the body, (2) regulation of osmotic pressure, (3) normal blood clotting, and (4) response to nerve stimuli.

True or false. Minerals perform many vital functions in the body.

True.

*

We get minerals from most of the foods we eat and even from drinking water. As has been mentioned, many minerals are essential for optimum health. These nutrients are

51

most likely to be provided in a diet regularly consisting of a wide variety of foods because no one food contains all the minerals needed by man. Some foods, such as milk and dairy products, will be high in some minerals, such as calcium and phosphorus, and low in other needed minerals; other foods will complement their mineral contribution by providing important amounts of other minerals.

A variety of food is/is not important to be sure that necessary _____ are available in the diet.

Is, minerals.

*

Although the quantity of minerals we need seems rather small, it is still essential to have adequate amounts of them in the diet. A gram or more of calcium and phosphorus may be needed each day while other minerals, like vitamins, are needed in milligram or trace amounts.

Minerals are needed in larger/smaller quantities than are proteins.

Smaller.

*

Since calcium and phosphorus are needed to form bones and teeth, it is not surprising that the recommended allowance for these two minerals is greater for persons who are growing than for those who have reached physical maturity. However, even adults need these two minerals to maintain their bones and teeth. Milk is an outstanding source of these two minerals and should be consumed by adults as well as children. Protein-rich foods and cereals are high in phosphorus; legumes and leafy green vegetables, except spinach, contribute some calcium.

If you do not drink milk, can you get enough calcium by eating other foods?

You can, but it is quite difficult. For instance, an adult would need to eat either four ounces of cheddar cheese, four cups of cottage cheese, seven cups of Brussels sprouts, or four and one-half cups of baked beans to meet his calcium needs for one day. Children would need still more.

*

Iron and copper are essential minerals for the formation of hemoglobin, the red pigment in blood that carries oxygen in the blood stream. When there is too little iron or copper available, not enough hemoglobin is formed to supply the tissues with the normal amount of oxygen. This

52

causes you to feel tired and listless and makes your skin pale. People with too little hemoglobin in their blood have a condition called *anemia*.

To make hemoglobin, we need ____ and ____.

Iron, copper.

*

Meats, eggs, whole-grain cereals, leafy green vegetables, dried fruits, and legumes are sources of iron and copper. Oysters and clams are also high in copper. It is important to eat foods rich in these minerals to avoid becoming anemic.

If you eat many whole-grain cereals, leafy green vegetables, and legumes, it is unlikely that you will develop ____, which results when you lack hemoglobin in your blood.

Anemia.

*

Iron is used over and over again in the body, but eventually it will be excreted and will need to be replaced. Anytime bleeding occurs, iron will be lost, thus increasing the body's need for this mineral. For this reason, women need to be particularly careful to eat foods rich in iron. Growing children also have a high iron need. An iron deficiency is one nutritional problem that can be found in any part of the United States.

True or false. Women and children have a greater need for iron in their diets than men do.

True.

*

Humans need iodine to enable the thyroid gland to produce its hormone, thyroxine. Thyroxine regulates energy metabolism and also is necessary for normal growth. When there is not enough iodine available in the diet, the thyroid gland enlarges in a futile attempt to make more thyroxine. This enlargement is easily seen at the front of the throat and is called a goiter. We can obtain iodine from fish or shellfish from the sea and also from fruits and vegetables grown on iodine-rich soil. Often drinking waters absorb enough iodine from the land to give an adequate supply. However, the northern states from Ohio through to the Pacific Ocean are in what is called the "goiter belt"; the incidence of goiter in this area used to be quite high compared with other parts of the country. Although it is not required by law, producers of table salt have made iodized salt available throughout the country. Simply using iodized salt rather than plain salt will provide sufficient iodine to prevent goiter in normal healthy individuals.

True or false. If nobody in your family has ever had goiter, you need not be concerned about developing one.

False. Goiter is due to a dietary deficiency of iodine and is not a hereditary condition.

*

The value of fluorine in reducing the number of dental caries in children has been demonstrated in several scientific studies. As a result, many community water supplies now have fluorine added to maintain the level of this mineral at one part per million. There is also some evidence to indicate that fluorine may be helpful in maintaining strong bones in older persons.

Fluorine has been shown to reduce/increase the incidence of dental caries in children.

Reduce.

*

REVIEW

1. *True or false.* We have examples of life surviving without food.

   False.

2. We use food to satisfy our nutritional needs, but food is also important to meet our _____ and _____ needs.

   Social, psychological.

3. *True or false.* It is almost always necessary to use diet supplements.

   False.

4. Enzymes *are/are not* important in digestion.

   Are.

5. Most nutrients are absorbed in the:
   A. Mouth.
   B. Stomach.
   C. Small intestine.
   D. Large intestine.
   E. None of these.

   C. Small intestine.

6. The acid in the stomach that aids digestion of some carbohydrates is _____.

   Hydrochloric acid.

7. Normally, food stays in the stomach about *three/seven* hours; the type of food that stays in the stomach the longest time is *fats/carbohydrates.*

   Three, fats.

8. When the body does not burn fat completely, _____ develops.

Ketosis.

9. Besides water, the chemical substances in food are: _____, _____, _____, _____, and _____.

Carbohydrates, proteins, fats, minerals, vitamins.

10. *True or false.* It is possible for us to change weight rapidly because of a change in the amount of water in the tissues.

True.

11. Carbohydrates are broken down in digestion and absorbed into the body in the form of *monosaccharides/polysaccharides*.

Monosaccharides.

12. We *can/cannot* digest cellulose.

Cannot.

13. Name a food that is almost 100 per cent carbohydrate.

Granulated sugar.

14. Match the item in column A with the best answer in column B.

A	B
1. Incomplete protein	A. Lacks amino acids needed to support life
2. Complete protein	B. Has all the amino acids needed to support life
3. Protein of high biological value	C. Comes from animal sources
4. Essential amino acid	D. Cannot be manufactured in the body

1. A.
2. B.
3. C.
4. D.

15. The largest number of Calories in your diet comes from: (a) carbohydrates, (b) fats, (c) protein, (d) vitamins, or (e) minerals.

Carbohydrates.

16. The second largest source of Calories in your diet comes from: (a) carbohydrates, (b) fats, (c) protein, (d) vitamins, or (e) minerals.

Fats.

17. Most adult men need *more/less* protein than do adult women.

More.

18. Fats *are/are not* burned in the body for heat and energy.

Are.

19. *True or false.* Vitamins help to regulate body functions.

True.

20. A lack of vitamin C (ascorbic acid) will cause _____.

Scurvy.

21. *True or false.* Most vitamins cannot be manufactured in the body, but sunlight can cause body fats to change into vitamin A.

False.

22. The number of Calories available from one gram of protein is _____, from one gram of carbohydrate is _____, and from one gram of fats is _____.

Four, four, nine.

23. Which will have the most Calories, a pound of bread or a pound of tomatoes?

Bread.

24. You need *fewer/more* Calories than did your forefathers.

Fewer.

25. Identify the function(s) of vitamin A in the body:
    A. Helps you to see in dim light.
    B. Keeps hair from turning grey.
    C. Gives resistance against infections.
    D. All are functions of vitamin A in the body.   *A and C.*

26. Carotenes *are/are not* the same substances as vitamin A.   *Are not.*

27. Thiamine can cure the disease called _____.   *Beriberi.*

28. _____ is destroyed by sunlight.   *Riboflavin.*

29. Pellagra is cured by the vitamin called _____, one of the vitamins of the _____ complex.   *Niacin, B.*

2

Cooking

Although we eat a number of foods raw, the variety and interest at meals may be greatly increased by using various cooking procedures to prepare foods. Here are just a few examples to show how cooking can improve palatability, digestibility, and appeal. Foods such as macaroni and rice are much easier to chew, have a more pleasing taste, and are more easily digested when they have been properly cooked. Most people find meat is more appetizing and palatable when cooked. Cakes are easier to eat after they are baked, because heat changes them from liquids to solids. Flavors in foods often blend better as the ingredients are cooked together. Cooking also helps to retard food spoilage by destroying some harmful microorganisms.

Food may be cooked to:
A.   Increase palatability and digestibility.
B.   Increase nutritive value.
C.   Give it a desirable texture and appearance.
D.   Destroy harmful microorganisms.
E.   Blend flavors.

All except B but, in some cases, B is also correct. Usually cooking reduces nutritive value slightly, but in some cases a cooked food is actually more nourishing than the raw food because digestibility is improved.

*

THE COOKING PROCESS

Cooking means to subject food to heat. Heat energy can be used to bring about desirable changes in foods. Some useful changes are: (1) the softening of structural materials in fruits and vegetables, (2) the thickening (gelatinization) of starch-containing foods such as gravies and puddings, (3) the firming (coagulation) of protein foods including egg mixtures and baked products, (4) the changing of color as in cooking meat or browning cookies and rolls, and (5) the leavening of foods as when we use baking powder to leaven a cake.

Heat energy can cause _____ in food that are _____.

Changes, desirable.

*

One important lesson in food preparation is to learn to control heat. Too much heat can cause undesirable actions, as when a custard separates, cocoa scorches, or meat is overcooked.

True or false. Fast, uncontrolled heating of food often results in unpalatable changes in food.

True.

*

Food prepared on conventional ranges must have heat to cook. We bring heat to food mainly by conduction, convection, or radiation. Heat can be transmitted through substances such as water, fat, or metal. Examples of this means of transferring heat include boiling vegetables in water, frying chicken in deep fat, or baking a pancake on a metal griddle. This method of heat transfer is *conduction*. Heat flows or is conducted from a warmer area to a colder region when there is some substance through which it can travel.

True or false. Conduction of heat is the flowing of heat through a substance from a hot area to a colder one.

True.

*

When gases or liquids such as water or oil are heated, they rise and push through the colder material. The colder material then sinks, is heated, and then rises. The cycle is thus established and continues to repeat and repeat during cooking. This movement causes currents to flow. When currents carry heat from one place to another, we call this *convection*. When water circulates in a pot or fat in a fry kettle or air in an oven, heat is carried to the food by convection.

Can both convection and conduction occur at the same time in a pot of water?

Yes. The heat is conducted through the water (conduction) and the movement of the water in carrying heat to the food is convection.

*

An object that is so hot that it glows radiates heat. Radiated heat is a wave of energy much the same as light is. It does not need matter for transfer such as is needed in the conduction or convection of heat. It can pass through a vacuum. The sun's energy is transferred to us here on earth in little more than eight minutes through a vacuum. This is radiated heat.

Heat from the sun is transmitted by **conduction/convection/radiation.**

Radiation.

*

Radiated heat cannot warm something until it is absorbed. Light or bright objects reflect rather than absorb radiated heat. Dark or dull objects absorb heat. For this reason, a cake will bake more rapidly in a dark pan than in a bright aluminum one. When you broil or toast food, you are using radiated heat.

Match the method of heat transfer in column A with the cooking procedure in column B.

	A		*B*		
1.	*Conduction*	*A.*	*Baking in an oven*	*1.*	*B.*
2.	*Radiation*	*B.*	*Boiling carrots*	*2.*	*C.*
3.	*Convection*	*C.*	*Toasting bread*	*3.*	*A.*

*

Heat must be created from other energy. The heat used in cooking is usually obtained by combustion or friction. Combustion is rapid oxidation of a fuel to develop a large quantity of heat. Probably today's most common fuel for combustion is gas, but charcoal, wood, oil, and kerosene are also used. These substances have stored energy that can be released to cook food.

Gas is the most common fuel today for heating by _____.

Combustion.

*

Electricity flows along a wire conductor much as water flows through a pipe. If you create a resistance against this electrical flow, you will develop friction; this friction can be so great that it causes a large quantity of heat. We can bring electricity to a heating element. Here the element sets up so much resistance to the electrical flow that it becomes very hot. This heat can be used to cook food.

59

The heat from an open flame results from a **chemical/ physical** *reaction while the heat in an electrical element results from* _____ *set up against the flow of electricity.*

*

Recently there has been increasing use of another means of heating. In this method, food is heated by *microwaves* (sometimes called radar waves). These waves penetrate the food and their energy is transmitted to the food molecules in the form of heat. Such heat does not brown food. This manner of heating is very rapid and has become important in the food-vending business to warm packaged sandwiches and other foods.

Heating by microwaves is very **slow/fast** *but* **does not brown/burns easily.**

*

The speed with which heat is conducted depends upon the substance through which it is passing. Some metals, such as copper and aluminum, conduct heat rapidly. Iron is also a good conductor, but not as good as copper and aluminum. Stainless steel is a poor conductor and develops hot spots where food sticks and burns. You will find that stainless steel pans with a copper coating across the bottom or an inner core of iron or aluminum between two sheets of stainless steel will heat more evenly than will pans made entirely of stainless steel.

What other advantages would aluminum have for cooking utensils? What disadvantages?

*

When you heat water, you can continuously raise its temperature until it begins to boil. As soon as water begins to boil, the temperature will not go higher. If you put more heat into it, the water merely boils faster and forms more steam.

Does food cook faster in rapidly boiling water than it does in gently boiling water?

*

When enough heat is applied to water, it vaporizes into steam which, in turn, condenses again to water. However, when melted fat or an oil is heated until it vaporizes, it

60

does not return to its original form when it is cooled as water does. The high temperatures cause fat to break down and change into different compounds. This is an irreversible chemical change.

If you overheat butter and cause some of it to vaporize, it will/will not return to its original form when it cools.

Will not.

*

The temperature at which water and many other substances will boil depends upon the atmospheric pressure. At sea level a pressure of approximately 14 pounds per square inch is pushing down on the surface of water. Water will boil when it has enough heat energy to just exceed this downward push. At 212°F, water has enough energy in it to vaporize or change to steam and boiling occurs. However, in the mountains where the elevation is higher, atmospheric pressure is less than at sea level. This means that less force is pushing down on the water. Consequently, less heat energy in the water is necessary to overcome the downward pressure and the boiling point is lower than at sea level. The pressure of the atmosphere also changes as the weather changes, so you will find that the boiling point of many liquids used in cooking will vary according to the weather. You will find that elevation or the weather is important when you cook sugar syrups and make frosting or candies.

If you were boiling potatoes in mile-high Denver, would the boiling temperature be higher or lower than at sea level? Would the cooking time be longer or shorter? Why?

The boiling point is lower in Denver than at sea level because the atmospheric pressure is less at a high elevation. It would then take longer to boil potatoes in Denver because boiling water is not as hot as it is at sea level.

*

Heat is also used to cause useful chemical changes in food. It is heat (along with water) that enables baking powder to react and release carbon dioxide gas to leaven many baked products. When a sugar solution is heated in the presence of an acid, a chemical reaction takes place and different sugars are formed. This is important when you make candy or frostings.

Would you say that cooking involves much chemistry?

Chemical processes are constantly going on when we cook; in fact, a kitchen is one type of chemical laboratory.

*

Heat not only causes chemical changes but also effects physical changes in foods. It softens many foods: it changes solid fats to liquids; if caramels are warmed slightly, they become softer and easier to cut. Some other foods become thicker or firmer when heat is applied. A sugar solution cooked to 240°F will not be as hard as it would be if it were cooked to a higher final temperature. Eggs are another illustration of a food that thickens when heat is applied.

True or false. In cooking we are constantly using heat to cause desirable physical and chemical changes.

True.

*

You can regulate the total quantity of heat put into food by controlling the temperature of the heating source and the length of time food is cooked. It may be important to introduce a large quantity of heat quickly such as is necessary to produce the steam needed to leaven a cream puff properly. At other times more palatable foods are prepared by heating them slowly for a long time, as is done when a less tender cut of meat is braised.

We can regulate the total quantity of heat we put into food by controlling the _____ _____ or _____ _____.

Cooking temperature, cooking time.

*

RECIPES AND COOKING

No pharmacist attempts to mix a prescription unless he has a written formula. No carpenter attempts to build a house until he has a blueprint. No one who is successful in the kitchen attempts to cook without a recipe. Some cooks are able to make things without a written recipe in front of them because they have had considerable practice and experience. They know the proportions of ingredients to use and are familiar with the correct appearance at different stages of the preparation. Although it seems that they are working without a recipe, they actually are using a recipe that is based on extensive experience. Even with this background, there may be noticeable variation in a product from one time to another unless careful measurements are made each time. Recipes are important in controlling the quality of foods we prepare.

In the old days cooks were trained to cook without a recipe. How could this be done?

*People were **trained** because this was the only way available to teach cooking, but as recipes were standardized and more and more people could read, a better way was found. In the early days, when recipes were not used, there was a wider variation in quality than there is today.*

When we use a recipe at home or in quantity food preparation, we should always be alert to reduce work required and to simplify procedures. New ingredients or prepared foods are appearing constantly on the market and often may be used to advantage. When a product is finished, we should critically examine it and evaluate its qualities. Then it is wise to jot down modifications that might be incorporated in the recipe to improve the product. Recipes are vital things and constantly need revision to meet changing conditions. There is always a better way of doing something.

If you had a recipe that called for simmering four pounds of bones and a pound of lean beef shank for eight hours and then straining the broth and moving through another set of steps to make jellied bouillon, would you think it a good idea to change the recipe to the use of canned bouillon?

In this day and age, yes. The cost of labor and facilities must be considered when deciding the relative value of a commercially prepared product and one prepared in your own kitchen.

*

Recipes for quantity food preparation should contain information needed to plan menus and to prepare the food. A standardized recipe for quantity cookery is a statement of (1) the number of portions and their size, (2) ingredients and their quantities both in measure and weight, and (3) how these ingredients are used. Any additional information required to modify, change, or substitute items in the recipe is plainly noted. Panning weights, baking times and temperatures, and other required information are listed. It is desirable that portion and total cost also should be stated.

Which of the following can be omitted in a standardized recipe?
A. Ingredients.
B. Methods for combining ingredients.
C. Number of portions and their size.
D. Calories per portion.
E. Panning weights, baking times, and equipment used.

D, but there are times when it might be a good idea to list Calories.

*

A standardized recipe yields a known quantity of known quality. In quantity work it is desirable to know also the cost per portion. It is desirable to know how long it takes to prepare a recipe. Sometimes one may start a recipe only to find that the time required is beyond that available.

The time needed for various pieces of equipment is also important, especially in quantity work.

A standardized recipe yields a known____of____quality.

Quantity, known.

*

Recipes for home use should include much of the same type of information contained in a standardized quantity recipe. Specifically, home recipes should list the ingredients and their exact measures, precise directions for preparation including exact mixing times, size and type of pan, correct baking time and temperature, and number of servings. Recipes accurately stating all this information may be prepared time after time with comparable results.

Quantities of ingredients for home recipes are usually given by **weight/measure.**

Measure.

*

Recipes are easiest to use if they are in a book that opens out flat or on index cards. Mistakes in following a recipe can usually be avoided if the ingredients and their amounts are clearly listed apart from the directions. Numbering the steps in the directions also makes a recipe easier to use.

True or false. A clearly written recipe is a valuable aid in avoiding mistakes in the kitchen.

True.

*

While there are many acceptable forms for recipes, the descriptive form is perhaps the best and most commonly used for quantity recipes. The name of the food is at the top center with any file or code number at the right. Different colored cards for different types of foods make it easier to file and to find groups of recipes. Normally the total yield and number of portions are given under the name at the left of the recipe and the portion size is given at the far right. Ingredients *in their order of use* are listed at the left with weights following in the next column to the right and measures in a column to the right of weights. Methods for handling the ingredients follow in a column at the right. These procedures for handling the ingredients should be numbered by distinct steps. Frequently a line or space indicates this distinction between steps.

Using the descriptive form write the following recipe for 2½ cups of white sauce: ¼ cup flour, ⅔ cup instant dry milk, 1 tsp salt, 2 Tbsp butter or margarine, and 2 cups water. The water and dry milk are mixed together. In a saucepan the fat is melted and the flour and salt are mixed in. Then the milk is stirred in and the mixture is placed over low heat and stirred until thickening is complete. It is advisable to cook another minute to be sure gelatinization (thickening) of the starch is complete and to help remove any remaining raw starch flavor.

This white sauce recipe, in descriptive form, is given below.

*   *   *White Sauce*   *   *

2½ cups or 20 ounces; 10 portions

Each portion ¼ cup or 2 ounces

Water, tepid	1 lb	2 cup	1. Put water into a container and pour the milk over the water. Stir until dissolved.
Milk, dry, nonfat, instant		⅔ cup	
Margarine or butter	1 oz	2 Tbsp	2. Melt margarine or butter in a saucepan. Add the flour and salt and stir until well blended.
Flour, all-purpose	1 oz	¼ cup	
Salt	1 oz	1 Tsp	

3. Add the reconstituted milk, stirring well during the addition. Place over low heat, stirring continuously and reaching all areas of the bottom of the pan.

4. When the mixture has completely thickened, cook slowly one minute more to remove the raw starch flavor.

*

In the recipe for white sauce, specific ingredients are listed. The milk is *dry, nonfat, instant* and the flour is *all-purpose*. Just saying "milk" might confuse the cook. Is it fresh whole, fresh nonfat, condensed, or evaporated? If cake flour were used instead of all-purpose flour, the sauce would be thicker than desired because cake flour has more starch in it and, therefore, has greater thickening power than all-purpose flour.

If a recipe simply listed sugar, ground meat, chicken, cheese, and tomatoes, what questions might you have about the ingredients?

Is the sugar granulated, brown, or powdered? Is the ground meat cooked or raw and is it beef or some other meat? Is the chicken a broiler, fryer, or roaster? Would cottage, cream, Cheddar, Parmesan, or some other cheese be best, and should it be sliced, diced, or grated? Are the tomatoes raw or canned, whole or cut up?

*

Before using a recipe, read it over carefully, noting ingredients, equipment, utensils, times, and other factors necessary to produce the item. Organize the work, assembling ingredients first and doing the required pre-preparation. Eliminate the use of extra bowls, cups, measures, tools, and manipulations. Then start to produce the item, working in an area that is not over five and one-half feet long and two and one-half feet wide. Work quickly and with purpose. A recipe is most successful when it is moved on to completion with sureness and purpose.

Why is it important to read a recipe before beginning preparation?

To be certain that all ingredients, equipment, and necessary time are available. If anything is lacking, preparation should not start.

*

Accurate measurements are essential to good food preparation and can best be accomplished with suitable equipment. A glass measuring cup with fractions of a cup clearly marked is the best utensil for measuring all liquids. A nested set of measuring cups (Mary Ann cups) consisting of a one-fourth cup, a one-third cup, a one-half cup, and a one cup measure is your best choice for all dry ingredients and solid fats. For measuring less than one-fourth cup quantities of liquids and dry ingredients, a set of measuring spoons is most accurate.

Match the ingredient with the utensil which should be used to measure the food.

1. Flour.	A. Glass measuring cup.	1. B.
2. Whole milk.	B. Mary Ann cups.	2. A.
3. Salad oil.		3. A.
4. Shortening.		4. B.
5. Sugar.		5. B.
6. Honey.		6. A.
7. Rice.		7. B.
8. Water.		8. A.

*

Liquids are measured by placing the glass cup on a level surface and then pouring in liquid until it reaches the desired level. The greatest accuracy is obtained when you read the level of the liquid by bending down until your eyes are even with the cup and then reading straight across.

The most accurate measurement of liquids is made by looking **down on/straight across** *the surface.*

Straight across.

*

Dry ingredients should be lightly spooned into the appropriate-sized Mary Ann cup until the cup is more than full. Then you should draw the straight edge of a spatula firmly across the top of the cup to obtain a level measurement.

True or false. Dry ingredients should be leveled by shaking the measuring cup until the excess material is removed.

False. The straight edge of a spatula drawn across the cup gives a more accurate measurement.

*

Three special ingredients should be mentioned in a discussion of measuring techniques. Most types of flour (with the exception of granular, instantized, and no-sift) should be sifted and very lightly spooned into the cup to avoid any packing. Careless handling of flour will cause it to pack and you will then have more flour than should be used in the recipe. Butter, margarine, and other solid fats should be pressed firmly into the correct-sized Mary Ann cup to avoid any crevices in the fat. The third special ingredient is brown sugar. The quantity of brown sugar required in a recipe is based on the assumption that this type of sugar will be gently but firmly packed in the measuring cup. As with other dry ingredients, the three ingredients discussed here should be leveled after the cup is filled slightly beyond capacity.

*Flour usually should be **sifted/packed** before being measured; solid fats should be **gently spooned/solidly packed** into the cup; brown sugar should be **lightly spooned/gently but firmly packed** into the cup.*

Sifted, solidly packed, gently but firmly packed.

*

In cooking we use many terms which have very specific meanings. Unless you understand these meanings and then apply the proper technique to produce a food item, you will not be a successful cook. We have a language which gives exactly what technique should be used. Thus, when the term "fold" is used, it does not mean "stir." Braising is a different process than simmering. Chilling and cooling are two different temperatures in food preparation.

What do you think would be the result if you stirred egg white foam into a souffle rather than folding it in? (See Appendix A for definitions.)

The souffle made by stirring would have a smaller volume than the one in which folding was done because stirring would break up the foam more and air would be lost.

*

FOOD SAFETY

By destroying bacteria, molds, or other microorganisms that may be present, you can extend the safe storage period of foods. When you are boiling or heating foods, you are making them safer to eat. Ten minutes of boiling kill most microorganisms. Even when foods have been boiled 10 minutes or longer, they can easily become recontaminated with harmful microorganisms. These can multiply and cause food spoilage. If we reduce the temperature below boiling, the time required to destroy the microorganisms will be extended.

*If we wish to destroy microorganisms, we should use **moist/dry** heat; as we drop the temperature below boiling, it is necessary to **increase/decrease** the time.*

Moist, increase.

*

To destroy harmful bacteria in some foods, we pasteurize them. Milk is pasteurized by holding it 30 minutes at 145°F or for 15 seconds at 161°F. Milk also can be flash pasteurized by quickly bringing the temperature of the milk to 180°F and dropping it immediately. Rapid cooling to less than 50°F follows immediately in any of these pasteurization methods. Notice that the time required for pasteurization is much longer at a lower temperature than it is at the higher temperature.

Milk is pasteurized to make it **safe to drink/taste better.**

Safe to drink.

*

Eggs can be pasteurized by holding them for a long time at 138°F. Below 135°F it is difficult to destroy bacteria even by extending the time.

True or false. In pasteurization of foods, we destroy all bacteria present.

False. Pasteurization is actually partial sterilization, and some live microorganisms may remain. However, the temperatures used in pasteurizing milk do kill the disease-producing microorganisms that may be present. In sterilization, all bacteria are destroyed.

*

In commercial dishwashers the washing temperature is 160°F and the final rinse is 180°F for 10 seconds. The result is tableware that is fairly free from bacteria. Our home dishwashers wash dishes between 140 and 160°F and have a final drying period in which the temperature goes well above 180°F. Therefore, most bacteria will be destroyed in either type of dishwasher. If you do not use a dishwasher, you should clean dishes and silverware by thoroughly washing and then carefully rinsing them in very hot water. Just a few tablespoons of chlorine bleach in the rinse water will assist in destroying bacteria. It is better to air-dry than towel-dry tableware. The towel contaminates the clean tableware and spreads this contamination around.

If a _____ _____ is added to a rinse water for tableware, it will assist in destroying bacteria.

Chlorine bleach.

*

Sprays or powders designed to destroy microorganisms should be used only with extreme care in the kitchen because some are poisonous. Bleaches that contain chlorine can be used safely to sanitize objects. Approved sanitizing solutions are available which, when added to water, destroy bacteria, yeasts, and molds. Soaps and detergents remove microorganisms from objects and act as a deterrent to their growth. Some commercial foods today are irradiated to destroy insects, their eggs, and larvae.

Which of the following will not kill microorganisms?
A. *Chlorine or sanitizing solutions.*
B. *Irradiation.*
C. *Soap or detergent.*
D. *Sprays and powders containing poisonous substances.*

C. Most soaps and detergents are more bacteriostats (deterrents) than bactericides (killers).

*

Bacteria, molds, and yeasts are in the air and easily can contaminate food. Human beings, utensils, and other foods are also sources of contamination. Animals, sewage, insects, dirt, and even water are other sources of contaminants. Most foods have some bacteria, molds, and yeasts on them; under proper conditions these can grow and spoil food or make it dangerous to eat.

*Microorganisms can contaminate food from a **few/many** sources.*

Many.

*

Mixtures of protein and carbohydrate foods are particularly favorable media for growth of microorganisms. Custards, gravies, and similar food items are excellent media for the growth of bacteria. Acid foods are less likely to spoil than those that are neutral or slightly alkaline.

Would a dish such as macaroni and cheese or a beef stew be a good culture for food-poisoning bacteria.

Yes. These foods spoil easily since they are composed of a mixture of carbohydrate and protein.

*

Dried foods and foods that are extremely salty or very high in sugar do not favor growth of microorganisms. Consequently, such foods keep well, but eventually different molds, yeast, and bacteria may attack them and prepare them for invasion by others. Some molds grow on very sweet foods, even jam or jelly, and some yeasts grow in extremely salty foods. Some bacteria can grow well on quite acidic foods. Dried foods may gradually take up enough moisture so that microorganisms may be able to grow on them.

Which food would be likely to encourage growth of micro-organisms?
A. *A neutral one that is completely dry.*
B. *A neutral, moist mixture of protein and carbohydrate.*
C. *An acid, salty food.*
D. *An acid, yet sweet item.*
E. *A very dry, alkaline food.*

B. A neutral, moist mixture of protein and carbohydrate.

*

Most microorganisms grow best at temperatures between 40° and 120°F. Below this temperature, growth is slowed appreciably and above this range, many are destroyed. Some microorganisms actually grow best at temperatures below 40°F and some develop rapidly between 100° and 135°F. We call those that grow at low temperatures *cryo-philic* (cold-loving) microorganisms and those that grow best at higher temperatures *thermophilic* (heat-loving) microorganisms. Freezing does not destroy microorganisms, but frozen storage significantly slows their growth and can cause some types to die.

Why do some foods spoil even when refrigerated?

Cryophilic microorganisms thrive at refrigerator temperatures.

*

Some microorganisms can hibernate by going into a rest-ing, highly protected dormant form. Such forms are called *spores.* They are very resistant to heat and other forces that ordinarily destroy microorganisms. Spores can remain dormant for many months or even years; then, when con-ditions become favorable, they develop into active organ-isms and multiply rapidly.

The highly resistant, dormant form some microorganisms assume is called a _____.

Spore.

*

A violent food poisoning will result if you eat a toxin made by *Clostridium botulinum* (a kind of bacteria). These bac-teria are quite resistant to heat, especially when in a food not highly acidic, and can thrive in canned vegetables that have not been processed to a high enough temperature. Home-canned vegetables are the most likely sources of this type of food poisoning.

71

What are some home-canned vegetables that might contain the toxin produced by Cl. botulinum?

Peas, beans, beets, and corn are only slightly acidic and are representative of the types of vegetables to be examined carefully. Bulging cans or jar lids quickly tell you that the food is not safe to eat.

*

Some microorganisms will not grow in air. These we call anaerobes. Others need air and we call these aerobes. *Clostridium botulinum* likes no air, but can grow if the oxygen supply is limited.

Microorganisms that grow best in air are called _____; those that grow best without air are _____.

Aerobes, anaerobes.

*

Food poisoning may result from toxins developed by bacteria, as is the case when *Clostridium botulina* flourish in canned food. Other types of food-borne illness result from eating food containing *Staphylococci, Salmonellae,* and other harmful microorganisms. Dysentery, diphtheria, tuberculosis, scarlet fever, and other diseases are caused in this manner. Careful handling of food from the time it is harvested until it is eaten is essential to keep food safe to eat.

Illnesses caused by microorganisms in food are called _____ illnesses.

Food-borne.

*

Once food is contaminated and favorable conditions exist for growth, it is sometimes possible for enough microorganisms to grow in four hours to cause illness. Therefore, no food requiring refrigeration should ever stand at room temperature that long. If certain foods are unrefrigerated for less than four hours, some distress still may be caused. Flesh foods, milk, and prepared foods should be held at room temperature as short a time as possible.

The total time a food may be held safely at room temperature is less than _____ hours.

Four.

*

Care should be taken to see that foods are cooled quickly and refrigerated. If a food is used at more than one meal, you should be very careful that the total time unrefrigerated is kept to an absolute minimum. For instance, a turkey or roasting chicken that is cooked and allowed to cool at room temperature for a couple of hours before being refrigerated, and is then in and out of the refrigerator for varying lengths of time while being used as leftovers, may be very dangerous to eat because it has probably been unrefrigerated more than four hours.

True or false. A food such as turkey is safe to eat if it is never out of the refrigerator more than four hours at any one time.

False. The total time out of the refrigerator should not exceed four hours. The total time in this case is not stated, but it is likely considerably more than four hours.

*

Many cases of food poisoning can be avoided if food handlers are careful about personal habits. Food is frequently contaminated by human beings. *Staphylococci* are found on the hands, especially in cuts or bruises, and pus contains large numbers of them. Fecal matter carries large quantities of *Salmonella* bacteria. Thorough and frequent washing of hands with soap removes much of this hazard. You should be careful to never lick your fingers or handle your mouth and never put tasting spoons back into food. Persons with colds and other contagious conditions should not work with food.

A frequent way in which food is contaminated is by _____ _____.

Human beings.

*

Salmonellae sometimes are found in eggs, pork, and the intestinal tract of poultry. On rare occasions, food poisoning has resulted from other foods picking up *Salmonellae* from a table top on which chickens were cut up for cooking. Most cases of food poisoning would be avoided if food handlers at every stage of production would be more careful about keeping everything clean that touches food.

Salmonellae may be found occasionally in _____, _____, and _____.

Eggs, pork products, poultry.

73

Some foods are poisonous to humans. Solanine, the green in potatoes exposed to sunlight, is a poisonous glycoalkaloid, but nearly 18 pounds of green potatoes would have to be consumed to make a person ill. Rhubarb leaves, fava beans, and ergot, a substance that grows on moist or moldy rye, are all poisonous to us. Many mushrooms and toadstools are poisonous. Unless you are an expert—and even the experts get fooled—you should not eat varieties that are not grown from approved cultures. Eating mushrooms and toadstools is the most common way people become poisoned from natural foods.

Which of the following is the most common cause of poisoning in human beings: mushrooms, rhubarb leaves, moldy rye, green potatoes, or fava beans?

Mushrooms.

*

Trichinosis is an illness caused by eating meat containing live *Trichinae*. The meat most likely to contain this parasite is pork, and it only rarely is found to have these organisms. Sometimes bear and rabbit also are infected with them. The one sure way of avoiding the possibility of developing trichinosis is to heat these meats to an interior temperature of at least 170°F. Federal inspection of meat does not detect the presence of *Trichinae*.

Although most pork will not contain them, an occasional carcass may be a health hazard because it contains ____; such organisms are killed by cooking to an interior temperature of ____.

Trichinae, *170°F.*

*

Oysters, clams, shrimp, and mussels may be poisonous at times when they have eaten a poisonous plankton. Areas where shellfish are sometimes contaminated with this type of plankton are the Gulf of St. Lawrence, the Pacific Coast, and the Bay of Fundy. The U.S. Public Health Service watches such waters and, when the plankton appear, all seafoods from the area are condemned. The problem occurs in the summer months. There are also some poisonous tropical fish.

Some shellfish, including ____, ____, and ____, may cause serious poisoning at certain times of year; shrimp also may be toxic at times.

Oysters, clams, mussels.

Some metals such as cadmium, which is used on plated utensils, and antimony, which is used on enamelware, can be poisonous. Tin, lead, and zinc can contaminate food and make it poisonous. Copper can cause gastric upsets when acid foods or carbonated beverages come in contact with the metal. Frequently, acids react with metals and then the metal may be carried into the food. For this reason, plated utensils that are beginning to wear through the plating to the core metal should be discarded.

*Some metals **are/are not** poisonous.*

Are.

*

Some household poisons are very harmful, in fact they may cause death, if eaten. Rat poisoning and other sprays and poisons may accidentally get into food if they are stored in the wrong place. All items of this type should be stored out of the kitchen in some area remote from all food. Other materials that are used in the kitchen may also be harmful if ingested. A dishwashing compound may be mistaken for cornmeal; liquid detergents might look like cooking oil. Store these items away from food so there is no possibility of confusion.

Should any special precautions be taken about the storage of items which might be poisonous or harmful if they get into food?

Poisons should be stored in another part of the house away from the kitchen and all food, preferably in a locked cabinet. Kitchen cleaning items and detergents should have a special storage place in the kitchen, and no food should be stored in this area.

*

REVIEW

1. Heat can be transmitted by ____, ____, or ____.

Conduction, convection, radiation.

2. *True or false.* When you are deep-fat frying, you are transmitting heat by convection.

True.

3. Heat from the sun is transmitted by ____.

Radiation.

4. The temperature of boiling water at sea level is *higher/lower* than of water boiled at 8000 feet elevation.

Higher.

5. *True or false.* Baking powder is as effective dry as it is when moist.

False.

6. *True or false.* When you are cooking food in fat, the temperature is probably higher than it is when you are cooking in water.

True.

7. A standardized recipe yields a *known/unknown* quantity of known quality.

Known.

8. Quantities of ingredients in a home recipe usually are given by *weight/measure;* those for quantity work often are given by *weight/measure.*

Measure, weight.

9. *True or false.* Milk can be flash pasteurized by quickly heating it to 180°F and then immediately cooling it to less than 50°F.

True.

10. *True or false.* Eggs cannot be pasteurized without coagulating the protein.

False.

11. People *are/are not* a frequent source of contamination in food.

Are.

12. Foods containing milk and eggs are much *more/less* likely to spoil than acidic foods.

More.

13. Dried foods are *favorable/unfavorable* for the growth of microorganisms.

Unfavorable.

14. Foods high in salt are *favorable/unfavorable* to the growth of microorganisms.

Unfavorable.

15. *True or false.* Freezing is not a very effective means of controlling microbiological growth.

False.

16. The dormant form of some microorganisms is called a _____.

Spore.

17. *Clostridium botulinum* produces a toxin that often *is/is not* lethal.

Is.

18. Home-canned foods *high/low* in acid are the most likely source of *Cl. botulinum.*

Low.

19. To maintain the safety of foods, avoid keeping them unrefrigerated more than _____ hours.

Four.

20. _____ are sometimes found in uncooked egg products.

Salmonellae.

21. Solanine, the green alkaloid in some potatoes, *is/is not* likely to make humans ill.

Is not.

3

Food from Plants

Many years ago, Spanish explorers searched for a fountain whose waters kept one young forever. Belief that the Fountain of Youth was in Florida arose because the natives there displayed remarkable health, vigor, and long life. Although the Spaniards never found the long-sought fountain, the health, vigor, and long lives of the early Florida Indians probably were directly related to their diets being rich in the fruits and vegetables growing there. Many regions where North American Indians lived did not have such a liberal supply; because of this, they were less healthy and vigorous and their lives were shorter.

What did the vegetables and fruits furnish that made these Indians so healthy?

In Chapter 1 we learned that fruits and vegetables are important in the diet because they can furnish a liberal supply of many needed minerals and vitamins and also supply bulk for good digestion.

*

Fruits and vegetables, by providing variety in color, flavor, and texture, make meals more appetizing. They add zest to breakfast and delightful contrasts in our other meals. They can play a variety of roles in a meal: as a tempting appetizer, a crisp salad, a colorful complement to meat in the main course, or a refreshing dessert.

Do you think that you could plan a well-balanced menu in which there were no fruits and vegetables?

It would be difficult to do and the foods would lack the variety in color, texture, and flavor needed to make the meal tempting and appetizing.

Fruits and vegetables are most nourishing and appetizing when top-quality produce is carefully and skillfully prepared and served. Knowledge of the best ways to select, store, prepare, and serve fruits and vegetables will make it possible for you to use these foods to maximum advantage in meal preparation. Unfortunately, fresh fruits and vegetables are delicate items that can deteriorate rapidly; nutrient losses can occur and flavor, color, and textural change may happen all too quickly. These foods are still living, even though they have been harvested.

Why is sweet corn that is cooked almost immediately after being picked so much better than if it first stands several hours at room temperature?

Corn, like other vegetables and fruits, is still living and continues to change even after picking. In vegetables, there often is a conversion of some natural sugar to starch within a matter of hours. Naturally, this causes a change in flavor.

*

STORING FRUITS AND VEGETABLES

After harvest, fresh fruits and vegetables continue to take in oxygen and release carbon dioxide; enzymes continue to work. In fruits, there is a gradual change in the pectic substances to more soluble compounds. As a result, there is a softening of cell walls. There is also an enzyme-catalyzed conversion of starch to sugar. These enzymatic reactions eventually end in spoilage. Even in a short time, enzymes can cause marked deterioration; asparagus toughens and loses flavor in a day, and corn-on-the-cob loses its sugar in a few hours.

True or false. Enzymes catalyze important reactions in vegetables and fruits.

True. In ripening fruit, enzymes promote important flavor and textural changes. Unfortunately these changes do not stop when fruits and vegetables are ripe, but they continue and result in spoilage.

*

Packing too tightly in storage may cause fruits and vegetables to spoil more quickly. During storage they need air for respiration. For optimum storage, moisture of the air needs to be controlled. Some moisture in the atmosphere will reduce moisture loss, but excessively moist air encourages spoilage. Cool storage temperatures effectively slow down enzymatic and chemical changes and thus permit longer storage periods.

Which of the following conditions are important to retain quality during storage of fruits and vegetables?
A. Not tightly packed.
B. Closely packed.
C. Cool temperature.
D. Room temperature.
E. Moderate moisture.

A, C, and E are correct.

*

Vegetables and most ripe fruits should be stored at temperatures just above freezing (approximately 33-38°F), although tropical fruits are best when stored at room temperature. If you wish to ripen fruits, hold them at room temperature to hasten enzymatic and chemical reactions essential to the ripening process. Potatoes should not be refrigerated because some of their starch will change to sugar, a change that impairs their mashing, baking, and frying qualities.

Why do most refrigerators have closed containers at the bottom for storage of fruits and vegetables?

This region of the refrigerator is normally cool enough to hold the food well without freezing it. The closed container helps to keep the air damp but not too humid around the items.

*

SELECTING FRUITS AND VEGETABLES

Root vegetables can be stored satisfactorily for quite long periods, but many fruits and green leafy vegetables in particular lose quality rapidly. When you purchase fresh produce, keep the perishability of the food in mind and buy only the quantity you can use before loss of quality is apparent.

Which of the following foods may satisfactorily be purchased in large quantities and stored for later use: peaches, tomatoes, potatoes, lettuce, Hubbard squash, peas-in-the-pod, bananas, onions, cauliflower, and rutabagas?

Potatoes, Hubbard squash, onions, and rutabagas.

*

We can purchase fruits and vegetables fresh, dried, canned, freeze-dried, frozen, pickled, and preserved. Canning preserves food by using heat to sterilize the produce and inactivate the enzymes. Dried foods are so low in moisture that they cannot spoil. We use salt or sugar to preserve some items; foods containing a salt concentration greater than 8

79

per cent or a sugar concentration greater than 50 per cent can be stored safely for a fairly long time. Frozen fruits and vegetables last because deteriorative changes occur very slowly in the frozen state. Foods can be safely stored for long periods of time only when some processing method, such as those just mentioned, is used to either destroy microorganisms and enzymes or to retard their action greatly.

To preserve foods, we must destroy the ____ and ____ or retard their action greatly.	*Microorganisms, enzymes.*

*

Canned vegetables are usually packed in water with about one per cent salt added. Sometimes they are vacuum-packed in only a small quantity of brine. Fruits are commonly packed in either light, medium, heavy, or extra-heavy sugar syrup. Fruits packed in the more dense syrups are usually higher grade, but there are exceptions. Current interest in weight control has stimulated production of canned fruits packed simply in water, in juice, or in an artificially sweetened liquid.

Match the items in column B with those in column A.

A	B	
1. *Vegetable water pack*	A. *About one per cent salt added*	
2. *Light syrup pack*		
3. *Vacuum pack*	B. *Syrup commonly used with higher grade fruit*	
4. *Extra-heavy syrup pack*		
	C. *Syrup with low quantity of sugar*	1. A.
		2. C.
	D. *Only a small quantity of brine added*	3. D.
		4. B.

*

The amount of food we get in a can or package is important. Some packers put more food in a can or package than others, and the wise buyer knows this. You can shop more wisely if you learn to note can size and label information. The quantity of food in a can is directly influenced by the size of the food, its specific gravity or density, and the amount of liquid on this food. The law requires that labels state the net weight, but this does not give much information on the ratio of solids in the can to liquid. Canned fruits and vegetables should be about 65 per cent food and about 35 per cent liquid. Table 3.1 summarizes the most common can sizes and how much they hold.

What is the closest size to a No. 303 Can: No. 300, No. 2, or No. 2½?

No. 300 and No. 303 are quite close in size. Because of this, these two cans are sometimes mistaken for each other on grocer's shelves; it is always best to read labels.

*

Table 3.1 Summary of Can Size and Contents

Size Can	Weight* (in ounces)	Volume (in cups)
No. 300	13½	1¾
No. 303	15½	2
No. 2	20	2½
No. 2½	28	3½
No. 46	46	5¾
No. 10	104	12¾

*Weight varies with the density of the product and the style of the product; fruit cocktail in a No. 2½ can will usually weigh 28 ounces, while sliced pineapple will weigh about an ounce less.

The size of the items in the can or package is important. Peas are sized by sieve size, from No. 1 sieve to No. 6 (largest). String beans, lima beans, beets, asparagus, and other canned vegetables are also sized. The number of pieces of fruit in a can should be stated on the label.

It is helpful when labels tell ____ ____ ____ and ____ ____ ____.

Size of items, number of pieces.

*

Federal labeling requirements are that pictures and sizes of items on the label must bear a relationship to the item in the can. Thus, if a label on a can of ripe olives says "extra large" and a huge olive is shown on the label, this is misbranding and the packer is liable to fine and imprisonment. An extra-large olive has a diameter of five-eighths inch.

81

If you saw on a can label a picture of an asparagus spear about five-eighth inch wide and the term used was extra large and you opened the can to find that the spears were five-eighths inch wide, would you feel the label misrepresented the size by using the term "extra large?"

You might, but this is the size for extra-large asparagus spears; the buyer has to learn that sometimes terminology for size used in foods can be misleading.

*

Labels on foods must comply with certain federal regulations. Food labels must state the amount in the package either by weight, volume, or count. The packer or distributor's address must be on the label. If the contents are below weight or standard quality, this must be noted. The packing medium, style of pack, and variety of item must be indicated.

According to federal regulations, the quantity of food in a package must be stated by _____, _____, or _____.

Weight, volume, count.

*

Some food items are carefully defined by the Food and Drug Administration as to ingredients and proportions; consequently they may be marketed by name without listing ingredients. Most food mixtures must have a list of all ingredients with the substance present in the greatest quantity listed first, the second most abundant ingredient listed next, and then continuing in decreasing order until everything is listed. Even the presence of any artificial or imitation flavors or colors must be stated.

Suppose the following label appeared on a package of dry chicken-noodle soup mix: noodles, hydrolyzed vegetable protein, monosodium glutamate, salt, autolyzed yeast, chicken extract, chicken fat, chicken, and spices. How much chicken do you suppose you would be getting?

You would get little chicken. In fact, the label tells you that there is even more salt and yeast than chicken.

*

Quality in canned fruits and vegetables is indicated by the color, appearance, flavor, clarity of the liquid, and defects. Form is important. Broken or mushy vegetables are of low quality. The variety of the vegetable and the locale where it is grown influence the quality. For instance, Blue Lake string beans are noted for their soft tenderness, excellent flavor, and lack of strings. However, Oregon-grown Blue Lake string beans generally are superior in quality to any grown elsewhere in the United States. Sweet corn grown in Minnesota is usually top quality. The variety of a vegetable can be selected to give you the qualities you like. For instance, the wrinkled varieties of peas retain their skins better than do June or early peas, but they are less sweet.

How can you tell which variety and area are best for canned items?

Although there are books on the subject, you can learn a great deal by looking carefully at foods when you use them, noting quality factors, and then looking at the label. Then the next time you buy, you can purchase the item again or avoid it.

*

Frozen vegetables ordinarily are packed without liquid unless they are frozen in a sauce. Packing in a sauce or butter has been found to reduce flavor losses somewhat. The recent appearance of a variety of frozen vegetables with sauces no doubt is improving vegetable consumption, but the price of such products is appreciably higher than is that of the vegetable alone.

True or false. Vegetables frozen in a sauce are of higher quality than are the vegetables selected to be frozen alone.

False. The same quality vegetables can be used for either product, but many people enjoy the variety available in the vegetables with sauces. The vegetables in sauces are frozen in plastic bags which help retain flavor for a longer period of time in storage.

*

Frozen vegetables should be carefully cooked before judging their quality. After cooking, they are judged on the basis of color, flavor, and texture. You will recognize that good cooking practices are necessary to retain quality in a vegetable. Unfortunately, good cooking cannot improve an inferior frozen vegetable. Learn to select frozen vegetables of good quality and retain this quality by careful preparation.

The quality of frozen vegetables should be judged after **thawing/cooking.**

Cooking.

83

When buying frozen fruits, you will find that price is greatly influenced by the quality of the fruit that was frozen and by the ratio of juice to fruit. The use of a high ratio of sugar to fruit tends to draw water out of the fruit, with the result that there are a large proportion of juice and a shriveled fruit when thawing is complete. Despite this drawback, frozen fruits are commonly packed in sugar to improve their flavor and keeping qualities. As a buyer, you should understand that you are purchasing some sugar as well as fruit. The ratio of fruit to sugar varies; usually berries will be packed with four parts of fruit to one of sugar (4:1), peaches and apricots may be packed in a 5:1 ratio, and apples and rhubarb in a 7:1 ratio.

If you purchased a 12-ounce package of frozen strawberries for 49 cents, how much are you paying for the sugar and how much for the strawberries?

Of the five parts, four are berries and one is sugar. This is about 10 cents per part, so you are paying about 10 cents for 2½ ounces of sugar and about 40 cents for 9½ ounces of strawberries.

*

The quality of canned and frozen fruits should be judged on the basis of texture, color, and flavor. Frozen fruits are thawed but not cooked before they are judged. The variety of fruit and the area where it is grown are important guides to quality when selecting canned or frozen fruits. For example, Marshall strawberries grown on the Pacific Coast are superior to almost any other frozen strawberry. They are red all the way through rather than having the white heart found in other berries. They also have a superior flavor and texture. Apples are another fruit where variety is important. If one wishes to purchase canned applesauce, the summer varieties are preferred. Canned or frozen apple slices for pies should come from apples that stay firm yet cook to a tender product. A Northern Spy apple does this.

Compare brand A of peaches with brand B.

	Brand A	*Brand B*
Weight	*18 ounces*	*17 ounces*
Price	*38 cents*	*34 cents*
Form	*Even slices*	*Broken slices*
Color	*Rich orange yellow*	*Greenish yellow*

Which would be the better buy?

Brand A would be a better buy in spite of the slightly greater cost per ounce because of the distinctly better quality.

*

When buying fresh vegetables, you should look for a bright, fresh appearance. Very mature vegetables may be lacking in flavor and be tough. Vegetables should be crisp and full of juice. Fresh vegetables of high quality will be firm, plump, free of soft areas, and a good color. A soft, flabby texture is a mark of deterioration in many vegetables.

How would you check the quality of fresh corn-on-the-cob, green peppers, cucumbers, parsnips, or potatoes?

Look at them carefully to be sure that they are a good color, show no signs of spoiling, and are rigid rather than flabby when touched.

*

Other tests are used to judge quality in vegetables. For instance, egg plant should have a soft, velvety sheen to it. Dry onions, when shaken in a sack, give a loose, dry sound. A knife cutting through a beet held close to the ear will make a grating sound if the beet is tough. Deep leaf scars on parsnips, turnips, or rutabagas indicate age. Good quality in head lettuce and cabbage is indicated by a firm feel when the head is pressed gently.

Are experience and observation useful when you are buying fresh vegetables?

Very much so, but books provide important suggestions to help make you a more observant and discriminating consumer. In this way you can learn rapidly from your shopping experiences.

*

Appearance is an important indicator of quality in fresh fruits. A sunken, wrinkled, or dull appearance indicates old fruit. Berries should be firm and bright appearing. Look under the box to see if there is any mold or leakage of juice. Fruit with these signs has very poor keeping quality. Lift apples, oranges, lemons, grapefruit, and other items in the hand and see if they feel heavy for their size. They should because this tells you that they are juicy. Quality cranberries bounce when dropped.

Would you prefer to purchase fruits and vegetables by the dozen or by the pound?

By the pound would be preferable in most instances.

*

Usually canteloupe and pineapple have an aroma very similar to their desired flavor when they are ripe. The center spikes from pineapple should pull out readily when this fruit is ripe. A good ripe watermelon sounds hollow when tapped, but this is not an infallible test.

85

True or false. Odor is a useful guide to the ripeness of canteloupe and pineapple.

True.

*

Ordinarily we want the younger, just mature fruits and vegetables. Apples that are very large or very small frequently lack the flavor of the medium sizes. Size is usually not a mark of quality. Small asparagus can be as flavorful, tender, and delicious as large sizes. Open buds on cauliflower or broccoli indicate an older vegetable that is no longer at its best.

Are there tests for quality in fruits and vegetables other than eating them?

Yes. Odor, appearance, and feel are useful guides in the market, but eating is the final critical test.

*

Fruits and vegetables are seasonal items. Axiomatically, they are highest in quality and lowest in price at the peak of their season.

True or false. High-priced fresh fruits and vegetables are usually best in quality.

False. High price is more likely an indication of a poor crop or an out-of-season item.

*

Many fruits and some vegetables are not completely ripe when they are harvested. These are picked while green and still firm so they can be shipped without damage. Select fruits that will be ripe on the day you plan to use them. Bananas that are flecked with brown specks are just right for eating, but if they have to be held for several days, they will be too soft and perhaps spoiled. A pineapple that is ripe will not hold for more than a day or so; neither will a melon. Sometimes pears are put onto the market unripened. A few days at room temperature will be required before they are ripe enough to be juicy and sweet. Avocados also may be put on the market unripened.

True or false. Purchase fruits and vegetables only at optimal ripeness.

False. Purchase them so that they will be at optimal ripeness when you plan to use them.

*

It is important to select the type of fruit or vegetable that is best suited to the preparation you plan. A Delicious apple is fine to eat raw and in salads, but it bakes into a tough-skinned, shrunken product. A Roman Beauty apple is soft, delicate, and mellow in flavor when baked. Some Irish potatoes are particularly good for baking, mashing or deep frying. If you wish to prepare potatoes in these ways, select those high in starch and low in moisture. Such potatoes have a dry, mealy quality. When baked or mashed, they are light and fluffy. They also make an excellent French-fry which is done in the center by the time the outside is golden brown. Russets and Katahdin varieties are good potatoes for this purpose. These usually are grown on the Pacific Coast or in Maine and Long Island.

Russets or other mealy potatoes are well suited for _____, _____, and _____ _____.

Baking, mashing, French frying.

*

A potato that boils well or is well suited for potato salad and sautéing is more moist. The dry, mealy potatoes that are good for baking, mashing, or deep-frying break up easily when boiled or fried, while the more moist types do not. Irish Cobblers, Red Triumphs, and some others grown in the Midwest are waxy, moist potatoes. New or immature potatoes of either the moist or dry varieties are quite moist and are best for boiling, potato salad, or similar purposes where it is important to hold their shape. They do not bake, mash, or deep-fry well. A potato of specific gravity 1.08 or more will be good for baking, mashing, or deep-frying. This is a heavy potato with a high starch content; a potato with a specific gravity of 1.07 or less is a moist potato—it will float in a salt brine made of two ounces of salt to a pint of water.

Probably 1.08 or more. This potato would be good for baking, mashing, or deep-frying.

If the potato sinks in such a salt brine, what specific gravity would you say it had?

*

GRADES OF FRUITS AND VEGETABLES

Fresh fruits and vegetables that are of good quality are graded U.S. No. 1 or A; the remainder of the crop that is marketable usually is graded U.S. No. 2 or B. There are grades higher than U.S. No. 1. When the crop is superior, grades of AA Fancy, or Extra Fancy may be found. In poor years these are not on the market, but U.S. No. 1 usually is available. The grading for fresh fruits and vegetables may not be consistent.

87

Fresh fruits and vegetables of good quality will be graded
U.S. No. 1/U.S. No. 2 *or* **U.S. Grade A/U.S. Grade B.**

U.S. No. 1,
U.S. Grade A.

*

The federal government has established letter and market grades for canned and frozen fruits and vegetables. From highest to lowest quality, these are:

Fruits	Vegetables
U.S. Fancy or	U. S. Fancy or
U.S. Grade A	U.S. Grade A
U.S. Choice or	U.S. Extra-standard or
U.S. Grade B	U.S. Grade B
U.S. Standard or	U.S. Standard or
U.S. Grade C	U.S. Grade C

True or false. The grades for canned or frozen fruits and vegetables are the same.

False. They are the same except for the second quality; U.S. Choice is the grade for fruits in this quality and U.S. Extra-standard is the grade for vegetables in this quality.

*

The grades for dried fruits and vegetables may be U.S. No. 1 for top quality or U.S. Grade A. Sometimes no federal grade is used, but trade grades are used.

The grades for dried fruits and vegetables **are the same/ differ** *from the grades for canned and frozen fruits and vegetables.*

Differ.

*

PREPARING AND CLEANING
FRUITS AND VEGETABLES

Fresh fruits and vegetables need to be cleaned; usually some preparation is done before they are cooked and served. Soil and harmful pesticide residues may still remain. Both should be washed off. Some fruits can tolerate only gentle washing, but fortunately most vegetables can easily withstand a good scrubbing with a stiff brush under running water. Extremely dirty vegetables will be easier to clean if you soak them briefly in salt water (one tablespoon of salt per quart of water). Leafy green vegetables such as spinach should be given a good soaking, followed by thorough up-

and-down agitation in plenty of water. The greens should be gently lifted from the water, the dirty water drained away, and fresh water used to rinse them; this cleansing is continued until the water is completely free of any sediment or dirt. Greens also need to have the heavier stems and any undesirable portions removed.

True or false. A light rinse is sufficient to clean most fruits and vegetables.

This is true for some fruits, but most fresh produce needs more thorough cleaning.

*

If you are planning to store fresh fruits and vegetables, look carefully for spoilage and bruises. These are signs that spoilage may occur soon. Trim off nonedible portions, including all coarse stems and leaves. Occasionally Brussels sprouts, broccoli, and other fresh vegetables may contain small mites, worms, or other insects. These frequently can be persuaded to depart to the bottom of the sink by briefly soaking the vegetable in salted water. Then they can be rinsed away.

What method of cleaning in column A would you use for the products listed in column B?

A	B	
1. Rinsing in a colander	*A. Brussels sprouts*	
2. Soaking in salt water to remove mites	*B. Carrots*	*1. C.*
3. Lifting up and down in water to remove soil	*C. Strawberries*	*2. A.*
4. Scrubbing with a brush	*D. Spinach*	*3. D.*
		4. B.

*

A rather large proportion of some fresh fruits and vegetables is waste material to be discarded. Only about 35 per cent of a cauliflower may be left when the outside portions are removed and the head is ready to cook. The small amount left after shelling peas or Lima beans may surprise you. Keep this waste in mind when you purchase fresh items. Often it may be less expensive to use a canned, frozen, or other processed item than the fresh. If cost of labor for preparation is a factor, it will be less expensive in many cases to buy processed rather than fresh produce.

If you could purchase fresh cauliflower at 45 cents a head (20 ounces after trimming) and a 10-ounce package of frozen cauliflower for 24 cents, which would be less expensive?

The fresh is 2.25 cents per ounce and the frozen is 2.4 cents per ounce. When you add the value of your labor to the fresh, the frozen product is definitely less expensive.

*

To prevent the darkening of some fruits and vegetables after they are pared, cook such items as sweet potatoes, yams, salsify, and Irish potatoes in their jackets and then pare them. Heat destroys the enzymes which cause the discoloration. Another way of avoiding darkening is to pare the vegetable while keeping it completely immersed in water containing a little vinegar. Sometimes browning is prevented by paring and then putting the item immediately under water and holding it there until needed. This procedure is often used to prevent discoloration of potatoes pared and cut for French fries.

Salsify, often called oyster plant, discolors very easily. How would you pare it so the browning could be avoided?

Pare it in water containing some vinegar.

*

Fruits may discolor quickly after they are peeled. You can prevent this by moistening the surface with an acidic juice such as lemon or pineapple. Ascorbic acid is used sometimes to prevent browning, and it is also used as a dip for potatoes to prevent darkening without having to soak them. Some other antioxidants are on the market that prevent discoloration in fruits and vegetables.

List three ways by which you would prevent discoloration of either fruits or vegetables after they were pared.

1. Cook first.
2. Use acid either in water or on the item.
3. Use anti- oxidants.

*

Many valuable nutrients can be lost from fruits and vegetables when we prepare them. Most of the minerals and vitamins in vegetables are water soluble. Therefore, if you soak them, these nutrients are lost by leaching. Many fruits and vegetables concentrate their minerals and vitamins just below the skin surface. If you pare too deeply, you can lose a large portion of these. Some nutrients are destroyed easily by oxygen in the air. Therefore, if you dice, chop, or mince and allow the fruit or vegetable to stand, there will be a substantial loss by oxidation.

Match the method of cooking or preparation in column B with its probable effect on nutrients listed in column A.

A	B
1. *Increases nutrient loss*	A. *Mashing a banana*
2. *Retards nutrient loss*	B. *Squeezing orange juice and holding several hours*
	C. *Cooking potatoes in their jackets*
	D. *Dicing potatoes and soaking them*
	E. *Paring an apple for salad*
	F. *Dipping pared potatoes into ascorbic acid.*

1. A, B, D, E.
2. C, F.

*

COOKING FRUITS AND VEGETABLES

When fruits and vegetables are cooked, their color, flavor, and appearance are modified and their texture softens so they can be eaten more easily. Good cooking of vegetables and fruits will optimize nutritive value, natural flavor, aroma, texture, and appearance. The cooking method used should be selected with care to bring out the best qualities of the food. Unfortunately, a method may be best for certain factors but not for others. Then the cooking method selected is a compromise in which we may sacrifice some color, nutrients, or flavor.

When deciding on a method for cooking fruits and vegetables, you should consider ____, ____ and ____.

Color, nutrients, flavor.

*

Texture

Cellulose and pectins give structural strength or firmness to fruits and vegetables. Cooking softens cellulose and dissolves some pectins. Moisture must be present for this softening to occur.

*Moist heat **dissolves/softens** cellulose and **dissolves/softens** pectins.*

Softens, dissolves.

*

The time it takes to cook a fruit or a vegetable is related to its cellulose content. Artichokes have a high content of cellulose, thus requiring a cooking time of 45 to 60 minutes compared with tomatoes whose lower cellulose content permits them to become tender in only a few minutes. This is true also of the stalk of broccoli compared with its top.

*Vegetables with a high _____ content require **more/less** time to cook than do those containing less.*

Cellulose, more.

*

The amount of cellulose in a fruit or vegetable influences the choice of cooking method. Some vegetables cannot be baked, broiled, or fried because these cooking methods do not give added moisture to help soften their structure. Steaming and boiling may be the two methods that must be used. Sometimes a vegetable is parboiled to soften the cellulose; then it can be finished by one of these three dry heat methods: baking, broiling, or frying.

Vegetables with a high cellulose content are best cooked by _____ or _____; other fruits and vegetables may also be cooked by _____, _____, _____.

Boiling, steaming, baking, broiling, frying.

*

A fruit or vegetable is considered done when it is quite tender and easily cut. Overcooking causes mushiness and detracts from palatability. You can tell when the food is done by piercing it with a fork or by drawing the item to the side of the pan and seeing how easily it can be cut. Appearance may also be a guide to doneness. Authorities recommend that vegetables be cooked just until tender. This is done not only to give a better texture to the product, but also to minimize nutritive loss and retain better flavor and color. The marvelous flavor and texture of many vegetables that the Chinese cook result from leaving some "bite" or small core of crisp texture.

*When fruits and vegetables are overcooked, they become _____; the longer cooking also causes **more/less** loss of nutrients.*

Mushy, more.

The stems of some vegetables contain more cellulose than the leaf or tips. Obtaining doneness in the parts higher in cellulose without overcooking the more tender portions becomes a problem. The thicker portions of spinach stems can be discarded to solve this. You may pare the stems of vegetables such as broccoli or asparagus, removing the thickest portions of cellulose, or you may split the stems so the heat penetrates more quickly and cooking occurs more rapidly. It also works well to tie the vegetables together with string and stand this bunch in the pan with the tips above the water. Then, tip them on their sides so all of the vegetable is covered with water for the last five minutes. One other solution is to cut the stems from the tips, start cooking the stems first, and add the tips for the last five minutes.

Which of the following is **not** *recommended when boiling broccoli?*
A. *Cover all the broccoli with water and boil until the stems are done.*
B. *Pare stems to get rid of some cellulose.*
C. *Split stems to encourage more rapid cooking.*
D. *Cook stems first and then immerse and cook tips with stems.*
E. *Cut stems from tips and cook them first; add tips later.*

A. The broccoli flowers will be too done if they are boiled as long as the stems.

*

Acids delay softening of cellulose during cooking of a vegetable. If you cook dried beans or even some fresh vegetables in an acidic sauce such as tomato or with an acid ingredient such as molasses, the cooking time will be greatly extended. Salts such as calcium chloride or magnesium chloride delay softening of cellulose. Very hard water may contain so many mineral salts that the cooking time is extended.

_____ *and mineral* _____ *delay softening of cellulose.*

Acids, salts.

*

In commercial food processing, knowledge of the behavior of cellulose is used to produce better products. When tomatoes are canned commercially or pickles are made, calcium chloride may be added. This salt firms the tomatoes and retains the desired crispness in pickles.

Why is calcium chloride added during canning of tomatoes?

The calcium chloride prevents the tomatoes from breaking up when cooked and canned.

93

During cooking, water passes back and forth between the cooking liquid and the cells. If fruit is cooked in a heavy sugar syrup, water will be drawn out of the fruit and into the syrup, causing the fruit to shrink and shrivel. Long cooking at a hard boil accentuates the problem. Best results are obtained by simmering fruits in a moderately sweet syrup just until tender. After cooking, additional sugar can be added to the syrup if desired. Sometimes, to retain the shape of an item such as watermelon pickles during cooking, it is cooked in water first until tender and then it is cooked in syrup. This permits a more rapid entry of the sugar into the product and less shrinkage occurs. However, many fruits will become mushy and fall apart when they are cooked without some sugar in the water.

A heavy sugar syrup causes water to **come out of/go into** *the fruit and the finished product* **falls apart/shrivels and shrinks.**

Come out of, shrivels and shrinks.

*

If you add soda to water in which a vegetable is cooking, the water becomes alkaline. This causes a vegetable to become soft very quickly, and it is often unpalatable. Excessive alkali causes sliminess. In areas where the water is very alkaline, a pinch of cream of tartar added to the water may prevent excessiving softening of the vegetable.

Why does cream of tartar do this?

Cream of tartar is an acid and neutralizes the alkalinity of the water.

*

Cooking time is reduced if dried legumes such as peas, beans, and lentils are cooked in slightly alkaline water. Thus, in areas where the water is alkaline, but does not contain calcium or magnesium chloride, the cooking time is shortened. An eighth of a teaspoon of soda to a quart of nonalkaline water gives a favorable softening action for dried legumes. The slight alkalinity does not harm nutrients, but more soda will destroy the normally high thiamine content of legumes.

If the water is already alkaline, should you add soda?

No. Excessive alkalinity then would be a problem.

*

If dried legumes are presoaked, they will soften more quickly. The first steps are thorough cleaning and sorting, followed by a soak in fresh water. Use this soaking water to cook the legumes so nutrients leached into the water are retained. Dried legumes may be slightly undercooked if they are to be baked later in a sauce. "Converted" dried legumes that cook in a much shorter time than regular legume products are on the market. These have been treated with potassium phosphate and take only about 40 minutes from the dry stage to cook to a tender product.

Which will not shorten the cooking time of dried beans, peas, or lentils: (a) acid, (b) alkali, (c) presoaking, or (d) treating with potassium phosphate?

Acid.

*

Most dried fruits and vegetables should be presoaked and then cooked in their soaking water. Presoaking helps to produce a plump product. If dried fruits or vegetables are not presoaked, the final volume after cooking is less.

Why are dried foods cooked in the water used for soaking?

To retain the nutrients that may have been leached out in the soaking.

*

In addition to softening of cellulose, other things happen when fruits and vegetables are cooked. The starch in vegetables gelatinizes (swells by taking up water). A boiled or steamed potato gains weight when it cooks because the starch absorbs moisture as it gelatinizes. Dried beans enlarge because of the water they absorb. Although the effects of the coagulation of proteins in cooking vegetables or fruits are not obvious because these foods contain only a small amount of protein, this coagulation does help give firmness to the item. This is of particular significance in legumes, which are good sources of protein.

True or false.   The gelatinization of starch and the coagulation of proteins in the cooking of vegetables help to give some firmness to a vegetable.

True.

*

Color

The colors in fruits and vegetables are caused by pigments. Green comes from a pigment called chlorophyll. The yellow and orange colors come from carotenes. Most red colors come from anthocyanins such as in red cabbage, plums, cherries, beets, and others, but lycopene is the pigment in tomatoes and it is a carotenoid pigment. While we may not think of white as a color, the whiteness of some vegetables is affected by pigments called flavones.

Match the color with the pigment.

Pigment	Color	
1. Anthocyanin	A. Red (beets)	1. A.
2. Carotenes	B. Red (tomatoes)	2. C.
3. Flavones	C. Yellow (peach or carrot)	3. E.
4. Lycopene	D. Green (spinach)	4. B.
5. Chlorophyll	E. White (onion)	5. D.

*

Pigments are chemical compounds and these sometimes undergo chemical changes during food preparation. Heat, acid, or alkali can cause these reactions to take place. The result is a change in the color of the food.

The color in fruits and vegetables can be affected by:
A. *Heat*
B. *Alkali*
C. *Acid*
D. *All three.*   D. All three.

*

Chlorophyll (green) is easily altered by moist heat. When acid is present, the natural green color changes to an olive green. This olive-green color also develops when vegetables are cooked a long time. To preserve chlorophyll, cook the vegetable as short a time as possible and do not add any acid ingredient. Although alkali maintains the green color, you will not want to add it because it makes the vegetable mushy and destroys thiamine. Chlorophyll is not water soluble.

Chlorophyll is altered by ____ ____ and the change is speeded if ____ is present.   Moist heat, acid.

*

If green vegetables are dropped into a small quantity of boiling salted water and boiled without a lid until just tender, the chlorophyll will not change to an olive-green color. Vegetables contain volatile acids that can escape in the vapor if the vegetable is boiled in an uncovered pan. If the pan is covered, this acid will condense and return to the cooking water where it will speed the development of the olive-green color. Therefore, the pan is covered while the water is being heated to boiling, and the lid is removed when the vegetable is added to the boiling water.

*To boil green vegetables, place them in **boiling/cold** salted water **with/without** a cover.*

Boiling, without.

*

When vegetables are steamed, some of the acid volatilizes and comes in contact with the chlorophyll, and the color gradually changes. Under pressure, the temperature of steam is quite high and it is a still greater problem to maintain the desired green color. However, green vegetables can be steamed satisfactorily if they can be cooked quickly without packing in the steamer. Spinach packs so that steam penetrates poorly and uneven cooking occurs.

True or false. Steaming frequently causes some discoloration of green vegetables.

True.

*

The white pigments, the flavones, are found in vegetables such as onions, cauliflower, and potatoes. They are white in acid and yellow in alkali. Thus, onions cooked in hard water may turn slightly yellow. If we add a pinch of cream of tartar to such water, the vegetable will be white. Flavones are water soluble.

If you cook onions in hard water and the water and onions turn yellow, is this proof that the flavones are water soluble?

Yes.

*

Anthocyanins are red in an acid medium and blue or dirty purple in an alkaline one. Red cabbage, beets, berries, and cherries cooked in an acid medium remain bright red. If these foods are cooked in hard water, they can turn an undesirable color, but usually the acid in the fruits is sufficient to overcome the alkalinity of the water and keep the color red. Anthocyanins are water soluble.

Anthocyanins are _____ in an acid medium and blue or dirty purple in an _____ one.

Red, alkaline.

*

Red cabbage and beets are less acidic and may discolor in hard water unless a little acid is added. A little vinegar in the water retains the color of beets; red cabbage cooked with tart (acidic) apples has a pleasing red color.

97

Why does red cabbage cooked in extremely alkaline water turn green?

Red cabbage contains both anthocyanins and flavones. The anthocyanins in an alkaline medium turn blue and the flavones yellow; the combination gives a green color.

*

Lycopene, classified as a carotinoid, is very stable to heat, acids, and alkalies. It is not water soluble. Lycopene is closely related chemically to the carotenes, which we shall see are also quite stable to heat, alkalies, and acids.

The red color of some grapes used to make wine does not appear when the grapes are first crushed and fermentation starts, but the red color intensifies as fermentation proceeds. Could we deduce from this that lycopene, the red color in these grapes, is soluble in alcohol but not in water?

Yes. As alcohol is built up, the red color is dissolved and brought into the wine; rosé wines, which are a light pink, are made by permitting the skins to ferment in the wine only overnight before being removed. If they were not removed, more lycopene would be extracted into the wine and it would become a deep red.

*

Anthocyanins are not discolored by heat although long cooking may leach out enough of the pigment to cause some loss of color. Some compounds can combine with anthocyanins to give undesirable reactions. A red fruit punch may become a muddy, murky color when tannins from tea, iron, or some other substance from canned fruit juice combine with the red pigments of the fruit. There is little that can be done to correct this once it happens.

How does the iron get into the canned fruit juice?

Some fruit juices such as pineapple and cranberry may pick up iron salts from contact with machinery and equipment used in processing the juice.

*

Carotenes are a group of yellow or orange pigments found in vegetables and fruits such as squash, peaches, apricots, carrots, and rutabagas. They are also present in green vegetables, but the green masks their color and they cannot be seen. Carotenes are stable to heat, acids, and alkalies and are not water soluble. They are important nutritionally as precursors of vitamin A.

Complete the following chart where the dashes appear.

Pigment	Stability to Heat	Solubility in Water	Color	
			In Acid	In Alkali
1. Chlorophyll	Unstable	_____	_____	Bright green
2. Flavones	_____	Soluble	White	_____
3. Anthocyanins	Stable	_____	_____	Blue or purple
4. Lycopene	Stable	Not soluble	_____	_____
5. Carotenes	_____	_____	Yellow or orange	_____

1. Not soluble, olive green.
2. Stable, yellow.
3. Soluble, red.
4. Red, red.
5. Stable, not soluble, yellow or orange.

*

Nutrients

Some nutrients are lost when fruits and vegetables are prepared for cooking, and still more loss occurs during cooking. Leaching of water-soluble vitamins is a cause of vitamin loss in boiled vegetables. Heat and alkali cause appreciable destruction of ascorbic acid and thiamine.

Leaching of water-soluble _____ occurs when fruits and vegetables are cooked in water; heat and alkali destroy the two vitamins _____ and _____.

Vitamins; ascorbic acid, thiamine.

*

Nutrient loss can be minimized if cooking time is kept short because there is less time for heat destruction, leaching, and chemical reactions to occur. Baking, steaming, broiling, and frying are desirable cooking methods that cause less leaching of water-soluble vitamins than does boiling. However, the high temperatures of broiling and frying and the longer cooking time required for baking and steaming also cause vitamin losses. No cooking method eliminates vitamin losses. Thus, if potatoes are steamed under pressure, leaching loss is minimized, but the higher cooking temperature causes some losses.

99

True or false. To conserve vitamins in vegetables, cooking time should be kept as short as possible.

True.

*

Fruits and vegetables high in moisture can be cooked in only a small quantity of water because they can furnish the moisture needed to cook them. A heavy pan, a tight cover, and controlled heat are necessary to avoid scorching vegetables cooked with little added water. Greens such as spinach, turnips, and chard will usually have enough water clinging to them after washing to start cooking, and after that the greens themselves will furnish the moisture needed. Other vegetables need to have enough water added to just bubble over them when they are boiling. Vegetables that should be cooked without a cover to maintain color or improve flavor may need to have water added to offset evaporation.

Why is it desirable to cook vegetables in only a small amount of water?

To minimize loss of water-soluble vitamins.

*

If the cooking water from fruits and vegetables is used, the nutrients leached into the cooking water will be recaptured. Fruit juices can be used in punch, desserts, and gelatin products, and vegetable waters are used for gravies and soups. Some cooks save the liquids drained from fruits and vegetables and refrigerate them in covered containers for later use in recipes.

If you boiled some onions for a meal and were to make a gravy for meat balls, would it be a good idea to use the water from the onions as part of the liquid needed for the gravy?

Yes, it would add flavor as well as nutrients.

*

At times it is necessary to cook fresh vegetables ahead of service. The quality and nutrient loss would be too great if these vegetables were held at a serving temperature until needed. Therefore, it is necessary to cool them down and later rewarm them. The recommended procedure is to cook just until done, drain, and immediately plunge into cold water to cool rapidly and stop cooking. Then, the vegetable is drained and immediately spread on a shallow pan to cool rapidly in the air. Some of the water drained from the vegetable should be used for reheating the vegetable for service.

If a vegetable is precooked, cooled as recommended, and rewarmed in some of its cooking liquid, is the nutrient loss higher than when cooked and served immediately?

Yes, somewhat, and there is a small loss of flavor and color as well. This procedure is only followed when it is not possible to prepare the vegetable just before serving.

*

Flavor

The method used to cook vegetables and fruits should be suited to the flavor strength. Most fruits and some vegetables are very mild in flavor while others have strong flavors.

Would you cook green peas in the same way you cook onions?

No. The peas have a very mild flavor and the onions a strong flavor; with peas we want to hold all the flavor we can, but some weakening of flavor in onions is desirable.

*

Many flavors in fruits and vegetables are highly volatile organic compounds (esters, aldehydes, and ketones) that escape easily during cooking. Fats and oils readily absorb these compounds. If a vegetable is cooked in a sauce in which there is fat, flavors from the vegetable are trapped in the sauce. We frequently sauté onions, garlic, or other vegetables so we can extract their flavors into the fat and then add this to our other dishes for flavor.

Why is it helpful to add a tablespoon of butter, margarine, or salad oil to the water in which you are to boil mild-flavored vegetables?

These fats and oils help to capture the flavors of the vegetables and prevent their loss. In addition, butter and margarine contribute pleasing flavors of their own to augment the natural vegetable flavor.

*

Mild-flavored vegetables should be cooked in a minimum quantity of water so their flavor will be conserved. The use of a lid is recommended unless the cover would harm the color, as is the case with green vegetables. As short a cooking time as possible also should be used. This is accomplished by adding the vegetable to boiling water and by avoiding overcooking. Steaming is a good method of cooking mild-flavored vegetables if they do not pack.

Mild-flavored vegetables are best when cooked in a **small/ large** *quantity of water for a* **short/long** *time.*

*

Strong-flavored vegetables are boiled in a fairly large quantity of water to dilute the flavor. Some volatile flavors will escape, and the flavor will be more mild if no cover is used. If you dice or mince strong-flavored vegetables, more flavor will be lost during cooking. The time of cooking varies according to the type of vegetable. Most strong-flavored vegetables such as cauliflower, cabbage, turnips, rutabagas, and kohlrabi are cooked for as short a time as possible. This is done because cooking intensifies their strong flavors. Steaming can be used for cooking these vegetables, but the time of cooking must be watched very carefully to avoid a truly objectionable flavor; even at best the flavor is very strong. Members of the onion family behave in just the opposite way. Their pungent flavors become more mild with longer cooking.

All strong-flavored vegetables are cooked in a **small/large** *quantity of water, but members of the cabbage family are cooked as* **short/long** *a time as possible while members of the onion family are cooked for a somewhat* **shorter/ longer** *time.*

*

Members of the cabbage and onion families keep well in storage, but they do build up more and more of their strong flavors. Therefore, if you wish to avoid intense flavors in these vegetables, use those that have been freshly harvested. Since older vegetables usually have a stronger flavor than young ones, select young onions, cabbage, turnips, and kohlrabi if you wish the most mild flavor.

Select from each pair the more effective means of reducing the pungent flavor of garlic in a spaghetti sauce.
1. *A. Chop it. B. Mince it.*
2. *A. Add it to the sauce just before service. B. Add it early in the cooking period.*
3. *A. Use fairly young garlic. B. Use some garlic that is quite old.*
4. *A. Purchase fresh garlic that has just been harvested. B. Purchase garlic which has been stored a considerable time.*

Older vegetables and fruits usually have poorer flavor than younger ones. This is due to an increase in bitterness and other flavor developments with maturity; sweetness declines as starch is formed from the sugar in maturing vegetables. Texture differences may be marked and the older vegetables may, upon cooking, be rather unpalatable and too soft or too woody. However, fruits and vegetables that are too young may lack flavor. Some immature fruits may be highly acid and tough; the flavor of some vegetables that are quite young may be insipid. Other factors in addition to maturity may influence flavor. Some persimmons are not good until a frost has come. Parsnips are left in the ground until after a sharp frost, because this causes the parsnips to change starch to sugar and thus we have a sweeter parsnip.

Could we summarize this statement by saying that to obtain optimum flavor, it is wise to select fruits and vegetables that are neither too young nor too old but just at the peak of their flavor development?   *Yes.*

*

Cooking Methods

We can boil, steam, bake, sauté, deep-fry, or broil fruits and vegetables. The cooking method should be suited to the type of fruit or vegetable because all do not respond equally well to each cooking method. For instance, some varieties of potatoes boil well but are very poor for mashing, baking, or French-frying. Others boil less well, but are better for mashing, baking, or French-frying.

If you had to guess, which would you think would be the method suited for cooking most fruits and vegetables? Which can be used satisfactorily for only a limited number of fruits and vegetables?   *Boiling, broiling.*

*

The method of cooking is also varied because of the type of processing a fruit or vegetable has received. Frozen and canned vegetables are cooked differently from dried ones. Dried fruits are cooked much differently than fresh ones.

Cooking method is varied to suit the type of vegetable and the _____ it has undergone.   *Processing.*

103

Processed fruits and vegetables. Often canned fruits are served as they come from the can and frozen fruits simply are thawed before serving. However, these fruits may be cooked in a dessert or sauce. The usual practice is to drain the juice from them, thicken or otherwise prepare the juice as desired, and then add the fruit. Frozen fruits have a much more natural fresh flavor and better color when handled in this manner. A frozen loganberry or frozen cherry pie is much improved if only the juice is heated and the uncooked fruit is then added to the thickened juice, carefully folded in, and put into the pie for baking. Even though the fruit is cooked somewhat in the baking of the pie, the flavor and color are still good. Frozen fruits served without cooking have a more pleasing texture if served with just a bit of the frost still in them.

When you cook frozen or canned fruits, cook **the juice only and add the fruit/the juice and the fruit together.**

The juice only and add the fruit. There will be exceptions, and the recipe directions usually will indicate the preferred method.

*

As we have previously noted, dried fruits and vegetables are better when presoaked and then cooked. This gives a more plump, tender, and frequently more flavorful product. Soak dried fruits an hour and then simmer them in a covered pan until tender and plump. Dried legumes cook more quickly if they are first soaked overnight.

Why is a cover used when simmering dried fruits?

To retain volatile flavors.

*

Canned vegetables need only rewarming for service. However, it is advised that some imagination and care be taken so that the vegetables arrive at the table prepared in an interesting and flavorful manner. The method of just warming and serving canned vegetables is all too common and does not lead to a genuine liking for vegetables in many individuals.

True or false. Canned vegetables need only warming to be ready for service.

True, but they are more tempting if they are used imaginatively.

*

Frozen vegetables are usually broken into pieces and then dropped into boiling salted water and cooked quickly. With the exception of corn-on-the-cob and perhaps some greens, they are not thawed. Corn-on-the-cob needs thawing so the

104

center will be warm when the outside is cooked. Large blocks of greens may overcook on the outside before the inside is done if they are not thawed before cooking. The cooking time for frozen vegetables will be shorter than for fresh. Frozen vegetables have been blanched to inactivate the enzymes before freezing, and this partially cooks them.

Frozen vegetables are cooked very much like fresh vegetables, but frozen vegetables require a **longer/shorter** *cooking time.*

Shorter.

*

Boiling. The boiling of fruits and vegetables has been covered under other sections of this chapter and is only summarized here. Strong-flavored vegetables should be cooked in an excess of water without a cover to reduce their flavor. Mild-flavored vegetables are best cooked in just enough water to cover them. A lid is used unless the vegetable is green. Vegetables should be placed in boiling salted water and cooked until just tender. As we have noted, fruits are often simmered with additional sugar to retain their shape; sugar is withheld if you wish to have the fruit broken up as for applesauce. Gentle simmering rather than boiling also helps to retain shape.

Yes. Potatoes and some others will break up if the boiling is hard and vigorous.

Would you think that hard boiling would break up some vegetables?

*

Steaming. Vegetables can be steamed under pressure or at atmospheric pressure. At atmospheric pressure it takes longer to steam a vegetable than it does to boil it, but this method is often desirable because it reduces loss of vitamins due to leaching. As steam pressure builds up in a pressure cooker, the cooking temperature rises and cooking time is reduced. The loss of vitamins caused by high temperatures is offset by the shorter cooking time which conserves flavor and nutrients.

In steaming at atmospheric pressure, the cooking time is **longer/shorter** *than in boiling vegetables.*

Longer.

*

In steaming, one needs to be sure that the steam can penetrate evenly through the mass being cooked. Potatoes, corn, carrots, and similar vegetables are excellent when steamed.

105

Potatoes **do/do not** steam well, but spinach cannot be steamed satisfactorily because the steam penetrates **too quickly/unevenly**.

Do, unevenly.

*

When you use a pressure cooker, be sure it is in proper working order. These cookers are equipped with a pressure gauge that can be adjusted for the desired pressure. When this pressure is exceeded, a safety valve opens and allows the excess steam to escape. Both the gauge and valve should be checked to see that they work properly and that food or some other substance is not clogging them.

If you set your pressure gauge for 10 pounds, the cooking temperature will be **greater than/less than** *212°F.*

Greater than. The cooking temperature would be about 240°F at sea level if 10 pounds of pressure were used. At 4000 feet elevation, 12 pounds of pressure would be required to reach 240°F.

*

Baking. You can bake winter squash, potatoes, and other vegetables that are high in moisture and contain sufficient cellulose to retain their shape. Some fruit with a moderate amount of cellulose, such as bananas, Roman Beauty apples, pears, and peaches, also bake well. Fruits and vegetables high in cellulose do not bake well, because baking is a dry heat and there is not enough moisture available to soften the cellulose. Some items have so little cellulose that they collapse when baked and are unattractive to serve.

Would you expect tomatoes to bake well? Why.

Yes. They have enough cellulose to retain their shape if they are not overbaked, and yet they contain enough moisture to soften the cellulose.

*

Some fruits and vegetables, that do not bake well from the raw state, bake very well if they are cooked or partially cooked before baking. Thus, you can boil or steam onions until tender and then bake them in a sauce or butter glaze. Carrots, eggplant, cucumbers, beets, and artichokes can be baked if they are cooked first. Often you can precook an item, put it into sauce, and then bake it. We usually call such a dish *scalloped* or *escalloped;* if topped with crumbs and cheese, it may be called *au gratin.* Sometimes you may wish to cook a fruit partially and put it into a casserole, put syrup on it, cover it tightly, and bake it. This is more stewing than true baking.

If you put raw potato slices into a thin white sauce and bake them, would you have escalloped potatoes?

Yes. Baking food in liquid or a sauce is scalloping; the food need not necessarily be cooked before being baked.

106

Many fruits and vegetables need not be pared to be baked. Squash can be cut into portions with the skin on or baked whole with the skin on. Parsnips bake very well in their jackets. Corn-on-the-cob can be baked in its husk 30 minutes in a 350°F oven to give a product much like roasting ears baked on hot coals. The corn may also be shucked, buttered, seasoned, wrapped in aluminum foil, and baked. Apples may be cored and baked with the skin on. Sometimes basting during baking is recommended to reduce moisture loss and shrinkage.

Parsnips bake fairly well while carrots do not. From this fact, what can you deduce about the quantity of cellulose each contains?

Carrots have more cellulose than do parsnips.

*

For baked potatoes, a mealy variety, high in starch and low in moisture, is preferred. Select medium to large potatoes of uniform size, wash them well, and rub them lightly with oil or fat if you wish to have a tender jacket. Bake in a hot (425° to 450°F) oven for from 45 to 60 minutes. To test for doneness, pick up and squeeze or pierce with a fork; if soft, they are done. Immediately cut an X in the top and gently squeeze the potato open partway to allow the steam to escape. This gives a mealy, light quality to the potato.

Why are Russet potatoes preferred for baking?

This variety typically has the high starch content necessary to bake into a dry, mealy, light product.

*

Potatoes can be pared, left whole, and baked in an oven. If they are rubbed with oil or fat and oven-roasted from the raw state, they are called oven-roasted or *rissolé* potatoes. When they are boiled or steamed to cook them partially before they are coated with oil and baked, they are called *Franconia* potatoes. Quartered or halved pared potatoes are sometimes baked with a roast. When this is done, the potatoes are turned frequently so they will absorb meat juices in their baking.

True or false. Franconia potatoes are cooked from the raw stage and oven-roasted; rissolé potatoes are partially cooked and then baked.

False. Just the opposite.

*

Broiling. Not many vegetables broil well. Raw tomatoes can be broiled, and mushroom caps broil well if they are

107

at least six inches from the heat. Some quite moist fruits also can be broiled. Sometimes, if you partially cook a vegetable or a fruit and then put it under the broiler, you can get a satisfactory broiled item. Eggplant can be cooked in this manner. It is excellent when it is spread with mayonnaise before it is finished under the broiler. After being parboiled, zucchini, onion slices, boiling onions, carrots, and parsnips can be dipped in melted butter and broiled briefly, being turned as they brown. Fresh grapefruit halves sprinkled with brown sugar also broil well. Canned pineapple slices and peach halves broil satisfactorily.

True or false.   Broiling is a very rapid and drying type of heat that tends to dry out and overcook the surface of fruits and vegetables before they are done in the middle.

*

Sautéing or frying. Frying or sautéing is cooking in shallow fat. Most vegetables and some fruits sauté or pan-fry well. Irish potatoes are excellent prepared in this manner, as are celery, wax beans, mixed green vegetables, leafy vegetables, broccoli, cauliflower, cabbage, onions, and many other vegetables when they are thinly sliced. The Chinese are famous for cooking vegetables quickly in oil, stirring and tossing them in the pan until they are almost done. To speed cooking, they may add a few tablespoons of water and then allow the vegetables to steam a bit. This method of sautéing vegetables also may be called panning.

Why is it necessary to slice vegetables thinly before sautéing them?

*

Raw sliced potatoes sautéed in shallow fat are called "raw-fried" or "American- or German-fried" potatoes. If they are boiled first, cut into slices, and fried, they are called "country- or cottage-fried"; if they are cubed and browned well in shallow fat, they are called "hash browns." "Hash brown" is also the name used for potatoes that are grated when raw and then sautéed. Waxy potatoes do not break apart easily and are generally preferred for sautéing if potatoes are boiled before being fried. If the sugar content in waxy potatoes is too high, you will discover that they may burn before they are cooked throughout. Mealy potatoes are satisfactory for sautéing when you start with raw potatoes.

Why are waxy potatoes the preferred type for cottage-fried potatoes?

*

Some fruits sauté well. Raw apple slices or bananas can be sautéed. Canned peaches, pears, pineapple slices, and some other fruits also can be sautéed. A bit of brown sugar adds to the flavor.

If we put vegetables or fruits in shallow fat or oil and cook them in an oven, is this sautéing?

*

Deep-frying. Mealy potatoes (high in starch) deep-fry very well either as slices or in strips. They can be pared, cut into the desired form, soaked in water, dried well, and then deep fried. You may choose between two methods for frying them: (1) the one-stage method in which the slices are fried at 375°F in deep fat for about seven minutes or until tender and golden brown, or (2) the two-stage method which begins with prefrying at 325° to 350°F in deep fat until tender but not brown (about three or four minutes) and then browning in fat maintained at 375°F for about two minutes. The potatoes should be a golden brown and crisp when cooked by either method. Drain well, salt, and serve them as soon as possible because the quality of a deep-fried vegetable quickly deteriorates.

Deep-frying of potatoes may be accomplished by either the _____ or _____ -stage method. In the one, the cooking is continuous from the raw to the cooked product; in the other, the cooking is completed in **one/two** fryings.

*

Some other root vegetables also can be deep-fried. Raw onion rings or eggplant deep-fry well if batter-dipped, but usually vegetables are best parboiled and then finished in the deep fat. Carrots, parsnips, okra, sweet potatoes, and yams may be cooked by this method. Some other vegetables may be partially cooked, breaded or batter-dipped, and then deep-fried. Cauliflower that is parboiled, breaded, and deep-fried takes on new character when cooked in this manner. Many others respond well to such treatment. Corn-on-the-cob may be removed from its husk and deep-fried.

Would either sautéing or deep-frying of vegetables tend to leach many nutrients from them?

No, very little nutrient loss occurs except in parboiling the vegetables before frying.

*

Some fruits, if batter-dipped or breaded, can be deep-fried. Apple slices cooked this way are popular and are called apple fritters. Pineapple and banana slices also may be batter-dipped and deep-fried.

Would you think that a quartered and well-drained dill pickle that is batter-dipped and deep-fried would be a good accompaniment to corned beef?

Yes, it truly is good. Sometimes the dill pickle is cut into round slices, drained well, and then batter-dipped and fried.

*

Vegetables such as Irish potatoes, sweet potatoes, yams, parsnips, pumpkin, or squash can be puffed or souffled by frying them in hot fat. To souffle Irish potatoes, pare them, slice them into half-inch strips or fourth-inch thick slices, and soak them in cold water. Drain, and dry well, and deep-fry for about four minutes in fat at 350°F. Chill thoroughly! Then, drop the pieces into 425°F fat and cook for about two minutes. As the potatoes strike the hot fat, steam is generated in the potato, and the product puffs up. It cooks in this puffed shape to a crisp, golden-brown product.

We ordinarily think of a souffle as a dish made puffy by beating air into egg whites, but souffled potatoes are puffed up by _____.

Steam.

*

MAKING FRUIT GELS

Fruits can be kept indefinitely by making high-sugar gels. You are doubtless well acquainted with the various types. Jelly is a clear gel made from fruit juice. Jam is similar to jelly but contains pieces of crushed or finely chopped fruit evenly distributed throughout. Preserves are small whole fruit or uniform pieces of larger fruit in a thick syrup or gel. A marmalade is the special term used for citrus fruit preserves. Conserves are preserves to which chopped nuts have been added.

Match the gelled product in the two columns.

1. Jam
   A. Crushed or finely chopped fruit suspended in a gel

2. Jelly
   B. Small whole fruit or uniform pieces of larger fruit suspended in a gel or thick syrup

3. Marmalade
   C. Whole fruit and chopped nuts in a gel

4. Preserves
   D. Clear gel

5. Conserves
   E. Gel with slices of citrus fruit

1. A.
2. D.
3. E.
4. B.
5. C.

*

Jellies are clear fruit juice which has gelled because of the correct balance of pectin, sugar, and acid. If the mixture is to gel, it will need: (1) to be quite acidic (pH of about 3.0), (2) to contain pectin at a level of one-half to one per cent, and (3) to contain between 55 and 70 per cent sugar. Usually, to make a good jelly from fruit, we add two-thirds to one cup of sugar per cup of fruit juice.

To make a jelly, the sugar should be from _____ to _____ per cent, pectin _____ to _____ per cent of the total, and the pH should be around _____.

55, 70, one-half, one, 3.0.

*

Slightly green fruit will be more acidic and contain more pectin than will ripe fruit. When fruit ripens, the pectin (carbohydrate with gel-forming properties) changes to pectic acid and loses its gel-forming ability. If the fruit is too ripe, both pectin and acid may have to be added to make jelly that will set or gel. Pectin extracted from apples or orange rinds often is used because it sets well. Commercial pectins that are suitable for making high-quality jellies are useful when using fruits that do not naturally contain enough pectin.

What ingredients are required to form a gel?

Liquid, pectin, acid, and sugar.

*

To make jams and jellies, the ingredients are boiled rapidly to 220°-221°F. Test the mixture to see if it will gel by putting a small quantity on a cool plate and letting it cool. It will start to gel if it is ready to set. Another test is to remove a spoonful, let it cool a moment, and then pour it slowly back into the mixture. The first part of the mixture should pour off, but the last part should drop off in a mass or sheet.

111

How do you know whether a gelled product is ready to set when you are making it?

Use a thermometer, use a sheeting test, or put it on a plate and let it cool.

*

Cooking longer than 20 to 25 minutes may alter the pectin so much that a gel will not form. Jellies or jams should be poured immediately into sterile jars before they start to gel. If they are poured after they start to gel, they will not form as strong a gel. The product is then sealed with paraffin so that molds cannot grow on the surface.

To make a gelled item, you must control the quantity of pectin, acid, sugar, cooking time, temperature, and freshness of the juice and also see that you preserve the item by putting it into _____ jars to keep it from spoiling.

Sterile.

*

Sometimes sugar crystals form in gelled fruit products. This results when there is too much sugar in the finished product. Overcooking of the mixture is a likely cause of this problem. Overcooking also makes a darker, less attractive product.

When you see a dark-colored jam with sugar crystals evident in it, you know the jam was _____.

Overcooked.

*

COOKING CEREALS

Cereal products are high in starch and are also important sources of the B vitamins if the cereals are whole grain or enriched. The cooking of these foods is designed to gelatinize the starch and give a palatable product. Boiling is the preferred method of cooking. Rice, other cereals, and macaroni products are boiled in salted water until tender and the raw starch flavor disappears. Macaroni and other pastas are cooked in an excess of water which is drained after cooking. However, for all types of cooked cereals, just enough water is used to gelatinize the starch; consequently, these will not need to be drained. This cooking method conserves the water-soluble B vitamins in cereals.

What are some cereal grains commonly eaten in the United States?

Wheat, rice, corn, and oats.

*

To boil one cup of macaroni or other pasta, a quart of water containing one teaspoon of salt is brought to a boil and the product is stirred in. Long spaghetti and lasagne noodles are slowly pushed into the water as they soften. Some stirring is necessary to keep the pasta from sticking to the pan until boiling resumes. The food is gently boiled to keep the pieces moving in the container and to prevent them from sticking together.

Macaroni and other cereal products are **boiled/broiled** *to gelatinize the* ____.

Boiled, starch.

*

Good cooks watch their macaroni products very carefully because overcooking makes them soft and mushy. The Italian cooks say to cook these products *"al dente"* or "to the tooth" which means that they still should have some bite left in them but be tender enough to cut easily. This texture gives a better eating product. When this stage of doneness is achieved, the item is drained and quickly dipped in cold water. This stops cooking and also removes any loose starch which would cause the cooked pieces to stick together. Then it is ready to use.

True or false.   One important thing to watch in the cooking of macaroni products is to not overcook them.

True.

*

The quality of the cooked pasta will be affected by the kind of product used. Macaroni products should be made with quality semolina (durum) flour or from high-protein farina. The protein content of a good macaroni paste will be as high as 14 to 16 per cent. This amount of protein gives a good firm texture. If a macaroni paste is made with a poor-quality flour, it will be white and pasty and become overly soft and mushy when cooked. Good semolina products are very yellow and look almost as if eggs had been added to them.

For the best macaroni products, select those that come from **low-protein wheat/high-protein semolina (durum)** *flours.*

High-protein semolina (durum).

113

Several types of rice are available in the United States. If you prefer a light, fluffy rice with distinct grains, select a long-grain rice. Short-grain rice is preferred by people who like a sticky rice. Some excellent long-grain rice varieties are grown in Louisiana, Texas, and Arkansas. Short-grain rice is grown in California.

*Long-grain rice is **fluffy/sticky** and short-grain rice is **fluffy/sticky.***

Fluffy, sticky.

*

Parboiled rice is more nutritious than regular polished rice because it is steamed to drive the vitamins to the inside of the grain before the bran is removed. Enriched rice is another white rice that is higher in vitamin content than plain polished rice. Enriched rice has a premix of vitamins and iron added to it. These two types (parboiled and enriched) are cooked like regular polished rice. However, be especially careful not to have excess water remaining when the rice is done, because the added vitamins then will be lost into the excess water that is discarded.

True or false. Parboiled and polished rice both contain B vitamins.

False. Polished rice contains little but starch.

*

Also available in today's markets is minute rice, a precooked and dehydrated rice. The addition of hot water rehydrates it and it is then ready to serve. Such rice is a light, delicate product; avoid overcooking or letting it stand in excess liquid because these conditions may cause minute rice to break up. If it is to be used in dishes which later will be baked in a sauce, let it stand the minimum time and do not overbake.

Minute rice is rice that has been _____ and _____.

Cooked, dried.

*

Brown rice is a nourishing food because the vitamin-containing bran layer has not been removed .The high cellulose content of this bran layer slows the cooking, and brown rice requires 40 minutes to soften compared with 20 minutes for polished, parboiled, and enriched rice.

Brown rice requires _____ *minutes to become tender; parboiled rice cooks in* _____ *minutes.*

40, 20.

*

Wild rice is not a grain but is a grass seed picked by Indians in their swamplands in Minnesota, Michigan, and Wisconsin. It is scarce and therefore quite expensive. The strong flavor of wild rice combines particularly well with wild and domestic fowl. It is cooked like rice but requires 40 to 45 minutes to become tender.

Which would be the least expensive, regular or wild rice?

Wild rice is at
least 10 times more
expensive!

*

Sometimes rice is washed and lightly sautéed in a small quantity of melted fat, butter, or margarine until delicately browned. It is then boiled. The addition of oil, even without the sautéing, assists somewhat in keeping the rice grains separated after cooking.

If an oil or fat is added to rice before it is cooked, it sticks together **more/less** *easily.*

Less.

*

Cooked breakfast cereals will vary in the quantity of water required, according to their type. Cornmeal and other finely ground cereals will require about three cups of salted water to every cup of cereal. Part of this water should be mixed cold with these finely ground cereals before they are added to boiling water. Coarser cereal such, as oatmeal, cracked wheat, or coarse grits will take about two and one-half cups of salted water per cup of cereal. The cereal is sprinkled over the boiling water and slowly stirred until the water boils again. Good stirring is necessary to prevent lumps from forming. When boiling starts and thickening begins, only a minimum stirring should be done to keep the cereal from becoming sticky and pasty. It is lighter on the tongue if stirring is resticted to a minimum. The cereal is done when the raw starch taste disappears.

Stirring **is/is not** *recommended before a cooked cereal begins to thicken but* **is/is not** *recommended after that.*

Is, is not.

115

REVIEW

1. During storage, fruits and vegetables *do/do not* need air.

    Do.

2. *True or false.* Some moisture in the air during storage is desirable.

    True.

3. A salt concentration greater than _____ per cent prolongs the storage life of a food.

    Eight.

4. A sugar concentration greater than _____ per cent prolongs the storage life of a food.

    50.

5. Which is larger, a No. 10 can or a No. 303 can?

    No. 10.

6. *True or false.* Federal law requires that pictures and sizes of items on the label must bear a relationship to the item in the can.

    True.

7. In a list of ingredients on a package, the item present in the largest quantity will be listed _____.

    First.

8. Frozen fruits are commonly packed in a _____ pack.

    Sugar.

9. *True or false.* Very young fruit is preferable to just mature fruit.

    False.

10. For good mashed potatoes, select a *waxy/mealy* potato.

    Mealy.

11. For French-fried potatoes, select a *waxy/mealy* potato.

    Mealy.

12. If you are making scalloped potatoes, select a *waxy/mealy* potato.

    Waxy.

13. *True or false.* It is important to avoid storing fruits and vegetables with blemishes if you wish to keep them very long.

    True.

14. A vitamin that often is used to prevent browning is _____ _____.

    Ascorbic acid (vitamin C).

15. A fruit or vegetable is considered to be properly cooked when it is _____ and easily cut.

    Tender.

16. Asparagus spears can be cooked most effectively by *tying them together and standing them upright in the water for part of the cooking time/placing them in water to cover them completely for the entire cooking time.*

    Tying them together and standing them upright in the water for part of the cooking time.

17. The white pigments in vegetables are _____.

    Flavones.

18. The reds and blues in vegetables are the pigment _____.

    Anthocyanin.

19. Yellow and orange colors in vegetables are _____.

    Carotenes.

20. Chlorophyll is *green/olive green* in acid.

    Olive green.

4

Salads and Salad Dressings

SALADS

The definition of a salad is difficult and cumbersome because of the many variations possible in this category. Salads may be served hot, chilled, or frozen. They may contain any type of vegetable, meat, fish, shellfish, cheese, or nuts. Actually, almost any food may be used, either singly or in combination with other ingredients, to make a salad.

What foods may suitably be included in the preparation of a salad?

Practically any food, either hot or cold, may be used.

*

Before the days when people were so concerned about their weight, there was one common factor in almost all salads. They all had a dressing of some type: However, today weight watchers often eat their salads without a dressing, in an attempt to reduce their caloric intake. Consequently, even this distinguishing feature is not common to all salads.

True or false. Salads may be defined as follows: Any hot or cold dish of meat, fish, fruits, or vegetables, either singly or in combination, served with mayonnaise or other dressing.

False. This is a reasonably good definition, but salads are served rather frequently without a dressing.

*

Salads can be good sources of all the nutrients needed for good nutrition. Foods from any of the groups in the Basic

117

Four Food Plan can be included in salads. From milk and dairy products, you can select many types of cheese, or you might occasionally use whipped cream or whipped nonfat dry milk solids. A wide variety of meats, poultry, and fish can be used to represent the meat category. The food group most abundantly represented in salads is the fruit and vegetable class. Examples from the bread and cereal group are limited primarily to various styles of macaroni or to croutons.

Which groups in the Basic Four Food Plan are used: (1) in the largest quantity in salad preparation and (2) in the smallest quantity?

1. *Fruits and vegetables.*
2. *Bread and cereal.*

*

Salads may be used in a variety of ways. A small, brightly flavored salad is a pleasing appetizer for a meal. One that is served as an accompaniment to the main course needs to be planned to complement the color, flavor, and textures of the other foods served at the same time. A hearty salad focusing on a protein food such as meat, eggs, cheese, or beans can be a complete meal. Sometimes fruits, fruited gelatin, and frozen fruit salads are the dessert course.

A salad may be served as an _____, an _____, a complete _____, or a _____, but in one meal the salad would be used in only one of these ways.

Appetizer, accompaniment, meal, desert.

*

Now, let us examine a few types of salad that are enjoyed by many people. Greens, such as head lettuce, leaf lettuce, butter lettuce, endive, spinach, and romaine, may be used singly or in combinations of greens to make a tossed salad. Often wedges of tomato, marinated artichoke hearts, sliced radishes, carrot curls, flowerets of cauliflower, shredded red cabbage, and other vegetables are added to a tossed salad to heighten the color, flavor, and nutritive value. Occasionally, mandarin orange slices or other fruits may be added to greens to give variety.

Name three types of greens that are commonly available for salad making.

Any three of the following: head lettuce, leaf lettuce, butter lettuce, endive, spinach, and romaine.

*

Additional variety may be achieved in tossed salads by garnishing them with tiny shrimp, julienne strips of ham or beef, chopped or sliced hard-cooked eggs, crumbled bacon, or croutons. Such additions add interest to the appearance, texture, and flavor of a salad and also increase the nutritive value.

Can you think of some other foods that might be included in a tossed salad?

Rings of green pepper, cheese either diced or in strips, bits of pimiento, chopped chives, and onion slices are a few additional possibilities.

*

Cooked vegetables are sometimes chilled for use in a salad. Sliced beets topped with sour cream dressing and sprinkled with chopped chives comprise a colorful salad of this type. Other possibilities might be chilled cooked asparagus spears garnished with a dollop of mayonnaise, chopped hard-cooked egg, and a strip of pimiento. A nourishing, piquant combination of fresh onion rings, wax beans, kidney beans, garbanzo beans, and string beans marinated overnight in a tart vinegar and oil dressing is a popular cooked vegetable salad. Certainly, potato salad cannot be ignored when reviewing some cooked vegetable salads. This favorite is commonly served cold, but hot potato salad with onion, crumbled bacon, and a sharp dressing of bacon drippings and vinegar is a pleasing cold-weather salad.

Vegetables may be used in salads either _____ or _____; cooked vegetable salads are frequently served chilled, but occasionally a _____ potato salad is prepared.

Raw, cooked, hot.

*

Occasional use of less familar foods in salads can add considerable interest to a meal. Hearts of palm can be cooked and served chilled as a gourmet salad item. Nasturtium blossoms may be heaped into a mound and served with a sweet fruit dressing, or the nasturtium leaves may be used as a garnish or flavor accent. Other salads that may be unfamiliar include chilled artichoke hearts or a raw spinach salad.

Is it necessary to follow convention when preparing a salad?

No. The most pleasing salads often are made when you allow your imagination to roam the realm of foods.

*

Important color, flavor, and textural contrasts can be achieved in a meal by thoughtful planning and careful preparation of salads. This portion of the meal should be

119

planned to complement the total impact of the other foods being served. A mild guacamole and tomato salad is a colorful yet pacifying accompaniment to a spicy enchilada entree. A mild-flavored tuna and noodle casserole becomes more interesting when served with a tart grapefruit and pomegranate salad, a choice that also affords a color and texture contrast.

Important contrasts in _____, _____ and _____ can be added to a meal by careful selection of the salad.

Color, flavor, texture.

*

Salads that are good appetizers are designed to whet the appetite and raise abundant expectations of what is to follow. A crisp Caesar salad provides a provocative introduction to an Italian spaghetti dinner. A pleasing fanfare to a poached salmon entree is an avocado boat bearing a cargo of pineapple chunks, mandarin orange slices, and apple cubes sporting a bright red peel.

Would a jellied tomato aspic filled with tiny cooked shrimp and served on crisp greens be a good appetizer for a roast duck dinner?

Yes.

*

The aesthetic qualities of a meal are determined by the form and shape of the foods as well as by the color. Skillful and artistic cutting and arranging of salad ingredients can help the creative cook achieve a truly beautiful meal. Strive for a rhythmic uncluttered appearance. You may find that a stiff formal arrangement seems less inviting to eat than one that is casual but still attractive.

Salads enhance a meal if they are:
A. Planned to complement the rest of the menu.
B. Prepared using ingredients that have been carefully cut or broken into pleasing sizes and shapes.
C. Arranged to present an appetizing, uncluttered appearance.

A, B, and C. All these factors contribute toward the preparation of a delectable salad.

*

Salads are more pleasing to eat if they are prepared with the consumer in mind. Ingredients that are difficult to cut with a fork should be divided into bite-size segments when you are making the salad. Unity in the salad is achieved by using a leaf of greens as an underliner. The total salad should be framed by the plate or bowl in which it is served. Salads that protrude beyond the edge are unattractive and difficult to manage. This problem can be solved by selecting a dish of an appropriate size for the quantity of salad.

*When arranging a salad, select a plate large enough to frame the salad and arrange the ingredients so that **nothing/at least part of the salad** extends beyond the plate.*

Nothing.

*

The size of the portion depends upon the role of the salad in the meal. An accompaniment salad is normally between two-thirds and three-fourths cup if it is a tossed or mixed salad. The appropriate quantity for an accompaniment portion of a molded salad is about one-half cup. When the salad is the main course of a meal, the portion size is increased to between one and two cups.

What size serving would you plan for the following salads when served as accompaniment: (1) tossed, (2) mixed fruit, and (3) molded gelatin? What size for a chicken salad served as an entree?

1. Three-fourths cup. 2. Two-thirds cup. 3. One-half cup. One to two cups of chicken salad.

*

If a salad is to be a complete meal, it should contain foods that combine to make a filling dish and give a feeling of satisfaction. From one-third to one-half of such a salad should be meat or other food high in protein. Some foods suitable for this purpose include chicken, turkey, lobster, shrimp, crab, ham, beef, and cottage cheese. Chopped egg or diced cheese may be added to contribute to the nutritive value and general palatability of the salad.

Why do you feel more satisfied when you have meat in a salad than you do when you have a large salad of mixed greens?

Meat and other protein foods move sedately through the digestive tract and are slowly absorbed. Greens contain a large quantity of cellulose, which helps to move food through the body, but cellulose is not digested and absorbed.

*

Perhaps the most important thing to remember when preparing salads is the necessity to use greens that are crisp and chilled. Limp greens transform a salad from an appetizing inviting food to one with little appeal. Proper storage will eliminate this problem. Greens have a fragile wall that relies on water within each cell for the support necessary to impart crispness. If the water normally found in the cell is allowed to pass out of the cell, the cells begin to collapse, much as an inflated balloon collapses when the air escapes.

The cell walls of greens are supported by _____ inside the cell.

Water.

*

To keep greens crisp, it is necessary to store them in a manner that retains the maximum amount of water in the cells. If lettuce is allowed to stand at room temperature, water will pass from the cells and evaporate into the atmosphere. This evaporation will continue until all the free water in the cell has been drawn out, or until the air around the lettuce is saturated with water. It is apparent that one head of lettuce cannot possibly supply enough moisture to saturate all the air in a warm room. The lettuce will become very wilted because of the water lost from the cells into the atmosphere.

True or false. An unwrapped head of lettuce in a cold room will stay crisp somewhat longer than one in a warm room.

True. The rate of water evaporation is faster in a warm room than in a cold one. Consequently, the warm lettuce loses water more quickly and wilts rapidly.

*

It is possible to reverse this loss of water and actually cause water to flow back into the cells. To do this, reduce the temperature to about 34°F and moisten the greens. When the cells are once again filled with water, the greens will be crisp again.

The flow of water into and out of plant cells is a reversible reaction which is influenced by the _____ of the air and the amount of _____ in the air.

Temperature, moisture.

*

Cell walls are semipermeable membranes; that is, water can pass through them but salts and other substances in the cell are trapped in the cell and cannot pass in and out with the water. This passage of water through the cell wall is referred to as *osmosis*. Water moves both into and out of the cell. By varying the storage environment of the greens, it is possible to cause appreciably more water to pass into than out of the cell. Water accumulates in the cell when greens are placed in water or in a small chilled container with some moisture added to saturate the air surrounding the greens.

The passage of water through a _____ membrane is called _____.

Semipermeable,
osmosis.

*

To keep greens crisp and fresh during storage, it is best to store them in a refrigerator in a hydrator drawer, in a moist cloth, or in a covered container. As previously mentioned, the cold temperature retards evaporation and the limited amount of air around the greens quickly becomes saturated. This further slows evaporation of the moisture passing from the cells. Some air is necessary for the slow respiration of the greens during cold storage.

Greens should be stored:
A. On the top shelf of the refrigerator.
B. On the kitchen counter.
C. In the hydrator drawer of the refrigerator.
D. In a tightly closed kitchen drawer.

*C. In the hydrator
drawer of the
refrigerator.*

*

Water passes into the cells of plant foods when the food is placed in water because of the presence of dissolved salts in the fluid within the cells. These salts cause a pressure drop within the cell, and water is drawn in. Unfortunately, when you add salts to the liquid in contact with the cells, you can reverse the situation and cause the water to be drawn from the cell. Of course, this causes the cells with fragile walls to begin to collapse or wilt. You can observe this phenomenon when a dressing containing salt, sugar, or other seasoning is placed on a tossed green salad and allowed to sit several minutes. The salad will gradually become limp as the water is drawn from the greens. For this reason, it is very important to add the dressing for greens just before serving.

The wilting of greens that occurs after a dressing has been used to coat the leaves is due to _____.

Osmosis.

*

Making salads requires considerable labor. Frequently, there is also a need for last-minute preparation so the salad will be as fresh as possible. Therefore, plans for preparing the ingredients and assembling the finished product need to be carefully worked into the time plan for meal preparation. Greens often can be prepared ahead and then stored under proper conditions until the salad is assembled just before the meal. Garnishes also can be made in advance and properly stored. Gelatin and frozen salads must be made in advance if they are to be of proper serving consistency. When the work is well organized, the quality of salads and the pleasure in making them will be increased.

If you were planning to serve a crab salad as the main dish for a luncheon, when might you prepare the underliner greens?

They may be prepared the day before and allowed to crisp by wrapping them loosely in a dampened towel or placing the slightly moist greens in the hydrator drawer or in a plastic bag so they can crisp. Refrigerator storage in one of these ways will ensure perky, crisp greens.

*

Greens for use in salads need to be thoroughly washed to remove dirt and possible insecticides and fertilizers. Heads can be washed by holding them under a stream or spray of water. Spinach and other leafy greens are most efficiently and thoroughly washed by swishing them in a sinkful of cold water, draining the sink, and then repeating the process until the water in the sink is free of sediment. Excess moisture can then be shaken off, and the greens can be placed in a container and refrigerated to crisp them. Fresh greens may become suitably crisp in half an hour; some greens may require several hours of chilled storage to regain the necessary moisture.

True or false. Thorough washing of greens is an optional step in the preparation of salads because present marketing practices include washing of the greens before they are displayed in the store.

False. The heaviest soil may be removed from the produce before it is sold, but less obvious contaminants such as insecticides and fertilizers may be present. In addition, workers along the marketing chain have handled the produce and added human contamination.

*

Hot salads provide an occasional variation from the cold salads customarily served. For a change of pace, you may wish to try an oven-baked seafood salad, a hot potato salad, or a hot slaw. These salads, which would be served piping hot, can be heated by baking in the oven or by heating the ingredients together over a surface unit. Some hot salads, like hot potato salad, are made by stirring and heating together the seasonings, vinegar, oil, salt, and pepper until the mixture comes to a boil. At that time, the salad ingredients are added and thoroughly warmed preparatory to service. To add the gourmet character to a wilted spinach and wilted leaf lettuce salad, boil vinegar with bacon fat and bacon bits just prior to pouring it over the fresh greens. The heat of the dressing causes the wilting suggested in the names of these salads. Some chopped egg can be used to add the finishing flourish.

What would you think should be a precaution about serving hot salads and cold ones?

Serve hot salads hot and cold salads chilled thoroughly.

*

Care must be taken to make each salad a truly handsome creation. The ingredients should look sparklingly fresh, and the individual pieces should be completely intact. Fruit salads are particularly delicate and need to be handled even more gently than vegetables if they are to retain their form. Even tossed salads need to be handled with restraint. Just sufficient tossing to lightly coat the individual pieces with dressing is a suitable rule; vigorous tossing only bruises the leaves.

Fruits/vegetables *are particularly fragile and should be handled as little as possible to preserve the shape of each piece.*

Fruits.

*

The fitting climax to many salad arrangements is a small garnish to serve as a focal point. Such garnishes should be carefully selected for their simplicity and their ability to enhance the eye appeal and flavor of the salad. An edible garnish is definitely more appropriate than a paper flag or other inedible addition. Suitable edible garnishes include a sprig of mint, watercress, or parsley; a small cluster of sugared grapes; a green or ripe olive; a paper-thin slice of sweet onion; chopped egg; or a dash of paprika. Occasionally, you may wish to incorporate the garnish in the dressing. Chopped pickles, beets, anchovies, crumbled Roquefort cheese, or celery are suitable for this type of garnish. Caraway, celery, sesame, or poppy seeds also can add a flavorful touch to a salad.

125

The ideal garnish will be **edible/inedible** *and* **simple/elaborate.**

Edible, simple.

*

Gelatin salads provide a unique texture, and they can be varied in many ways to augment the flavor combinations in a meal. The convenience of molded salads also should be mentioned. You can make these a day before you plan to serve them. The busy cook will appreciate being able to prepare a salad when it is convenient rather than always having to take time during the last few minutes before the meal is served.

The busy cook **should/should not** *plan to prepare a gelatin salad three days before it will be used.*

Should not. As gelatin salads age, they gradually lose some of the water they contain and the gelatin begins to be rather tough and slightly chewy. Gelatin made within the 24 hours prior to service will have a more pleasing mouth feel than will older gelatin mixtures.

*

You will find two types of gelatin products available in the grocery store. The plain, unflavored gelatin is pure gelatin ground into moderately fine pieces. Flavored gelatins contain the pure gelatin but also contain sugar and flavoring. The gelatin particles in the flavored gelatin are significantly smaller than in the unflavored product.

Flavored gelatin has **smaller/larger** *particles than unflavored gelatin.*

Smaller.

*

To disperse the unflavored gelatin uniformly, it is appropriate to hydrate (soak in cold water) the dry particles for about five minutes so that all the particles become moistened thoroughly before the boiling water is added. If there is at least one tablespoon of sugar in a recipe, you can mix the sugar and dry gelatin together and then add boiling water to this dry mixture. The gelatin granules in the flavored product are so small that hydration is not necessary; the boiling water can be added directly to the dry gelatin powder. Stir the hot gelatin continuously until absolutely no particles of gelatin can be seen. Add cold liquid or ice to finish diluting the gelatin to the desired strength. To avoid a rubbery layer on the bottom of the gelatin, it is essential that all the particles be melted before the cold liquid is added.

True or false. *A rubbery layer on the bottom of a molded gelatin salad is caused by a failure to melt the gelatin completely before adding the cold liquid.*

True.

*

It takes one level tablespoon of pure unflavored gelatin to set a pint (two cups) of liquid. This amount of gelatin is contained in one envelope of this product. A much larger quantity of the flavored gelatin mixture is required to set a pint of liquid because the package contains considerable sugar mixed with the gelatin granules. In addition, the presence of sugar causes the gelatin salad to be somewhat more tender than it would be without sugar.

Flavored. The flavored gelatin is a mixture of pure gelatin and sugar, which means that it is necessary to use more of the product to obtain enough gelatin to set the salad. The sugar also has a tenderizing effect on the finished product.

To make gelatin salads of equal gel strength using unflavored and flavored gelatin, it would be necessary to use more of the **unflavored/flavored** *gelatin. Explain.*

*

Sugar is not the only additive that makes a gelatin salad less firm. A very small amount of fruit juice may make the gelatin more firm, but larger quantities (such as are often used in preparing gelatin) will help to tenderize the product. If a very sweet or tart gelatin salad were to be prepared, you would be wise to add more than one tablespoon of gelatin per two cups of liquid to make the salad firm enough to serve easily. You may also wish to add more gelatin to a salad that is to contain coarsely chopped, heavy pieces of food. The extra strength helps to support the food pieces throughout the salad.

Acids such as fruit juices in large quantities **weaken/strengthen** *gelatin salads.*

Weaken.

*

Sometimes gelatin is whipped to make a frothy salad or salad dessert. During the whipping of gelatin, the gelatin is stretched out into thin films that cover a wide surface. Much air is trapped within the whipped mixture. These foams become more stable when they are chilled enough to congeal the gelatin.

127

True or false. Variety can be given to gelatin salads by preparing a recipe that uses whipped gelatin to provide a stable foam.

True.

*

A few fresh fruits contain enzymes (protein substances) that prevent gelatin from setting. These enzymes are found in fresh pineapple, figs, and papayas. If these fruits are heated, the enzymes are inactivated. Hence, the addition of canned pineapple will not prevent the gelatin from setting.

Do you think that you could add frozen pineapple to a gelatin mixture and get the gelatin to set?

No. Frozen pineapple has not been heated so the enzyme is still able to prevent setting of the gel.

*

Many times you may wish to have fruit or vegetable pieces suspended evenly throughout a molded gelatin salad. You can accomplish this by adding the foods when the gelatin has cooled enough to become syrupy. Or you can add them before the mixture begins to set and give them a final stir when the gelatin becomes somewhat thick and syrupy. For a salad containing more than one layer of gelatin, it is necessary to set the first layer and then pour on the second layer when it is almost ready to set. If the second layer is still slightly warm, it will meet the first layer and the two will fuse together.

If you wanted to have a cottage cheese in lemon gelatin salad topped with a layer of chopped sour cherries and pineapple in red cherry gelatin, how would you proceed?

Put the lemon gelatin mixture with the cottage cheese into the mold and be sure to stir just before it sets so that the cheese will be well dispersed throughout the gelatin. Then, have the cherry gelatin mixture almost ready to set and pour this over the gelled lemon mixture. The chopped cherries and pineapple should be stirred into the cherry gelatin before being added to the mold to avoid disturbance of the bottom layer.

*

Gelatin salads need to be unmolded with care. A quick dip of the mold in warm water will melt just a little gelatin around the edges and permit the gelatin to slip out. Avoid

128

letting the mold linger in the warm water! It takes only a moment to melt the gelatin in direct contact with the mold. A longer time in the warm water will melt too much of the gelatin, and you will have a molded salad with a tired, indistinct outline. Usually, you can unmold gelatin salads by simply covering the mold with a plate and quickly turning the plate and mold over together. A sharp, abrupt shake is usually adequate to release the gelatin from the inverted mold, but resistant salads can be released by slipping a knife blade down the side of the mold to release the vacuum that is holding the gelatin.

*A gelatin salad can be removed from the mold easily and without destroying its appearance by **slowly/quickly** dipping the mold in warm water.*

Quickly. However, be careful not to immerse the salad below the edge of the mold or warm water will flood the surface and melt the gelatin.

*

SALAD DRESSINGS

To have a successful salad, you should carefully select the best dressing for the salad. A French dressing touched with garlic is excellent with tossed greens but is not as well suited to a fruit salad as is a tangy, honey-fruit dressing. A creamy French dressing will give body and smoothness to a crunchy tuna salad, but it may overwhelm a delicate watercress salad.

What type of dressing do you think might be served with an avocado salad?

A light, delicately seasoned French dressing would give some tang which would be a desirable complement to the natural softness and oiliness of the avocado. Mayonnaise would not provide this necessary contrast.

*

Dressings are classified as emulsions. This means they are mixtures of two liquids that are not easily mixed together. The two liquids used in salad dressings are vinegar (or lemon juice) and oil. Just like oil and water, these two fluids quickly separate when shaken together and the oil rises to the top while the vinegar sinks to the bottom. The oil in a dressing forms small drops when vinegar and oil are beaten together, but these may coalesce (go together) and form a layer of oil floating on the vinegar.

Why is it desirable to have the ingredients in a salad dressing remain in an emulsion?

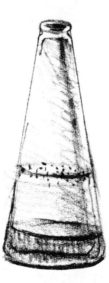

The full flavor of the dressing is obtained only when the ingredients are well blended and cover the salad uniformly. If the oil and vinegar separate, the dressing will be unevenly distributed. Some parts of the salad may then be too oily and other areas may have a very tart flavor of vinegar.

*

Salad dressings are made by forming an emulsion, that is, by separating oil into tiny droplets suspended in vinegar. The ingredients in some recipes are simply mixed by vigorous shaking, while in other recipes an egg beater or blender is used to separate the oil into droplets.

Name two ways of mixing salad dressing ingredients to form an emulsion.

1. Shake vigorously.
2. Beat with an egg beater or blender.

*

Some salad dressings separate into two distinct layers in minutes; others stay in a uniform dispersion (emulsion) for months. Emulsions that separate very rapidly are called temporary emulsions. The French and Italian dressings that begin to separate as soon as you stop shaking them are common examples of temporary emulsions.

Common examples of a temporary _____ include _____ and _____ dressings.

Emulsion, French, Italian.

*

These temporary emulsions separate because: (1) they have only a small amount of dry ingredients to keep the oil droplets from joining together, and (2) these dressings are very fluid. For these two reasons it is easy for the small droplets to come together and form two distinct layers.

Why is it necessary to shake Italian and French dressings vigorously just before they are put on the salad?

*

To get the total flavor of the ingredients in these temporary emulsions, it is essential to have the oil and vinegar completely mixed. Since these dressings separate quickly, you may be pouring only an oily dressing onto the salad if the dressing has time to separate.

Some salad dressing will stay blended in an emulsion for a few days. These dressings with somewhat greater stability are called semipermanent emulsions. Such dressings separate rather slowly because they are viscous enough to pile slightly when poured. They are as thick or thicker than honey. Some honey-fruit dressings and other sweet dressings containing a boiled sugar syrup are good examples of these semipermanent emulsions.

Why do semipermanent emulsions separate more slowly than temporary emulsions?

*

The oil droplets move with difficulty through the thicker dressing and, therefore, it takes a long time for the oil droplets to coalesce and form a separate layer.

A few dressings are very stable and rarely separate into two layers. Such dressings, classified as permanent emulsions, have one or more ingredients added to help keep the oil droplets separated. These ingredients coat the surface of each oil droplet, making it extremely unlikely that the oil droplets themselves can actually touch each other. Permanent emulsions are thick enough to pile well, a fact which also helps to stabilize them. Probably the best-known example of a permanent emulsion is mayonnaise.

*Mayonnaise, a _____ emulsion, is thick enough to **pile/flow** well.*

Permanent, pile.

131

The basic ingredients in dressings are simple. You need oil, vinegar, and some seasonings. The oil must be of excellent quality with absolutely no trace of a rancid flavor. Suitable oils include corn, cottonseed, soybean, olive, peanut, and safflower. The distinctive flavor of olive oil is preferred by some. With the exception of peanut oil, the other oils have such mild flavors that you will be unaware of any significant oil flavor in the dressing. The seasonings will obscure the delicate flavors. Occasionally, you may wish to make a hot dressing with bacon drippings replacing the oil. This dressing has a strong flavor of bacon and is particularly good in hot potato salad and wilted green salads.

*For most salad dressings, you would select an oil that has **very little/a distinctive** flavor.*

Very little.

*

Several vinegars are available to give just the desired flavor touch to a dressing. You may wish to select cider vinegar made from apples, malt vinegar with a nutty quality, fragrantly seasoned tarragon, or a zesty wine vinegar. Lemon juice can be used in place of vinegar. Fruit dressings benefit particularly from the use of lemon juice.

Could you use distilled vinegar in a salad dressing?

Yes, but distilled vinegar lacks the distinctive flavor characteristics of the vinegars mentioned here.

*

French dressing is a basic dressing made with vinegar, oil, and seasonings. The customary seasonings are dry mustard, paprika, cayenne, salt, and pepper. These particles coat the oil droplets when the dressing is shaken and thus help to keep the oil suspended briefly. They are vital to the flavor of a simple French dressing. The ratio of oil to vinegar influences the flavor of the dressing. The recommended ratio is two parts of oil to one part of vinegar.

A French dressing is _____ part(s) oil and _____ part(s) vinegar.

Two, one.

*

Pleasing flavor variations can be made by adding curry powder, diced hard-cooked egg, assorted herbs, crumbled Roquefort cheese, grated Parmesan cheese, chopped chives, or meat sauces to a basic French dressing. Chopped chutney added to the French dressing and served on a chicken and pineapple salad provides a pleasing sparkle to the combination of flavors.

True or false. When unusual ingredients are blended into a basic French dressing, this dressing is no longer considered a true French dressing.

False. These ingredients are used to make tempting variations, but the dressing is still considered a French dressing.

*

You sometimes may prefer to use a French dressing that is creamy and stable rather than one that separates very quickly. To do this, you can obtain gum tragacanth, gum arabic, Irish moss, or some other stabilizing agent from the drug store and simply blend this ingredient into the French dressing. These products will stabilize the ingredients and form a stable emulsion.

A French dressing which contains gum tragacanth is a **temporary/permanent** *emulsion.*

Permanent.

*

Imagination is a useful ingredient when concocting salad dressings. You have probably never tried currant jelly blended with chopped walnuts as a dressing on an apple and banana salad, but this is a pleasing flavor combination. Another suggestion is grated orange rind and angostura bitters on a melon ball salad. Other simple dressings can be made using sour cream, yogurt, cream cheese, or tomato paste as the base.

False. Cream cheese and tomato paste are but two examples of suitable starting materials for salad dressings.

True or false. Dressings can only be made by using vinegar and oil as the starting ingredients.

*

A dressing is intended to enhance a salad, not to obscure completely or disguise the ingredients. Enough dressing needs to be used to coat the ingredient lightly if it is to be tossed or mixed with the salad. Avoid adding so much dressing that a pool collects in the bowl and makes the ingredients at the bottom soggy. Occasionally you may wish to put a dollop of a more viscous dressing on the top of a salad to serve as a garnish as well as a dressing, but large quantities of dressing mask the pleasing appearance and flavor of the basic ingredients.

*If you find that a pool of dressing has drained to the bottom of a bowl of tossed salad, you know that you have added the   **correct amount/too much**   dressing.*

Too much.

*

Mayonnaise is a permanent emulsion that is stabilized by the presence of egg yolk which collects around the many small oil droplets in the dressing. To prepare mayonnaise, blend egg yolks, dry ingredients, and some vinegar together in a small bowl. Then, add oil dropwise while slowly blending with a rotary beater or in a blender. Slow addition of the oil is essential to the formation of a stable emulsion. If you add too much oil at one time, the emulsion will break and the oil will separate from the vinegar. The amount of oil added at each addition can be slowly increased as the volume of dressing increases. If you should add too much oil and the emulsion breaks, simply treat the broken emulsion as though it were oil. Start reforming the emulsion by beating an egg alone in another small bowl and then very slowly begin adding the broken emulsion. Gradually continue to add the remaining broken emulsion while slowly beating the new emulsion with a rotary beater. The result should be a mayonnaise that is pleasing in color and flavor and free of any oily surface layer.

*If a mayonnaise separates while you are making it, treat the broken emulsion as   **oil/vinegar**   and slowly   **add egg yolk to it/add it to a fresh egg yolk.**

Oil, add it to a fresh egg yolk.

*

When you shake a dressing or beat it with a rotary beater, you are helping to distribute the oil in small droplets throughout the vinegar. In a mayonnaise or any other type of salad dressing, the oil droplets are surrounded by the vinegar phase. An emulsion of this type is called an oil in water (o/w) emulsion because the oil is distributed in the water (vinegar in the case of salad dressings) .

*Mayonnaise is an example of   **an oil in water/water in oil** emulsion because the   **oil/water**   is distributed in droplets suspended in   **oil/water.**

Oil in water, oil, water.

*

Mayonnaise, like French dressing, is a good basic dressing that can be varied in many ways to complement a particular salad. Any or all of these ingredients may be added to make mayonnaise variations such as Thousand Island dressing: chopped egg, minced onion, chili sauce or catsup, chopped green pepper, snipped parsley, chopped olives, and pimiento.

Is a mayonnaise-type dressing always yellow to creamy yellow?

No. Variations made with catsup or other tomato products are a light-tomato color.

*

Cooked salad dressings are lower in calories than is either French dressing or mayonnaise because they contain significantly less oil. A cooked dressing is thickened with a starch (usually wheat flour or cornstarch) and either egg yolks or whole eggs. The maximum viscosity and best overall product of this type is obtained when you follow this procedure. (1) Heat the starch paste to a boil while stirring constantly. (2) Remove from heat and stir about one-fourth cup of the hot paste into the beaten yolks. (3) Stir the egg yolk-hot paste mixture carefully into the remaining hot paste. (4) Heat over boiling water for five minutes to thicken the egg. (5) Remove from heat and stir in the lemon juice or vinegar to complete the dressing.

When is the acid ingredient (lemon juice or vinegar) added in the preparation of a cooked salad dressing?

The acid is added after all heating ceases; therefore, this is the last step in preparing a cooked salad dressing.

*

Although cooked dressing may be used any time a recipe calls for mayonnaise, it is a particularly good choice for dressing potato salads, cole slaw, and fish and meat salads. Golden dressing is a variation of a cooked dressing that is particularly good with fruit salads. In this dressing one-half cup pineapple juice, one-half cup orange juice, one-fourth cup lemon juice, and one-fourth teaspoon salt are heated together before being stirred into four egg yolks that have been beaten vigorously with one-third cup sugar. This mixture is heated in a double boiler until it thickens. This cooked mixture then is folded cautiously into a meringue (four egg whites beaten vigorously until one-half cup of sugar gradually has been added and the resulting peaks almost hold their form). This foam then needs to be chilled. Just before serving, beat one-half cup heavy cream and then fold it into the cooked mixture.

135

In the recipe the one-half cup cream is measured **before/after** the cream has been whipped.

Before. Measurements made before the cream is whipped into a foam are more accurate than those attempted afterwards. The volume of whipped cream may vary considerably according to the conditions under which it is whipped, but whipping cream (unwhipped whipping cream) before it is whipped varies only slightly in its volume.

*

Here is a novel type of dressing that refuses to stay in any category discussed here. It is a fast dressing which is made by adding one-half cup finely chopped pecans to 1 pint lemon sherbet. This dressing is excellent served on fruit or cabbage salad.

This lemon-nut dressing should be served **the next day/quickly when it is ready.**

Quickly when it is ready. Sherbet loses its fresh, pleasing quality when it gets too soft to coat the salad effectively and no longer piles well.

1. To define a salad *is/is not* simple.   *Is not.*

2. Water can be drawn into a cell or withdrawn from it by
   _____.   *Osmosis.*

3. Use a salad to obtain contrasts in _____, _____, _____, _____,   *Color, flavor, texture,*
   and _____ with other foods.   *form (or shape),*
                                                                                 temperature.

4. A garnish should be both _____ and simple.   *Edible.*

5. Is a half cup of mixed greens enough to make a good portion?   *No, about half again*
                                                                                 as much is needed.

6. *True or false.*   You can toss many vegetable salads, but most
   fruit salads tend to break up during mixing.   *True.*

7. It takes _____ tablespoon (s) of pure granulated gelatin to
   set a pint of liquid or _____ cup (s) of gelatin dessert.   *One, one-half.*

8. If an acid is added to a gelatin mixture used for salad, use
   more/less gelatin and if sugar is added, use *more/less*
   gelatin than for a plain gelatin.   *More, more.*

9. *True or false.*   Some enzymes in fresh fruit digest gelatin;
   cooking destroys these enzymes.   *True.*

10. *True or false.*   Hot salads are always prepared by pouring   *False; they may be*
    hot liquids over them.   *tossed in a pan over*
                                                                                 heat with the
                                                                                 seasonings or baked
                                                                                 in an oven.

11. Three basic salad dressings are: _____, _____, and _____.   *French, mayonnaise,*
                                                                                 cooked.

12. A French dressing is _____ oil and _____ liquid such as   *Two-thirds, one-*
    lemon juice or vinegar.   *third.*

13. The variations for mayonnaise and French dressings *are/*
    are not   numerous.   *Are.*

14. *True or false.*   Most salad dressings are made from oils that
    have a definite flavor.   *False.*

15. By adding a stabilizer to a French dressing, you *can/cannot*
    make it into a stable dressing.   *Can.*

16. Mayonnaise is an *oil-in-water/water-in-oil* emulsion.   *Oil-in-water.*

17. *True or false.*   To make mayonnaise, add the oil rapidly at
    the start.   *False.*

18. *True or false.*   The way to make a mayonnaise is to add some
    boiling water to it early in the mixing.   *False.*

19. The thickening agents used in a cooked dressing are usually   *Eggs (or egg yolks),*
    _____ and _____.   *starch.*

5

Beverages

Good beverage preparation is important. Beverages are an important part of any meal and, in addition, are frequently served at snack time. The body's usual need for five or six glasses of liquid daily can be met by drinking beverages as well as by drinking water. Coffee, tea, hot chocolate, and fruit punches are often used to meet this need. These and other beverages add considerable interest and variety to our diet.

It is recommended that you drink _____ or _____ glasses of liquid daily.

Five, six.

*

COFFEE

Coffee is sometimes called the universal beverage because it is used so widely in many countries around the world. The source of this beverage is the coffee bean, which is actually the roasted seed of the cherry-like fruit from coffee trees. The origin of this tree is not known with certainty, but apparently some of the first coffee trees grew in Arabia.

Coffee is the beverage prepared from the roasted **seed/ leaf** *of the coffee tree.*

Seed.

*

Within the cherry, which is the fruit of the coffee tree, two seeds are found with their flat sides back to back. These seeds are removed from the pulp and carefully dried in the sun on large drying floors. It is these dried coffee beans (seeds) that are exported to the various coffee-consuming countries.

Coffee beans are shipped **before/after** *being dried in the sun.*

After. The dried beans can be shipped without becoming spoiled or moldy.

*

The quality of a particular coffee is influenced by the variety of coffee tree, the altitude, and the general climate where it is grown. The coffee available in the United States is blended from several sources and may be from such diverse parts of the world as Africa, the East Indies, Turkey, Hawaii, the West Indies, and South America. Some of the preferred coffees of the world are grown in the mountains that boast a semitropical climate; coffee can also be grown in the lowlands. The coffee from the plantations along the rivers is known as "rio" or "river" coffee.

Why do you think coffee is not grown in the continental United States.

The best coffee is grown in a semitropical climate—a climate not found anywhere in the United States except Hawaii.

*

Coffee from each area and each variety of coffee tree has a specific characteristic flavor profile. To obtain a flavorful beverage, several types of coffee beans from selected regions are blended together. The Brazilian Santos coffee contributes richness and flavor; mountain-grown Central and South American beans provide fineness of flavor and body; a small amount of the rios adds a slight astringency or bitterness. The blending of the many available beans to obtain a product with just the desired flavor characteristics is a discriminating, exacting art.

Is it possible to tailor a coffee to obtain one with exactly the characteristics preferred by many consumers in an area?

Yes. In fact, it is common practice to make a somewhat stronger, more bitter blend for sale in the South than is marketed in the West.

*

The characteristic flavor and aroma of coffee are developed during the roasting process. Green, blended beans are rotated in a chamber maintained at from 385° to 500°F until the desired light, medium, or heavy roasting is completed. As the bean changes in color from green to a brownish black, the bland aroma gives way gradually to a full, pleasing, rather caramelized odor. Accompanying these changes is the development of the full coffee flavor. Roasting is carefully controlled to obtain the desired flavor for a particular coffee-consuming public.

True or false. The extent of roasting has considerable effect on the flavor characteristics of a coffee.

True.

*

Although coffee is enjoyed all around the world, the degree of roasting and the way in which the beverage is served vary from country to country. Arabs and Turks choose a dark-roasted, finely pulverized coffee that is boiled and served, sweetened with sugar, in a very small cup. This coffee is brewed in a small brass pot called an ibrik. When the coffee is poured from the ibrik, most of the grounds remain in the pot, and some foam should be poured with the coffee into the cup. This foam is called "face" and is said to be where the expression "to lose face" (to lose dignity) originated.

How do you think the dark roasting of Turkish coffee affects the flavor?

The long roasting develops a strong, bitter flavor which has sometimes been described by the uninitiated as being similar to scorched shoe leather.

*

The French also serve an after-dinner coffee made with darkly roasted coffee. This is brewed as a dark, strong-flavored beverage traditionally served in demitasse (very small) cups. Variations of this beverage are sometimes made by adding brandy, cardamom, cloves, or cinnamon either alone or in combinations.

Would coffee made from lightly roasted beans have as strong and bitter a flavor as is typical of demitasse coffee made with a darkly roasted coffee?

No.

*

In Europe, coffee may be made with milk as the Spanish do, or milk may be mixed with brewed coffee in approximately equal amounts. This half-and-half mixture is called *"café au lait"* in France and *"café cappacino"* in Italy. In Vienna, coffee frequently wears a whipped cream hat.

Is coffee always combined with milk or cream when served in a Continental manner?

No. When served as a demitasse, no milk is added. The Italians also like "espresso," a beverage made by forcing steaming hot water through very fine particles of darkly roasted coffee.

140

The flavor of coffee is a blend of caramelized sugars, caffeol, and other organic substances. The most prominent flavor compound identified in coffee appears to be caffeol. Caffeine is a stimulant found in coffee, but this substance seemingly does not influence the flavor bouquet of coffee.

*The one compound most important in the flavor bouquet of coffee is **caffeine/caffeol.***

Caffeol.

*

The flavoring substances in coffee are very volatile; that is, they escape as gases into the air. You can easily detect the presence of these substances by their coffee-like aroma. Maximum coffee flavor is retained if these volatile substances are trapped rather than permitted to escape into the air. These compounds are readily lost after the beans are ground because grinding greatly increases the surface area available for evaporation to take place. Coffee in the bean will keep its flavor about 20 days, but ground coffee will lose 20 per cent of its flavor in three days at room temperature. Immediately after roasting and grinding, coffee is packaged in vacuum-sealed cans to halt the flavor loss.

This procedure stops the evaporation of volatiles and conserves flavor compounds.

Why is coffee packed in vacuum-sealed cans?

*

The flavor of coffee in an opened can needs to be protected by proper storage procedures. The can should be tightly capped at all times. Just remove the cover long enough to measure the coffee needed and immediately replace the cover securely. This precaution helps to keep the volatiles from escaping. Storage in the freezer or refrigerator also reduces volatile losses because of the cold temperature.

*Coffee should be **tightly/loosely** covered after the can is open to retain the volatile _____ compounds.*

Tightly, flavor.

*

Most people will agree that the single, most important factor in a cup of coffee is flavor. From the previous discussion, you are well aware of the importance of using coffee that has been carefully processed and properly stored. However, good preparation procedures are equally important if you are to make good coffee. The best flavor is extracted from the grounds into the finished beverage when the temperature of the water is between 185°F and 203°F during the brewing period. In this temperature range, the

141

maximum amount of flavor (caffeol) is extracted; at higher temperatures the bitter substances (tannins) also are extracted and cause the beverage to be bitter.

During preparation of coffee, the water should be held **between 185°F and 203°F/at the boiling point** *to* **maximize/minimize** *extraction of the bitter tannins and to* **maximize/minimize** *caffeol extraction.*

Between 185°F and 203°F, minimize, maximize.

*

The flavor of coffee is influenced by the quantity of grounds used to brew the beverage. Two tablespoons of coffee grounds per beverage cup (three-fourths measuring cup) are used if you wish to make a strong beverage without bitterness. One to one and one-half tablespoons per cup will produce the milder beverage that some people prefer. It is important to measure both the quantity of the coffee grounds and of the water used to prepare the beverage if you are to consistently make good coffee. A pound of coffee makes from one and three-fourths to two and one-half gallons of good coffee.

To prepare a cup of strong yet mellow coffee, use **one tablespoon/two tablespoons** *of coffee grounds for each* **three-fourths cup/one cup** *of water.*

Two tablespoons, three-fourths cup.

*

The length of time coffee is brewed and the size of the grind have important effects on the flavor of the finished product. The longer the extraction period, the more bitter is the coffee. To obtain a beverage with a mellow rather than a bitter flavor overtone, a short extraction time is recommended. A short brewing time is particularly necessary when a fine grind of coffee is used because water-soluble materials are extracted more quickly from these small coffee particles than they are from larger fragments of the coffee bean.

Name five factors that influence the flavor of a cup of coffee.

The degree of roasting, the blend of coffee beans, the quantity of coffee used for each cup of beverage, the brewing time, and the size of the grind are all factors that influence the flavor of the brewed coffee.

*

To brew a truly fresh-flavored, aromatic cup of coffee, you need to use a clean odor-free pot. Coffee contains oils that tend to cling. This oily film accumulates on the interior of the pot and gradually develops a strong, rather rancid character that is readily detected in the coffee. This problem is easily eliminated by simply scrubbing the pot thoroughly with hot soapy water and then giving it an adequate hot rinse after each use.

It is necessary to carefully wash a coffee pot after each use to avoid **off flavors in the coffee/accumulating caffeine in the pot.**

Off flavors in the coffee.

*

Commercial establishments often make coffee in equipment that requires cloth bags to hold the coffee grounds. Such bags need to be given special attention when they are washed to be sure that the oils are removed as well as possible. Sometimes flavor problems are minimized in these bags by soaking the washed bags under water when not in use. The water keeps the air from any oils that may be clinging to the bags and thus helps to slow down the development of an off flavor.

Off flavors develop more easily at **room/high** *temperatures in coffee; rancidity in coffee develops when the oils present are* **oxidized/removed.**

High, oxidized.

*

Glass or ceramic coffee makers are desirable because they are easy to clean and do not themselves contribute to the flavor of the coffee. However, they can be broken, which is an obvious disadvantage. Unbreakable metal coffeemakers are also available. Of the metals, stainless steel is the best choice because it does not contribute a noticeable metallic taste to the beverage.

If you were selecting a coffeemaker for use in a restaurant, what material might you select?

Stainless steel would probably be your best choice because the flavor of the beverage prepared in it should be good and breakage will be reduced.

*

There are three basic types of coffee pots: dripolator, percolator, and vacuum pot. Although you will notice various design features in coffee pots, any pot operates basically as one of these three types. Coffee is ground especially to be brewed in a particular type of pot. You will get the best results if you use drip grind coffee when preparing coffee in a dripolator, regular grind in a percolator, and fine grind in a vacuum pot.

What are the three types of coffee pots? What grind should be used in each?

Dripolator—drip grind; percolator— regular grind; and vacuum pot—fine grind.

A dripolator is considered to be the best pot to use if you wish to make coffee with a minimum amount of bitterness. A dripolator is designed with three separate compartments. The bottom section houses the finished beverage, the middle one holds the coffee grounds, and the top one holds the boiling water. To make coffee in this type of coffeemaker, you measure grounds into the center section (using filter paper if desired) and then assemble the pot. Water is boiled in a separate container and then the correct amount is poured into the top compartment; the lid is placed on top to retain the heat. After all the water has dripped through the grounds, the two upper sections are removed and the lid is placed on the bottom section containing the finished beverage.

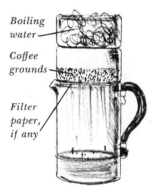

Boiling water

Coffee grounds

Filter paper, if any

Diagram a dripolator and show where the boiling water and coffee grounds are placed.

*

Dripolator coffee is flavorful and usually free of bitterness because: (1) the water is in contact with the grounds only briefly and (2) the water passing through the grounds is within the best temperature range (185°F to 203°F) for brewing a mellow coffee. The one disadvantage of dripolator coffee in the home is the need to heat the water in a separate container. Automatic dripolators for commercial use are now on the market. Coffee made in a nonautomatic dripolator can be held at serving temperature by simply placing the pot containing the prepared beverage on a heating element.

Dripolators **are/are not** *an appropriate choice for making mellow, flavorful coffee.*   *Are.*

*

A percolator is designed as a single container which houses a covered basket resting near the top of a hollow stem; the stem extends from the bottom to the top of the container. Place the measured amount of water in the percolator, insert the stem through the center of the basket containing the correct amount of coffee grounds, and place this assembly in the percolator. Make certain that you replace both the cover on the basket and on the percolator itself. Now heat the water rapidly until it begins to flow upward through the stem and sprays against the glass in the center of the lid. Reduce the heat to just the point where the percolating action is maintained steadily but not extremely vigorously. As soon as water begins to come up through the stem, start to time the percolation process. Most people find that a six- to eight-minute percolation period produces coffee of the desired strength.

Diagram a percolator showing where the coffee grounds and water should be placed.

*

The longer the percolation time, the stronger and more bitter will be the beverage. Even with the shorter brewing time, percolator coffee will be more bitter than dripolator coffee because the water is hotter and it passes repeatedly through the grounds rather than simply having a single brief contact.

Dripolator/percolator coffee will be more bitter and strong because the water that comes in contact with the grounds is above/within the temperature range shown to brew the most flavorful, least bitter coffee.

Percolator, above.

*

As soon as percolator coffee is brewed to the desired strength, remove the basket containing the grounds. This is necessary in both automatic and nonautomatic pots because the coffee may become hot enough to begin to percolate again. Of course, any such extension of the percolation process will increase the bitterness of the beverage.

True or false. The basket of coffee grounds should be removed as soon as the desired strength of beverage has been brewed.

True.

*

Automatic percolators are popular among many homemakers because they can simply assemble the pot, containing the water and coffee, and plug in the appliance. Such percolators have properly controlled heat and timing cycles to ensure regulation of both temperature and time throughout the brewing period. At the end of this time, the pot automatically cycles to a keep-warm temperature setting so that the coffee can be held for a considerable time without cooling below a serving temperature.

Automatic percolators have automatic controls regulating the _____ and _____ cycle.

Temperature (or heat), timing.

*

The third basic type of coffeemaker, the vacuum pot, has a lower container into which the water is measured. Tightly fitted into the top of this chamber is a second upper container with a funnel that protrudes almost to the bottom of the lower vessel. A snug, thick, rubber gasket provides the necessary close fit between the two portions of the vacuum pot. The coffee grounds are measured into the upper container and are held there by a metal filter securely fastened across the opening to the funnel. A lid is then placed on the top of the assembled pot.

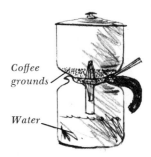

Coffee grounds

Water

Diagram a vacuum coffeemaker, showing where the coffee grounds and water are placed.

*

Heat is applied to the pot throughout the brewing process. First the water is heated until so much pressure develops in the bottom container that the boiling water is forced up through the funnel and into the upper chamber, where it is then in close contact with the grounds. Usually, the heat is maintained for approximately three minutes after the water enters the upper chamber. At the end of that time, the beverage is strong enough for most people. The finished beverage is returned to the lower container at the desired time by simply reducing the heat applied to the vacuum pot. This reduces the pressure and permits the beverage to return through the funnel to the lower pot, while the coffee grounds are retained in the upper portion by the metal filter. The upper assembly is then removed and the lid transferred to the lower pot.

The coffee will filter back to the lower container when the pressure from the bottom pot is reduced. This is accomplished by reducing the heat.

How do you make the beverage return to the lower pot when you are making coffee in a vacuum pot?

*

Automatic vacuum pots are convenient to use because you can simply assemble the ingredients and the pot and then let the machine take over. To use a vacuum pot, you must be sure that the heat switch is set on high. This will provide the heat necessary to heat the water and build up the pressure required to force the water into the upper cham-

146

ber. At the end of the correct brewing period, the heat switch automatically shifts to the low setting. This low setting permits the coffee to return to the lower container and then continues to hold the beverage at serving temperature.

When you make coffee in an automatic vacuum pot, set the heat switch on   **low/high.**

High.

*

Coffee made in a vacuum pot, like coffee made in a percolator, is brewed at a temperature a bit higher than is recommended. This higher temperature, combined with the intimate prolonged contact between coffee grounds and water, is responsible for the slightly bitter character of coffee made in a vacuum pot.

Will vacuum coffee be more bitter than coffee made in a dripolator?

Yes.

*

Sometimes you may need to make coffee when a regular coffee pot is not available or is too small to hold the amount of coffee you need. Steeped coffee, often referred to as boiled coffee, is your answer. To make steeped coffee, measure the correct amount of water into a kettle or old-fashioned coffee pot and bring it to a boil before adding the coffee grounds. It is best to have the grounds (regular grind) tied in a muslin bag so that they can be removed at the end of the steeping period. When the coffee is added to the boiling water, reduce the heat so the brewing beverage is simmering rather than actually boiling. This temperature control is important. Coffee will be bitter if it is allowed to boil. Usually a 10-minute extraction period brews coffee of the desired strength.

_____ *coffee can be prepared by tying* _____ *grind coffee in a muslin bag and steeping it in* **boiling/simmering** *water until the beverage is the desired strength.*

Steeped (boiled),
regular, simmering.

*

To estimate how much coffee to prepare for a group, you can calculate from 20 to 25 cups (5½ ounce size) per gallon of coffee prepared. Usually it is wise to plan on 1½ cups of coffee for each person. Approximately 1 pound of coffee should be used to prepare 50 cups of coffee.

147

How many gallons of coffee should you make to serve 100 people?

Make 7½ gallons (20 cups per gallon) and about 3 pounds (or slightly less) coffee.

*

Some electric coffeemakers have automatic controls that regulate the time the coffee will be in contact with the water. Some are also adjustable so a weaker or stronger brew can be made. This adjustment merely extends or reduces the contact time between the water and the grounds. The mild coffee flavors come from the grounds first in extraction, the heavier flavors coming with extended extraction time. These heavier flavors are composed largely of astringent substances such as tannins, caramelized substances, and stronger aromatic compounds. The job in making good coffee is to extract the desired mild flavors and just the right quantity of the heavier ones to give a flavorful, balanced brew.

If your coffee is too strong, what might be the reasons?

1. *Too much coffee in proportion to water is being used.*
2. *Too long a brewing time is being used.*
3. *Too fine a grind is being used,*
4. *The brewing temperature is too high.*

*

TEA

Probably more people in this world drink tea than coffee. The tea bush was originally found in China; today the peoples of the Orient and Near East drink considerable quantities of tea. The plant grows best in a semitropical climate, and the best tea is grown at high altitudes.

Are the ideal growing conditions for tea similar to those for coffee?

Yes.

*

The growing season is also important to the tea quality. The best teas are picked in the spring, when leaves are tender and young and the plant is growing vigorously. If the tea is picked during the rainy season (fall in many tea-growing areas), the flavor is apt to be marked by wine-like overtones. Quality is also dependent upon the size of the tea leaves that are plucked. The tip leaf and the two leaves just below this terminal bud are called a standard pluck.

148

If more leaves are picked, the tea has a coarser, blunter flavor.

Which of the following is not true of tea?
A.   *The size of leaf affects quality of the tea.*
B.   *The best tea comes when the plant is growing vigorously.*
C.   *The best flavor is obtained from leaves plucked during the rainy season.*
D.   *The terminal bud and two leaves are a standard pluck.*   C.

*

All tea leaves undergo processing to modify the color and flavor of the finished beverage. The first step is a slight withering of the leaves before they are rolled to release some of their juice. It is at this point that the processing is varied to develop the desired finished product. The rolled leaves can be treated in any one of three ways to make green, oolong, or black tea.

True or false. By selecting the appropriate processing method, a standard pluck can be made into green, oolong, or black tea.   True.

*

The type of tea produced depends upon the length of time the rolled leaves are allowed to ferment. Tannins, the astringent substances in the fresh leaves, are slowly oxidized during the fermentation period. Tea leaves are fermented at room temperature by spreading them on mats and turning them occasionally. Black tea is fermented the longest; it is usually left on the mats 18 to 20 hours. Oolong tea is said to be semifermented because it is on the mats only 3 to 4 hours. Green tea is not fermented at all. All fermentation stops when the leaves are fired (gently heated) to dry them and fix the flavor and color.

*The differences between green, oolong, and black tea are due to the amount of **fermentation/boiling** that the leaves undergo during processing.*   *Fermentation.*

*

Some of the finest teas are handrolled and handfired. On large tea plantations, machines roll over the leaves, causing them to roll up; but when fine teas are wanted, native girls sit and hand roll each leaf. They also may basket-fire them, a process in which the leaves are placed in a basket and held

149

over glowing coals while the native women lift the leaves and gently drop them back into the basket. The heat coming through the basket slowly dries the leaves.

An Indian maharajah is said to have paid $30,000 once for a kilogram (2.2 pounds) of rare tea. Why do you think this tea was so valued?

It must have come from only the finest of terminal buds, picked in the spring and at some of the highest altitudes tea can be grown. Probably it also was hand rolled and hand fired.

*

Tea shipped to this country is graded by tea tasters. The size of leaf or bud is the major factor in deciding grade, but they also sample teas to see if the flavor is equal to the grade or size of leaf or bud. They also reject teas that do not meet sanitary standards.

True or false. The Tea Board decides grades of tea solely on the size of the leaf or bud and only concerns itself with grading.

False. Flavor must correlate with size of leaf to meet a grade, and sanitary standards are also imposed.

*

Like coffee, teas are blended to give the desired flavor. Some tea bags may contain a blend of 30 different teas to have a proper flavor balance. Darjeeling teas from the Himalayas in India or high-grown Ceylon teas are excellent, but even these teas may be blended with some other grades for the more acid or pungent flavor preferred by some consumers.

Would a real tea connoisseur prefer to drink unblended tea or would he be more apt to select a blend?

He might like a fine Darjeeling or Pengelengen (Indonesian tea) without a blend of other flavors, because these flavors are so distinctive and fine that they can stand on their own. However, he may wish to build flavors by blending together several different kinds of tea for his beverage.

*

The best black tea grades, based on size of leaf, are (beginning with the top quality) : orange pekoe, pekoe, and souchong. There are also grades for broken tea leaves, fannings (small particles), and dustings (very fine particles). Since the quality of the brew depends on the speed or time of extraction, just as it does in coffee, the broken grades are often marketed, and tea leaves may even be broken up purposely to improve the quality of the brew made from the leaf.

Do you think that you could examine some tea and get an estimate of its grade? What would you think if you saw coarse leaves, stems, or a great deal of broken leaf materials?

Yes. The finer, tighter rolled leaves would indicate a high-quality tea; coarse leaves, stems, or broken leaves receive a lower grade.

*

Frequently, oolong tea is graded the same as green tea. Green tea grades, from best to poorest quality, are: gunpowder, young hyson, hyson, and imperial. Normally, the only two grades seen on the market will be gunpowder and young hyson. Gunpowder was a term given by the English because the size of the leaf and the size of the gunpowder used in the English ship's cannons were almost the same. The term "young hyson" comes from the sound of the word "spring" when said in Chinese. The Chinese indicated the fine quality of this tea by calling it spring tea.

Why is spring tea of excellent quality?

It is made from the tender young leaves produced in the spring when the tea bush is growing vigorously.

*

The expression "all the tea in China" means little to us today because we import very little tea from China. Most of our tea comes from India, Ceylon, Sumatra, Java, and Africa, although some green and oolong teas are imported from Okinawa and Japan. Some areas in southern Russia also produce excellent teas.

Check the countries that are the biggest exporters of tea: China, India, Ceylon, Sumatra, Java, Africa, Russia, Okinawa, and Japan.

Check India, Ceylon, Africa, and perhaps Sumatra and Java. Russia, Okinawa, Japan, and China ship tea to the democratic nations of the world in about that order.

*

To make a good cup of tea, use one level teaspoon to three-fourths cup of water. The water should be freshly boiled so that it will retain a maximum amount of oxygen to contribute to the brisk quality of the beverage. Boiling water is poured over the tea and then allowed to steep three minutes in the covered tea pot. The entire steeping operation is done without boiling the tea at all. The temperature usually remains between 190°F to 210°F. The leaves are removed after three minutes to prevent continued extraction of the bitter substances.

Would you preheat a pot if you were going to steep tea in it?

Yes. *A cold pot would reduce the temperature of the water below that at which best extraction of flavors occurs.*

*

At a temperature between 185°F and 210°F the desired flavor compounds will be extracted into the beverage during the three-minute steeping period. If tea is boiled, oxygen will be lost and the flavor will be flat. However, boiling harms the flavor of tea more than can be explained simply by loss of oxygen. Such a high temperature extracts tannins (also called polyphenols) in too great a quantity. Polyphenols are the astringent bitter compounds in tea. If the water is between 185°F and 210°F and the steeping time is three minutes, the desirable flavor components will be at a maximum and the tannin extraction will be minimized.

The temperature of the water for brewing tea should be between _____ and _____°F; the extraction time should be _____ minutes.

185, 210, three.

*

To make tea in quantity, bring the measured amount of water to a boil; stir in the correct amount of tea and allow it to stand covered for three minutes without further heating. Remove the leaves. Use a cup of tea for every two and one-half gallons of water. This will give 50 cups of tea.

To make 100 cups of tea, use _____ gallons of water and _____ cups of tea.

5, 2.

*

Black tea has a heavier body, darker color, and a more mellow flavor than oolong or green tea because of the fermentation it undergoes. Oolong is a tea in between green and black tea. The infusion made with black tea is an amber reddish-copper color (much like a bright copper penny) while oolong tea is a bit lighter with a trace of green in it; green tea is a pale greenish-yellow color.

Green tea can be distinguished from oolong and black tea by its flavor and _____.

Color.

The quantity of unoxidized tannins remaining in the tea leaf influences flavor as well as color, because these water-soluble substances leach out to become a part of the beverage. Green tea is more bitter than the other teas because it contains the largest quantity of tannins; oolong is somewhat bitter and black tea is quite mellow.

Match the tea in column A with its description in column B.

A	B	
1. Black tea	A. Semifermented, moderate tannin content	1. C.
2. Oolong tea	B. Unfermented, highest in tannins	2. A.
3. Green tea	C. Fully fermented, lowest in tannins	3. B.

*

Tea will have a better flavor if made from soft or barely alkaline water rather than hard water. The alkaline hard-water salts precipitate the tannins to produce a cloudy, dull-looking tea with a film on the surface. The film is quite objectionable because it coats tea cups and pots as the beverage cools. You can reduce this problem by adding a few drops of lemon juice. The flavor of the tea also seems to become more brisk when a small amount of lemon juice is used.

Hard water produces a dull, dark tea with a scum on top which can be removed by adding just a bit of ____ ____.   *Lemon juice.*

*

When metals come in contact with tea, they give a metallic flavor to the beverage. Therefore, tea tastes best when made in a glass or pottery tea pot.

Would tea made in a ceramic tea pot taste better than tea made in a stainless steel pot?   *Yes. The ceramic pot does not give tea a metallic taste.*

*

The best tea is made when the leaves are loosely encased in a container that allows good circulation of water through the leaves, even after they swell with water. At the end of the steeping period the leaves can be removed easily from the beverage. A thin parchment paper is an excellent device for holding tea leaves because it is strong enough not to tear easily, even when wet, and has the added advantage of not giving the beverage a metallic flavor. An aluminum tea ball may give a slight metallic taste.

True or false. A parchment paper bag containing the tea leaves is an excellent device for steeping tea, because water passes through the paper readily and yet the leaves can be removed easily.

True.

*

Tea may be judged by the following standards. It should have a clear sparkling appearance with no trace of film. The color should be typical for the type of tea used, i.e., amber for black tea, amber touched with a trace of green for oolong, and pale greenish-yellow for green tea. The flavor and aroma should be delicate and characteristic of the type of tea, with no suggestion of astringency or bitterness.

Preparation of tea should be judged on the basis of the ____, ____, ____, and ____ of the finished beverage.

Clarity, color, flavor, aroma.

*

Sometimes tea may be blended with fruit juices to make a punch. Remember that if iron salts are present in the fruit juice (canned fruit juices will certainly contain them) the iron may join with the polyphenols to form insoluble salts that cloud the punch. This happens especially when the fruit juices contain the red pigments known as anthocyanins.

If you blend tea with cranberry juice when making a punch and you get a cloudy precipitate, what has happened?

The polyphenols, anthocyanins, and iron have combined to form a precipitate.

*

Spices, orange or lemon peel, dried flower blossoms such as jasmine, or other substances can be added to give tea a special flavor. Fresh or dried mint leaves may be used similarly to give a special flavor contribution. A Russian tea is a sweetened product seasoned with orange, lemon, and pineapple juice and then just lightly flavored with a touch of cinnamon stick. Spiced tea is made by seasoning tea during its brewing with lemon and/or orange peel, whole cloves, and stick cinnamon.

*Tea **lends itself/does not lend itself** to flavoring by spices.*

Lends itself.

*

Good iced tea can be made by using an extra quantity of leaves and keeping the infusion time to strictly three minutes. The extra leaves are necessary to compensate for the dilution of the tea that takes place as the ice melts. Hard water will be more apt to cloud iced tea than soft water. The hard-water salts very readily precipitate any extracted polyphenols (tannins) causing the brew to become very murky.

Iced tea is made by using **more tea leaves/a longer extraction time** *to make a stronger beverage to allow for dilution by melting ice.*

More tea leaves.

*

To make good iced tea easily and successfully, the 1-2-3 method is recommended. One quart of tea is made by the 1-2-3 method as follows:

> 1 cup of boiling water poured over
> 2 heaping tablespoons of tea

Steep for three minutes while stirring, strain, and blend into

> 3 cups of cold water

Add ice and serve.

   For a gallon of iced tea use:

> 1 quart of boiling water over
> 2 ounces of tea.

Steep three minutes while stirring, strain, and blend into

> 3 quarts of cold water.

Add ice and serve.

The 1-2-3 method refers to pouring boiling water over _____, allowing it to steep three minutes, and then straining and adding **cold/hot** *water.*

Tea, cold.

*

Iced tea should be perfectly clear. As you have just learned, hard water can give iced tea a murky appearance. However, this cloudiness can happen even with soft water if the extraction temperature is too high or if the tea is steeped too long. Either of these conditions causes excessive extraction of the polyphenols and the danger of clouding is greater. This clouding is apt to occur if green or oolong tea is used (which is not done too often) or if a poor quality, highly astringent black tea is used, because these teas contain large quantities of unoxidized polyphenols.

The precipitation of polyphenols in a tea brew is increased by a **short/long** *extraction period and/or a* **low/high** *steeping temperature.*

Long, high.

155

If a tea brew is made and then placed into a refrigerator to chill, it is apt to cloud. Freshly made tea (especially black teas low in polyphenols) made slightly stronger than regular tea and poured over cracked ice usually will not cloud. Instant tea, made in much the same manner as instant coffee, makes a pleasingly clear iced tea.

For a good glass of iced tea, make the tea slightly stronger than desired and pour it **warm/chilled** *over cracked ice.* Warm.

*

Frequently, iced tea is served with a simple sugar syrup instead of granulated sugar because the granulated sugar goes into solution so slowly in the cold tea.

Powdered sugar is finely ground and would go into solution much more quickly than granulated sugar. Why isn't it used in iced tea? It contains starch which would cloud the tea.

*

COCOA AND CHOCOLATE

Cocoa and chocolate are popular beverages, especially with children or teenagers. The large quantity of milk in these beverages makes them highly nutritious. Hot chocolate is richer (contains more fat) than cocoa; cocoa is made with breakfast cocoa which contains 22 per cent cocoa butter, while hot chocolate contains bitter chocolate with a fat content of more than 50 per cent.

Hot chocolate is a **richer/less rich** *beverage than cocoa.* Richer.

*

Chocolate and various other chocolate-flavored products are made from cacao beans. These beans are harvested from cacao trees that grow in semitropical countries.

What other beverage products come from these regions? Tea and coffee.

*

The beans develop inside the large pods, which are the fruit of the tree. These pods are harvested and allowed to ferment so that the seeds can be removed easily from the pulp in the pod. The beans then are roasted to develop their

156

flavor. (Remember, coffee is also roasted to develop and fix flavor.) The chaff or skin on the beans is removed and the beans are cracked into particles called "nibs." These nibs are ground and the chocolate liquor that is released is processed into bitter chocolate.

The flavor of chocolate is developed during **roasting/ grinding.**

Roasting.

*

Several products are made from the chocolate liquor. Bitter chocolate contains 50 per cent or more cocoa butter, but breakfast cocoa contains only 22 per cent cocoa butter.

If a cocoa contained less than 22 per cent cocoa butter, do you think it could be called "breakfast" cocoa?

No, it could not. The federal government restricts the name "breakfast" to cocoas containing 22 per cent or more cocoa butter.

*

If milk solids are added to bitter chocolate, the product is called milk chocolate. Often this is sweetened. If only slightly sweetened, it is called semisweetened milk chocolate; if sweetened enough for use in most candies, it is called sweetened milk chocolate.

Do we have semisweetened and sweetened chocolates that contain no milk?

Yes, these are used often to coat dark chocolates.

*

If chocolates or cocoas are treated with alkali during their processing, they are called Dutched or Van Houten chocolate or cocoa in honor of the Dutchman who discovered the process. This treatment, which changes the color as well as the flavor and solubility of these products, is thought by many to give a superior product. However, the natural chocolate products often are said to have a truer flavor. It seems that preference for natural or Dutched chocolate is a matter of personal taste. Certainly, it is readily apparent that Dutched chocolate products have a milder flavor and lighter color than the natural chocolate.

The Dutch or Van Houten process uses **alkali/acid** *in the processing of the chocolate.*

Alkali.

157

The stimulants in chocolate products are theobromine and caffeine, mostly the latter.

True or false. Coffee contains the same stimulants as chocolate.

False, coffee has only caffeine in it.

*

Storage is difficult because chocolate products lose flavor when they get too hot or are in an area too high in moisture; under these conditions, bloom develops and discolors (turns brownish gray) the surface. Cocoa also can lump badly if left open in a damp place. Because of its fat content, chocolate products absorb flavors from other products easily; they should be protected from air containing other odors.

If you say that the chocolate or cocoa has a bloom, this means:
A.   A loss of flavor.
B.   A loss of texture.
C.   The development of lumps in cocoa.
D.   A change in color on the surface of the product.
E.   An increase in aroma.

D. A change in color on the surface of the product.

*

In cooking, it is sometimes necessary to substitute cocoa for chocolate; use three to three and one-half tablespoons of cocoa and one tablespoon of fat for each ounce (one square) of chocolate. The fat added may be butter, margarine, or shortening.

If you wish to use cocoa in place of chocolate, use _____ to _____ tablespoon(s) of cocoa and _____ tablespoon(s) of fat for each ounce of chocolate.

Three, three and one-half, one.

*

When chocolate or cocoa are cooked in milk or water, their starch gelatinizes and causes some thickening. Chocolate contains 8 per cent starch and cocoa 11 per cent. Therefore, in making hot chocolate or cocoa, you will find it wise to boil chocolate or cocoa briefly with water to gelatinize the starch before adding milk. This step assures you that the starch is gelatinized without overheating the milk. By heating the milk for a shorter time, you avoid forming considerable scum on the top of the beverage.

Precooking cocoa or chocolate with water before adding milk for cocoa or hot chocolate _____ the starch and shortens the heating with milk; this reducing the amount of _____ formed.

Gelatinizes, scum.

*

Cocoa is made by using either the: (1) syrup method, (2) paste method, or (3) dry-blend method. The dry-blend method is simplest but not the most satisfactory. In this method the dry ingredients (cocoa, sugar, and milk powder if used) are blended together and then vigorously stirred into the hot liquid. It has the disadvantage of not heating the starch sufficiently to gelatinize it. The result is a starchy flavor and a slightly grainy texture.

*In the **dry-blend method/syrup method,** all dry ingredients are mixed together and then stirred vigorously into very hot liquids.*

Dry-blend method.

*

The syrup method begins by boiling cocoa or chocolate with sugar and water for a few minutes to gelatinize the starch. Then, this syrup is added to hot milk.

Do you think that we could substitute dried milk or evaporated milk in proper ratios for fresh milk when making cocoa or chocolate?

Certainly.

*

The paste method is very similar to the syrup method. The first step is cooking cocoa or chocolate with water until a paste is formed. Subsequently, the sugar is blended in well, and the hot milk is added last.

*In the **syrup/paste** method the sugar is added before the cocoa and water are boiled; in the **syrup/paste** method the sugar is added after the cocoa and water are boiled.*

Syrup, paste.

*

Hot chocolate and cocoa need to be heated carefully to avoid scorching them. Avoid boiling them to minimize scum formation. Formation of scum on the top of these products can also be controlled if you whip them to create a foam, or add a whipped cream or marshmallow foam to cover the surface.

*Hot chocolate or cocoa **should/should not** be boiled.*

Should not.

Well-prepared cocoa or hot chocolate will have a pleasing flavor and will not settle out quickly. There will not be a scum. Careful heating and gelatinization of the starch are the keys to success.

Cocoa **should/should not** *separate readily.*

Should not.

*

To make French chocolate, blend melted chocolate with a slight bit of warm milk and then blend this into sweetened whipped cream. Add a quantity of this to a cup and pour in very hot milk, stirring gently to blend.

Would this be a type of product to serve to calorie watchers?

Decidedly not; it's loaded with calories but certainly good.

*

Instant cocoa mixes, containing all the dry ingredients for the preparation of hot cocoa, are now available. This is possible because the starch in the dry cocoa has been pregelatinized. It is only necessary to add hot liquid to this type of product.

Instant cocoa mixes contain cocoa whose starch has been _____.

Pregelatinized.

FRUIT AND VEGETABLE BEVERAGES

Fruit juices are used as beverages, either by themselves or combined with other liquids. Vegetable juices may also be used, but these often are not combined with other liquids. Both, besides furnishing liquid, also may furnish valuable nutrients.

What nutrients do you think might be provided in fruit and vegetable beverages?

Ascorbic acid, pro-vitamin A, and several trace minerals are present in selected fruit and vegetables juices.

*

The juices of some fruits such as limes, lemons, oranges, pineapple, or others having considerable acid may be diluted with water and sweetened with sugar to make ades. These are refreshing and still may contribute some ascorbic acid, even though they are diluted. The juices of some other fruits may be used to make ades if lemon or some other tart fruit juice is added to give a sufficient quantity of tartness and flavor.

True or false. Ades are fruit juices diluted with water, sweetened, and then served chilled.

True.

*

Punches are fruit drinks made by combining various fruit juices. Spices and tea may be added for flavoring.

Would a punch be very similar to an ade?

Yes. Both are very similar products, but the juice is diluted in an ade.

*

Fruit nectars are fruit juices. The term is frequently reserved for special types of fruits such as apricots and peaches. A fruit drink is one in which water and some artificial flavoring have been added to fruit juice. The term "juice" indicates that it comes from the item without dilution. The term "drink" does not indicate this.

True or false. If you purchase a nectar, you are getting a juice, but if you purchase a beverage labelled as a "drink," you are not getting the pure juice.

True.

*

REVIEW

1. *True or false.* Rio coffee is a high-grade coffee.

   False.

2. *True or false.* The highest quality tea and coffee are grown at high elevations.

   True.

3. Coffee comes from: (a) beans, (b) roots, (c) leaves, or (d) meat of a fruit.

   Beans.

4. *True or false.* Heat is used in processing tea, coffee, and cacao beans.

   True; roasting in coffee and cacao and firing in tea.

5. *True or false.* Roasted coffee has more flavor than has green coffee.

   True.

6. In which situation would ground coffee in a tightly closed container remain freshest: (a) at room temperature, (b) in the refrigerator, or (c) in the freezer?

   In the freezer.

7. As soon as the vacuum seal on a can of coffee is broken, the flavor will gradually *get stronger/be lost into the air.*

   Be lost into the air.

8. The grind of coffee having the largest particle size is _____; that having the smallest is _____.

   Regular, fine.

161

9. Drip grind is *finer/coarser* than that used for percolated coffee.

Finer. Percolated coffee is made with regular grind.

10. Many of the flavor compounds in coffee are *volatile/non-volatile.*

Volatile.

11. If you had a choice in selecting equipment for sending tea or coffee to hospital patients, would you select pottery or aluminum pots?

Pottery. The flavor would be better and the beverage would stay warm a long time if the pots were preheated.

12. Percolated coffee takes *a longer/less* time for brewing than does vacuum coffee.

A longer.

13. Hard water *improves/decreases* the clarity of coffee.

Decreases.

14. Coffee quality is judged by _____, _____, _____, _____, _____, and _____.

Color, clarity, taste, aroma, body, temperature.

15. To brew tea, use _____ teaspoon (s) for every six ounces of water; pour boiling water over the tea and allow to steep _____ minutes.

One, three.

16. The precipitate in cloudy tea is caused by precipitation of the _____ in the tea.

Polyphenols (tannins).

17. *True or false.* Coffees are blended to give desired flavors, but teas are not.

False.

18. Which tea is fully fermented? Semifermented? Not fermented?

Fully fermented—black; semi-fermented—oolong; and not fermented—green.

19. Darjeeling *is/is not* a name associated with good-quality tea.

Is.

20. *True or false.* Hard water may cause tea to be dull and dark; this may be corrected by adding a few drops of lemon juice to the water.

True.

21. *True or false.* The scum on tea is caused by hard water salts joining with polyphenols to form an insoluble substance.

True.

22. Chocolate has *more/less* cocoa butter than has cocoa.

More.

23. Chocolate contains *more/less* starch than does cocoa.

More.

24. The nibs from chocolate are:
   A. Chaff from the bean.
   B. The entire cacao pod.
   C. The beans after the pod has been removed.
   D. Chaff-free bits from the roasted beans.

D. Chaff-free bits from the roasted beans.

25. Dutched chocolate is *the same as/different from* Van Houten chocolate.

The same as.

26. Hot cocoa can be made by three methods. These are the: (1) _____, (2) _____, and (3) _____ methods.

Dry-blend, syrup, paste.

27. Which are the same thing: fruit juices and nectars or fruit juices and fruit drinks?

Fruit juices and nectars. They are both prepared without dilution.

6

Milk and Its Products

Milk and milk products are important foods for people of all ages. Although young children have a greater nutritional need for milk than adults, you will benefit from milk at any age. Milk contains significant quantities of almost all the nutrients needed by humans. The protein value of milk is high because casein (the chief protein in milk) is a complete protein that supplies all the essential amino acids. One pint of milk supplies almost one-third of the protein needed daily by a young woman. Protein from milk is well utilized by most people. Many of the foods you eat are also more palatable and appealing because of the milk used in them.

Milk   is/is not   one of the most healthful foods we eat.   *Is.*

*

MILK

Cow's milk is composed of 87 per cent water, 4.9 per cent carbohydrate, 3.9 per cent fat, 3.5 per cent protein, and less than 1 per cent minerals. The fat and carbohydrate in milk are good sources of energy; in addition, the fat is important as a carrier of the fat-soluble vitamins A and D. Milk contains seven minerals in significant amounts. These are calcium, phosphorus, potassium, sodium, chlorine, sulfur, and magnesium. The two nutrients that are definitely low in milk are ascorbic acid and iron.

The two nutrients that are inadequate in milk are ____ ____ and ____.

Ascorbic acid, iron.

*

Calcium and phosphorus are important because they are needed to build and maintain our bones and teeth and to regulate some body functions. Both of these minerals are particularly abundant in milk, and they are utilized efficiently by the body when they are supplied in milk that is fortified with vitamin D.

The three nutrients from milk that are important for growth and maintenance of bones and teeth are ____, ____, and ____.

Calcium, phosphorus, vitamin D.

*

Both fat and water-soluble vitamins are contained in milk. One quart of milk contains 400 I.U. of vitamin D, the amount recommended for children. Over one-fourth of the recommended allowance of vitamin A for adults is present in one quart of milk. The B vitamins are adequately supplied in milk. Niacin itself is present in only a very small amount, but tryptophan is available in milk and is converted in the body to niacin. Riboflavin content of milk is excellent when milk is marketed in brown glass or cardboard containers. Clear glass bottles, however, readily transmit the light waves that destroy the vitamin activity of riboflavin.

Milk is packaged in **clear glass/brown glass or cardboard containers** *to preserve its ____ content.*

Brown glass or cardboard containers, riboflavin.

*

The carbohydrate in milk is largely lactose, a sugar. It is less soluble than the table sugar you ordinarily use in food preparation; this characteristic is important when you are reconstituting dry milk. If you mix dry milk several hours before you use it, the product is better than if it is used immediately after mixing. This difference is due to the slow solution of lactose.

Milk that is allowed to stand 24 hours after mixing tastes **sweeter/less sweet** *than milk that is mixed and used immediately.*

Sweeter.

165

The insolubility of lactose in milk can be seen occasionally when you use evaporated milk. The percentage of sugar in evaporated milk is elevated because more than half the water has been removed. This greater concentration of sugar may cause lactose to precipitate and form the long needle-like crystals sometimes seen in evaporated milk. If lactose crystallizes out in ice cream and frozen desserts, the texture of the product may resemble broken glass.

If you found something that appeared to be broken glass crystals in a frozen dessert, would you dismiss it and say it was nothing but crystallized lactose or would you ascertain first whether or not it might be glass?

Better check first.

*

Milk Sanitation

The milk industry is regulated primarily through state and local codes, and these regulations are based largely on the U.S. Public Health Service's Grade A Pasteurized Milk Ordinance. State and local controls are enforced to maintain desirable sanitation standards throughout the processing of milk. The federal government does, however, establish the identification standards for dairy products. For example, the term "ice cream" is defined to be a product containing, among other specified quantities of ingredients, a minimum of either 10 per cent milk fat (less than 2 per cent flavoring added) or 8 per cent milk fat plus at least 2 per cent flavoring ingredients.

True or false. Compliance of the milk industry to the Grade A Pasteurized Milk Ordinance is regulated and enforced by the federal government.

False. Local and state authorities use the code as a model for writing their own enforceable code. The federal government does exert some control through its ability to define the identity of an item.

*

Very special attention must be given to maintaining high standards of sanitation throughout the production cycle of milk. Disease-free milk cows need to be milked under sanitary conditions to keep the milk as free of microorganisms as possible. All people working around dairies should have a clean bill of health, because milk is an excellent medium for transmitting infectious diseases. Growth of microorganisms can be discouraged by maintaining milk at storage temperatures below 45°F. One other important precaution is thorough washing of all equipment that comes in contact with the milk.

True or false.   *Milk is a very important food in the diet, but it does need to be handled carefully throughout processing and storage to minimize growth of microorganisms.*

True.

*

The safety of milk for human consumption is achieved when milk is pasteurized. This heat treatment effectively destroys microorganisms that might cause illness. Effective pasteurization is accomplished by one of two methods: (1) holding milk at 161°F for 15 seconds or (2) holding milk at 145°F for at least 30 minutes. The high temperature-short time method is commonly used today. This method safeguards the healthfulness of the milk and yet is sufficiently mild to protect its nutritional value.

_____ *is the heat treatment given to milk to make it safe to drink. It*   **does/does not**   *seriously reduce the nutritive value of milk.*

Pasteurization, does not.

*

In some regions of the United States, it is still possible to buy raw milk. Raw (unpasteurized) milk may contain live microorganisms that can cause tuberculosis, undulant fever, and other serious human illnesses. Because of this potential health hazard, the consumption of raw milk is not recommended.

_____ *milk has not been heated to destroy microorganisms and is a calculated health hazard.*

Raw.

*

Sometimes you may see milk that is labeled "certified." Such milk can be certified raw milk or certified pasteurized. Certified means that the milk was produced under strict sanitary conditions to control bacterial count at less than 10,000 per cubic centimeter. Such a count is distinctly low; the problem is that there is little assurance that none of the microorganisms is pathogenic unless the certified milk has been pasteurized.

True or false.   *Certified raw milk has been treated in such a manner that you can be assured that it contains no harmful microorganisms.*

False. Certified raw milk has a low bacterial count, but there is not an absolute guarantee that the milk is safe.

*

Certain milk grades are recommended in the federal code, but local or state regulations vary somewhat. Usually, raw

167

milk can be marketed as Grade A raw milk if the bacterial count does not exceed 50,000 per cubic centimeter; Grade A pasteurized milk cannot have a bacterial count exceeding 30,000 per cubic centimeter. Many codes also limit severely the number of coliform bacteria that can be present. Ordinarily, milk that does not meet the standards for Grade A is not sold in retail markets.

True or false. The bacterial count limitations for Grade A raw milk and Grade A pasteurized milk are the same.

False. Grade A pasteurized milk has a lower maximum bacterial count than does Grade A raw milk.

*

While the number of bacteria allowed per cubic centimeter in milk and other foods may sound high, the counts listed above actually are not high. All foods contain bacteria and these can be present in milk in the quantities listed above and not cause harm. Contamination is indicated when the counts go up into the hundreds of thousands or millions.

*The allowed bacterial counts in milk are **high/low** for those normally found in foods.*

Low.

*

Types of Milk

When you shop for milk, you are confronted with a reasonably wide selection of products. Let us see just what each one really is. Remember that an eight-ounce glass of fluid whole milk contains 8.5 grams of fat and 8.5 grams of protein.

In many states, whole milk, by definition, must contain a minimum of 8.25 per cent milk solids-not-fat and not less than 3.25 per cent milk fat. Milk marketed as whole milk can have vitamin D added; no other substances can be added. In some areas, fortified milks containing additional vitamins, minerals, and milk solids are available.

Match the type of fluid whole milk in column A with its definition in column B.

A	B	
1. Certified	A. Heated to destroy bacteria	
2. Raw	B. Produced under strict sanitary controls.	1. B.
		2. C.
3. Pasteurized	C. Not heated to destroy bacteria.	3. A.

Homogenized milk is usually made from whole milk. In the homogenization process, milk is forced, under considerable pressure, through a very small opening. This physical force breaks up the fat aggregates into tiny droplets that stay suspended in the milk rather than getting together and rising to the top of the container. Most consumers like homogenized milk because the fat is uniformly distributed in the milk. This process eliminates the problem of overly creamy servings of milk from the first part of the quart and too little richness of flavor toward the end. A side benefit of homogenization is the somewhat improved digestion of milk protein due to a modification of the protein.

Homogenization causes a change in the size of _____ globules and also modification of the _____.

Fat, protein.

*

For those who desire a less rich milk that is lower in calories yet almost as nourishing as fluid whole milk, many dairies are now marketing two per cent milk. An eight-ounce glass of this type of milk contains only 5 grams of fat instead of 8.5. Two per cent milk still retains a reasonably rich flavor, but the calorie value is 128 for eight ounces rather than the 159 contained in whole milk.

*Two percent milk is **higher/lower** in fat and calories than is _____ milk.*

Lower, whole.

*

Skim milk has had the milk fat removed. The fat content of skim milk is reduced to 0.25 gram and the protein content is increased slightly to 8.9 grams per eight ounces. These changes reduce the caloric value almost one-half—from 159 calories to 89 in an eight-ounce serving. The only significant nutritional loss is vitamin A. This fat-soluble vitamin is removed along with the fat.

Is skim milk as good for you as whole milk?

If you are too heavy, skim milk is really better for you than whole milk because it contains slightly larger quantities of most nutrients. The vitamin A value that has been lost can easily be obtained from other sources.

*

Fortified skim or fortified whole milk may have one or more nutrients added. Vitamin D is commonly added to these milks at a controlled level of 400 I.U. per quart, the

169

amount recommended daily for children. Often nonfat dry milk solids are added to increase the protein content. Vitamin A, minerals, and lactose (milk sugar) are also optional additives. The nutrients added during fortification are specified on the container.

Fortified milks will have 400 I.U. of _____ added to each **pint/quart.**

*

Nonfat dry milk has been processed to remove the fat and the water. This product is pasteurized to kill the microorganisms. One asset of this type of milk is that it keeps very satisfactorily at room temperature. This feature makes it convenient to keep in the kitchen. Dry milk can be reconstituted easily to fluid skim milk or it can be added in powdered form to many foods. Federal standards of identity exist for nonfat dry milk, evaporated milk, and condensed milk.

Nonfat dry milk solids **must/need not** *be stored in the refrigerator.*

*

Evaporated milk is prepared by placing pasteurized milk in a vacuum chamber and removing water by evaporation until the volume is reduced to just less than half the original amount. Vitamin D is then added and the cans are sealed and sterilized for 15 minutes. Unopened evaporated milk can be stored for months at room temperature without spoiling; opened cans need to be refrigerated to prevent spoilage and should be used within a couple of days.

Suggest two ways you might use evaporated milk in food preparation.

*

Condensed milk also has undergone evaporation, but the end product is distinctly different from the evaporated milk just described. Condensed milk contains between 40 and 45 per cent sugar. This additive, as you would expect, makes the flavor much sweeter than any other milk product. The sugar also is an effective preservative, thus making sterilization after canning an unnecessary step. Cans of condensed milk can be stored unopened at room temperature for months. However, the high sugar content does cause a very slow browning reaction between the sugar and the protein in the milk. Perhaps you have seen condensed milk that has undergone some browning. The most common use for condensed milk is in desserts. Its high sugar content makes it inappropriate for a beverage or in main course cookery.

A large quantity of sugar is added to **evaporated/condensed** *milk to act as a preservative.*

Condensed.

*

Cultured milk products are available in many areas. Buttermilk used to be available as a by-product of churning butter. Today's buttermilk is a skim milk product fermented mainly by *Streptococcus lactis.* During this fermentation period, an acidic flavor is developed. Some dairies add an aesthetic touch to their product by adding a few flecks of butter.

Buttermilk is a _____ milk product produced by culturing with Streptococcus _____.

Cultured, lactis.

*

Yogurt is another cultured product made from low-fat milk from which a small amount of the moisture has been removed. *Streptococcus thermophilus, Plocamo-bacterium yoghourtii,* and *Bacterium bulgaricum* are microorganisms used for fermenting the pasteurized and homogenized milk. During fermentation, yogurt acquires an acidic tangy flavor and becomes almost custard-like in consistency. Yogurt is now available in a variety of flavors.

Do you think that the tangy flavor of yogurt would combine well with fruits?

Yes, this flavor combination is pleasing. You may wish to make a mixed fruit salad and add a dollop of yogurt as the dressing.

*

Fresh milk is almost neutral, but as milk ages it loses carbon dioxide and becomes slightly acidic. Milk also becomes acidic when bacteria digest lactose (milk sugar) and turn it into lactic acid. When milk becomes sufficiently acidic, it clabbers and slowly the whey separates from the curds. Naturally, the flavor of this mixture is tart.

As milk ages, it becomes **more/less** *acidic.*

More.

*

Several substances in foods can cause milk to clabber. Tomatoes and other acidic foods will cause this reaction. Salts

171

used to cure meats, the calcium chloride added during canning to keep tomatoes firm, and common table salt all encourage curdling. Heating also hastens this undesirable change in cooked milk products.

The factors that encourage curdling of milk include: (1) _____ _____, *(2)* _____ _____, *(3)* _____ _____, *(4)* _____ _____, *and (5)* _____.

Acid foods, curing salts, calcium chloride, table salt, heat.

*

Different kinds of milk have different tolerances to curdling. Evaporated milk does not curdle easily; next in stability is fresh milk and the least stable is dry milk.

The kind of milk least likely to curdle is _____ *milk; the one most likely to curdle is* _____ *milk; the stability of* _____ *milk falls between these two extremes.*

Evaporated, dry, fresh.

*

If properly treated, milk and its products need not curdle. The following precautions greatly reduce the tendency to curdle: (1) reduce the quantity of salt, (2) heat at moderate rather than high temperatures, (3) use the shortest practical cooking time, and (4) avoid contact with calcium chloride or curing salts.

Why do you wish to avoid curdling the milk used in food preparation?

The appearance of curdled milk is not pleasing.

*

If you add a milk product into an acid food, you encourage curdling. The first part of the milk will be in an acid strong enough to curdle the milk. If you reverse the process by adding the acid food slowly to the milk, you keep the milk from being surrounded by the acidic food. This definitely reduces the tendency to curdle.

If you are making a cream of tomato soup, how would you combine the white sauce and the tomato mixture?

Slowly add the tomato mixture to the white sauce while you are stirring.

*

You may have noticed that scalloped potatoes are often marred by curds of milk. If you avoid too high an oven temperature and too long a cooking time, the milk is less likely to curdle. Cooking time with the milk can be reduced by parboiling the potatoes. You can use some of the cooking water to dilute evaporated milk for a thin white sauce and finish baking the scalloped potatoes. The use of evaporated milk, combined with the short cooking time, makes this method quite reliable.

If you cook scalloped potatoes and ham together using all of these suggestions, you still get a curdled mixture. Why?

The ham contains curing salts which make the milk protein extremely unstable.

Milk can easily scorch and develop a strong, objectionable flavor. The brown color comes from the caramelization or burning of lactose, a sugar. Often it is wise to cook milk in a double boiler to avoid scorching. You also can use a heavy pan and stir while heating gently.

Why is milk often heated in a double boiler?

To avoid scorching the milk.

Milk and its products should be handled carefully. Contamination may mean souring or even result in illness. To retard spoilage, milk should be stored in a refrigerator where the temperature is between 32° and 40°F. Spoilage is encouraged by allowing milk to stand unrefrigerated.

For optimum storage life and best flavor, milk should be stored: (a) in the freezer, (b) in the refrigerator, (c) at 40° to 50°F, (d) at 50° to 75°F, or (e) in a warm place.

In the refrigerator.

Sour milk is sometimes used as a liquid in food preparation. If you need sour milk for cooking, add one tablespoon of vinegar or lemon juice to one cup of milk and allow the milk to stand briefly. The acid in the vinegar or lemon juice will cause the milk to look a bit curdled. This clabbering of the milk protein is the cue that the milk is sour enough to use.

To prepare sour milk, add _____ tablespoon of _____ or _____ _____ per cup of milk.

One, vinegar, lemon juice.

BUTTER AND MARGARINE

Butter is made by whipping or churning cream. The churning action brings together the fat globules in the cream. These separate as butter, leaving behind the buttermilk which is almost free of fat. Although either sweet or sour cream can be used in the production of butter, sweet cream

173

is generally preferred because butter made from it retains its flavor particularly well. Customarily, butter is salted to please the tastes of most people. However, unsalted butter also is available. Butter contains approximately 81 per cent milk fat, about 12-15 per cent moisture, 2 per cent salt, and 1 per cent milk curd plus other subtances.

True or false. Butter is a pure fat and contains no substances but fat.

False. Butter is only about 81 per cent fat. The remainder of the content of butter is primarily water plus small amounts of milk curd, vitamins, and added salt.

*

Butter is graded AA (93), A (92), and B (90). The number after each grade indicates the score that the butter must obtain during evaluation to receive a particular grade designation. Seldom will you see Grade B or lower butter on the retail market. The grade a butter receives is based on flavor, body, salt, color, and packaging. Good butter shows a rough edge when broken, has a pleasant flavor, and is a good color. Butter (and also cheese) can be colored with annato (a yellow) coloring without this being marked on the package. Carotene may be used for coloring, but if used, it must be stated on the label. The butter should have no streaks.

If a pound of butter shows distinct variation in coloring on the surface and tastes strong, would you expect this butter to be Grade AA?

No. The butter would in all probability score lower than B.

*

Butter is used to fry foods and to add richness of flavor to cooked vegetables and some desserts. In cakes and some breads, butter is used to give a soft yellow color and to increase the tenderness of the baked product.

In food preparation butter is used to improve _____, _____, and _____ of various products.

Color, flavor, tenderness.

*

One of the most successfully formulated dairy-like products is margarine, which is made to resemble butter. Margarine is made from many different vegetable oils by hydrogenating (adding hydrogen) oils to change them into solids. Just enough hydrogen is added to give the margarine the desired spreading ability. The oils commonly used are corn oil, soybean oil, cottonseed oil, and safflower oil. Margarine, like butter, is about 81 per cent fat; the remainder is moisture, salt, and flavoring compounds that make it taste like butter. Vitamin A (in the form of its precursor, carotene) is added to margarine to improve the nutritive value and to change the color from white to yellow.

True or false. For many purposes in food preparation, butter and margarine can be used interchangeably in a recipe.

True, although some people with a keen sense of taste can distinguish the difference in flavor between the two products. This flavor distinction may soon disappear as flavor research helps margarine manufacturers more closely duplicate the flavor of butter.

*

The fat in butter and other dairy products readily takes on the taste of strong odors and flavors of foods stored in the same refrigerator. By storing these foods in air-tight containers or wrapping snugly in an air-tight wrap, you can avoid picking up off flavors in fat-containing foods.

*Strong flavors are **easily/not easily** absorbed by butter and other _____ containing foods.*

Easily, fat.

*

CREAMS

Creams containing various percentages of milk fat are presently marketed. Of these, the product lowest in fat is half-and-half, which contains 11.7 per cent fat as contrasted with the 3.5 per cent fat content of whole milk. Half-and-half is often used on cereals. It is useful in making stirred custards and some ice creams.

Half-and-half contains _____ per cent fat.

11.7.

*

Light cream is much richer than half-and-half. The high fat content (20.6 per cent) of light cream makes it desirable to serve with coffee. It is also used in making caramels and rich ice creams.

Half-and-half/light cream contains almost six times as much fat as whole milk.

Light cream.

*

Whipping cream contains 37.6 per cent fat. This high quantity of fat is important for preparing a relatively stable whipped cream foam. Whipped cream is a favorite dessert topping. It is also folded into refrigerated and frozen desserts as well as sweetened fruit salad dressings. The term "whipping cream" is used to designate the liquid cream. "Whipped cream" refers to whipping cream that has been whipped into a foam.

When cream containing 37.6 per cent milk fat is whipped into a foam, the product is called **whipping cream/ whipped cream.**

Whipped cream.

*

The foam formed from whipping cream is really many bubbles of air encased in thin films of milk protein supported with the butter fat. It is important to keep the cream cool enough to keep the fat firm during whipping. If the fat becomes warm, it will become quite fluid. Then it tends to drain from the foam rather than helping to give rigidity to the whipped cream.

What can you do to help make cream whip more quickly and to form a more stable foam?

1. Chill the bowl and beaters used for beating.
2. Keep the cream refrigerated until you are ready to whip it.

*

Whipping cream is an oil-in-water emulsion in which the fat is dispersed or scattered throughout the milk (aqueous) portion. This form is maintained during beating until, when beaten too much, the fat suddenly clumps together into a thick buttery mass and the aqueous portion is squeezed out. This is actually a reversal of the original system and a water-in-oil emulsion is formed. This is the process by which butter is made. However, most people intend to use this heavy whipping cream as whipped cream and have no intention of making butter. To prevent the formation of butter, it is necessary to watch the cream as it is whipping and stop beating when the whipped cream piles softly.

*Whipped cream should be beaten until it is thick enough to **stand rigidly/pile softly**. If you overbeat heavy cream, it will form **ice cream/butter**.*

Pile softly, butter.

*

FROZEN DESSERTS

We have a vast array of dairy products called frozen desserts. Ice cream is one. Federal regulations say it must contain not less than 10 per cent milk fat if vanilla and not less than 8 per cent milk fat if flavored. Some specialty ice creams contain as much as 22 per cent milk fat.

*Vanilla ice cream must contain **8/10** per cent milk fat.*

10; flavored ice creams must be not less than 8 per cent milk fat.

*

If a frozen dessert contains less than 10 per cent milk fat if vanilla or 8 per cent if flavored, it cannot be called ice cream. Sherbets are about 4 per cent milk fat. Some frozen desserts, milk desserts, or others have about 4 per cent milk fat content. An ice or snow contains no milk fat but is made solely from fruit juices or other nonmilk liquids.

*If a frozen dessert contains less than 8 per cent milk fat, it **can/cannot** be called an ice cream.*

Cannot.

*

Philadelphia ice cream is made from cream, milk, sugar, flavoring, and a stabilizer. Often the stabilizer is gelatin, but agar-agar and other stabilizers approved by the federal government may be used. If eggs are added, the dessert assumes new richness in color and flavor. This is marketed as French vanilla ice cream. Ground vanilla may be added to make the popular French vanilla ice cream. Some ice creams are made by first making a custard of eggs and then freezing this mixture. A mousse is a rich frozen dessert made from whipped cream.

The simplest ice cream is Philadelphia ice cream; if you add eggs to this, you have made _____ ice cream.

French.

*

The addition of sugar to ice cream mixtures lowers the freezing point below that of water. A moderately sweet ice cream mixture will freeze at around 24°F. As freezing be-

177

gins, the ice cream should receive a good, vigorous whipping action to incorporate air and increase the volume. This increase is called overrun. Ice creams should not have more than 100 per cent overrun; this means they should not more than double their original volume. If whipped to a greater volume, they become too light and fluffy and lack flavor. If not whipped to about 80 per cent overrun, they feel heavy on the tongue and seem to have a poorer flavor.

Overrun is an increase in _____ caused by whipping and freezing. In ice cream the overrun should be between _____ and _____ per cent.

Volume, 80, 100.

*

Sherbets have from about 35 to 50 per cent overrun and ices have from 25 to 35 per cent overrun.

*Ice cream usually has **more/less** overrun than sherbet and sherbets have **more/less** overrun than ices.*

More, more.

*

Frozen desserts made in home freezers or refrigerators are chilled to the freezing point in a bowl or container and then whipped vigorously. This procedure of chilling and whipping is repeated until you have a light, whipped mass that is partially frozen. Pour the partially frozen mixture into trays and freeze.

True or false. When frozen desserts are made in a home freezer or freezing compartment of a refrigerator, you do not get any overrun.

False. You may get less than you would in an ice cream freezer, but the vigorous whipping action does create a foam.

*

You can make frozen desserts by chilling a dessert base until it is near the freezing point and then folding in whipped cream. This dessert is called a mousse. You may wish to stabilize your frozen dessert by adding gelatin. To do this, a half teaspoon of gelatin per pint of liquid can be lightly dusted over the surface during the beating and mixed in.

*A mousse is a frozen dessert made by _____ whipped cream into a mixture and then **freezing/whipping** it.*

Folding, freezing.

*

If thin desserts are frozen quickly and whipped well during freezing, small crystals form and the dessert has a smooth texture. When the mass is completely frozen but still mobile, it is advisable to whip it vigorously one final time and then let the dessert finish hardening undisturbed. The frozen dessert should be held at low temperatures to retain the desired small crystals. If held at temperatures very close to the freezing point of the ice cream, the crystals slowly enlarge and the dessert gets coarse.

Have you ever seen coarse crystals develop in a frozen dessert held in a refrigerator freezing area? Why does this happen?

The storage temperature is too high. The original small crystals melt and recrystallize as larger crystals that feel coarse.

*

Some additives in ice cream help to prevent this undesired crystal growth. These substances are called stabilizers. In home-made ice cream, gelatin is frequently used as a stabilizer. Agar-agar, alginates, and other stabilizers are suitable for commercial use. Usually stabilizers are used at approximately the one per cent level.

Why are stabilizers added to ice cream?

To promote a smooth texture by retarding crystal growth.

*

CHEESES

For centuries cheese has been an important food. According to an ancient tale, the first cheese was made unintentionally one day when a shepherd boy carried some milk in a bag made from the stomach of a sheep. By evening, to his amazement the milk had changed to a white thickened mass and a thin liquid. Now it is known that this change in the milk was caused by an enzyme in the stomach. This enzyme is called rennin.

One means of clotting milk is to add the enzyme _____ to it.

Rennin.

*

Cheese may be defined basically as a concentrated milk product resulting from the clotting of milk protein by acid or enzyme. Variations in production techniques result in cheeses with widely differing flavor and texture. In fact, there are more than 2000 named varieties of cheese. These many varieties actually can be classified into 10 basic types.

179

There are _____ basic types of cheese and more than _____ variations of these.

Ten, 2000.

*

Milk can also be clotted or clabbered by making it more acidic. At home you can do this by adding acids such as lemon juice or vinegar. Commercially, lactic acid-producing bacteria are added to milk to speed the tempering of milk for cheese production.

By adding vinegar, lemon juice, or _____ _____-_____ bacteria, milk becomes acidic enough to clot.

Lactic acid-producing.

*

Cheese is an important food because it is a concentrated source of calcium, phosphorus, and protein. It is also a good source of the B vitamins. The cheeses made from whole milk are high in fat; they are 32.2 per cent fat as compared with only 3.5 per cent fat in whole milk. Cottage cheese and others made from skim milk, however, are much lower in fat. Uncreamed cottage cheese has only 0.3 per cent fat and creamed cottage cheese has 4.2 per cent fat.

True or false. All cheeses contain approximately the same amount of fat.

False. Cheese made from whole milk contains almost 8 times as much fat as creamed cottage cheese, and creamed cottage cheese has 14 times as much fat as plain cottage cheese.

*

Manufacturing Cheese

There are basically two ways to make cheese. One is to allow acid-producing bacteria to sour milk. This causes the protein to coagulate and precipitate as a curd. The curd is then separated from the liquid, called the whey. Most cheeses are made from this curd, but a few are made from the whey by using heat and acid to coagulate the proteins in the whey.

Most cheeses are made from the _____, but a few are made from _____.

Curd, whey.

*

Three well-known cheeses are made from the curd that precipitates when milk is coagulated by acid. These acid-coagulated cheeses are cream cheese, Neufchatel, and cottage cheese. The most well known of the acid-coagulated whey cheeses is ricotta, which is commonly used in lasagne.

Cream cheese, cottage cheese, and ricotta cheese are made using _____-coagulated milk.

Acid.

*

Cheese is more commonly made by adding rennin to slightly acidified milk. This enzyme quickly causes protein (casein) in milk to clabber or clot. Many cheeses are made from the curd formed by adding rennin. Familiar examples of such cheeses include Parmesan, bleu, Roquefort, cheddar, and Edam. This means of clotting milk protein to make cheese is particularly good from the nutritional viewpoint, because most of the calcium is held in the curd. Acid-clotted cheeses retain less calcium in the curd than do rennin-clotted cheeses.

Milk can be clotted to make cheese by adding _____ or _____.

Rennin, acid (or lactic acid-producing bacteria).

*

Usually, before the curd and whey are separated, the whole mass is heated gently to give a firmer curd. Some cheeses, such as Swiss, may be heated to around 100°F; if a dry hard cheese is wanted, the temperature is higher than this. Parmesan, a hard cheese that is usually grated, is heated to 124°F. Cottage, cream, and other soft cheeses are just barely warmed to set the curd and then are drained from the whey.

*"Cooking" is a method in which curds and whey are subjected to temperatures of around 100°F; the higher the temperature, the **firmer/softer** the curd will be.*

Firmer.

*

The amount of moisture left in a cheese greatly influences its keeping qualities. Hard cheeses for grating, with a moisture content of only from 30 to 45 per cent, can be kept a year or more; semisoft and soft cheese contain from 45 to 80 per cent moisture and are significantly more perishable. The desired percentage of moisture is achieved during storage in a cold room with controlled humidity. This drying is done very slowly. Parmesan cheese, with a moisture content of 30 per cent can be kept as long as two years. Most of the familiar natural cheeses, such as cheddar, Swiss, and camembert, range in moisture from a low of about 38 per cent to a high of 52 per cent. Cottage cheese has a moisture content of almost 80 per cent.

What familiar cheese has a very low-moisture content? Very high?

Parmesan, cottage.

*

Cheese made by the acid-setting method is seldom aged. Cottage cheese will be moved onto the market immediately after being made because it loses quality rapidly. Cream, Neufchatel, and other similar cheeses also will be moved through the marketing system soon because they can spoil quickly.

Cannot. If they are aged, they are extremely strong and bitter.

*We **can/cannot** age cottage and other similar cheeses.*

*

Cottage cheese is usually made from skim milk. If the label on the container says "creamed," then the cheese is 4 per cent or more milk fat. Cream cheese is 37 per cent milk fat and Neufchatel is 22 per cent.

Match the cheese in column A with its milk fat content.

A	B	
1. Cottage cheese	A. 22 per cent	
2. Creamed cottage cheese	B. 37 per cent	1. D.
		2. C.
3. Neufchatel	C. 4 per cent or more	3. A.
4. Cream	D. Less than 4 per cent	4. B.

*

Rennin-coagulated cheese such as cheddar, Swiss, Provolone, Roquefort, brick, Limburger, Liederkranz and others are aged. As they age, they develop a mellow flavor and become more acidic. Rennin-set cheeses also become softer and more pliable during aging. A well-aged cheese should have a waxy quality to it and should roll into a ball in the fingers, whereas a new cheese is quite rubbery.

*As a cheese ages it develops **more/less** flavor and becomes **more/less** rubbery.*

More, less.

*

Most rennin-coagulated cheeses are made by what we call the "cheddar" process. After the cheese is cut and cooked, the whey is drained and the curd is a solid rubbery mass about two to four inches thick. This is cut into strips about eight inches wide. These strips are piled one on top of the other about three or four deep. The pressure of the strips on top forces moisture from those below. The strips are alternated in position from top to bottom so all get equal pressing. This piling of the strips is called "cheddaring."

Cheddaring is a process of **cutting the curd/piling strips of curd on top of each other/draining away the whey from the curd.**

Piling strips of curd on top of each other.

*

The strips are then chopped into small pieces (a process called milling), and the curd is salted. Most cheese will have about a 0.2 per cent salt content to improve the flavor and to control undesirable microbiological action during aging.

Milling occurs **before/after** *cheddaring.*

After.

*

After milling and salting, the cheese is hooped or packed into large rounds or shaped as desired. As the cheese is put in for hooping, special molds or bacteria are added. Roquefort and some other blue cheese have *Penicillium roqueforti* added. Camembert has *Penicillium camemberti* added. Bacteria that develop gas in the cheese are added to Swiss cheese to make the large holes and the special flavor. During aging, these molds or bacteria multiply in the cheese, giving the cheese special characteristics. Aging is a very critical and difficult process; it takes a master to know how to handle the cheese to achieve the desired results.

It is **easy/difficult** *to age cheese.*

Difficult.

*

Cheese that is aged only a short time or not at all is called "current" or mild cheese; cheese that is cured or aged for several months or more may be called medium or sharp cheese; and cheese that has been aged a considerable length of time is called "aged" or extra sharp cheese. An aged cheddar is cured five or more months and aged Swiss is cured longer than six months. Because milk used for the making of cheese may contain harmful bacteria, any cheese aged less than 60 days must be made from pasteurized milk. After that time, aging destroys harmful bacteria in cheese.

True or false. Cheddar cheese aged for six months will have a sharper flavor than one aged six weeks.

True.

183

Many large gas (sweet) holes in a cheddar are an indication of a lack of quality, yet Swiss cheese should have large holes and surfaces surrounding the holes should be shiny. Small holes and dull surfaces in Swiss cheese indicate a lack of quality. A Swiss cheese that does not develop holes is called a "blind" and one with quite small holes is called a "nizzler."

Large holes in **Swiss/cheddar** *cheese are an indication of quality.*

Swiss.

*

The texture of a good cheese is firm, smooth, and flexible. It is not rubbery. When rolled up in the fingers, it should have a waxy quality and ball up. A sticky, dry, or crumbly texture or horizontal cracks that look like fractured glass indicate poor quality. A slight shortness or mealiness is not harmful to quality; it may indicate aging. A good cheese has a slightly translucent quality. Cheddar and some other cheeses lighten in color as they age. Swiss cheese may take on a lighter color but should not lose its creaminess.

Which of the following is **not** *a mark of quality in cheddar cheese?*
A. Solid, compact structure with few gas (sweet) holes.
B. Waxiness.
C. A number of wide gas (sweet) holes.
D. A sticky, crumbly, or dry texture.

C and D.

*

When milk is clotted and made into cheese, the product is called a natural cheese. The cheese can be processed differently, as indicated previously, to give a wide variety of cheeses commonly used in food preparation. In the United States, much of the cheese produced is not natural cheese but instead is pasteurized process cheese, pasteurized process cheese food, or pasteurized process cheese spread. These cheese products are made by mixing natural cheeses together, adding an emulsifier, and heating this mixture to pasteurize it and stop the curing that is a part of the finishing of natural cheese.

Cheese may be divided into two distinct categories: _____ and _____ _____.

Natural,
pasteurized process.

184

Pasteurized process cheese frequently is very similar to its natural counterpart, while pasteurized process cheese food has less fat (23 per cent minimum) and a maximum of 44 per cent water. Pasteurized process cheese spreads have a minimum of only 20 per cent milk fat and between 44 and 60 per cent moisture.

A food called a pasteurized process cheese spread must contain not less than _____ per cent milk fat.

Twenty.

*

We have some cheeses which are made from concentrated whey. The famous Scandinavian Primost, a light brown cheese that is slightly sweet, is made from whey. It is buttery and slightly grainy in texture. The sapsago cheese of Switzerland is made partially of whey; it contains a special clover called Alten Krauter which gives it a greenish color and a special flavor. Some cheeses may be a blend of the curd of milk and the concentrated whey. Usually they are sweeter than regular cheeses because the lactose or milk sugar goes into the whey rather than into the curd, and cheeses made with some whey in them are sweeter in flavor.

Why would Primost cheese be slightly grainy in texture?

We stated earlier that lactose had limited solubility. Therefore, when whey is concentrated, the lactose precipitates in crystals and causes this graininess.

*

Cheese Cookery

When you are cooking any food containing cheese, remember that cheese is a protein food. Natural cheeses, whether aged or not, should be heated as short a time as possible and at as low a temperature as is practical. Cheese heated at too high a temperature or for too long a time will become stringy and rubbery to chew. The fat will be squeezed from the rest of the cheese. These difficulties can be avoided by simply heating the cheese in a double boiler or in an oven at 325°F or lower just until the cheese is melted and then serving the food. You can improve the quality of cheese-containing casseroles by protecting the top layer of cheese from the heat with a cover of such insulating items as buttered crumbs or a tomato sauce.

Why is cheese in a pizza usually stringy and greasy?

Pizza is usually baked at a very high oven temperature, which toughens the protein and squeezes out the fat.

*

Although natural cheeses require gentle treatment during food preparation, they contribute important flavors to many food dishes. Aged (ripened) natural cheese blends more readily with other ingredients and imparts a more intense flavor than does the slightly aged natural cheese.

Aged/slightly aged   cheese blends more readily with other ingredients and adds more flavor than does unripened cheese.

Aged.

*

The cooking characteristics, as well as the flavor of cheese, are altered in the cheeses that have been made into pasteurized process products. These products melt readily into a smooth creamy mass with little tendency for the fat to separate. Process products are easy to work with in cooking but often lack the distinctive flavors of natural cheeses.

Natural/pasteurized process   cheeses are easy to use in cooking but may not add a distinctive flavor.

Pasteurized process.

*

Cheese is often served at a meal as an accompaniment for a fruit or with crackers for dessert. Although there is no heating involved in preparing this type of cheese dessert, there is a trick you should know. Cheese will have the best flavor and texture if you will let it warm up almost to room temperature before serving it.

*Slices or wedges of cheese for desserts or snacks should be served   **chilled/at room temperature.***

At room temperature.

*

FOAMS

Evaporated or dry milk can be whipped to a foam. While these foams are less stable than whipped cream, they give greater volume and far fewer calories. To whip undiluted evaporated milk, chill it thoroughly in a bowl until ice crystals begin to form. Whip vigorously. When a slight foam appears, add one tablespoon of lemon juice and continue to whip to a thick foam. Add flavoring and sweetener as desired.

*Evaporated milk whips to a   **larger/smaller**   volume than an equal quantity of whipping cream; it has   **more/fewer** calories.*

Larger, fewer.

186

If you wish to make a foam using nonfat dry milk solids, blend three-fourths cup of regular dry milk solids (whole or nonfat) or one and one-third cups of instant dry milk solids into one cup of ice water in a chilled bowl. Beat to a soft peak, add one-fourth cup of lemon juice, and beat again until a stiff foam forms. Fold in the sugar and flavoring desired. Such a foam is not as stable as that made from evaporated milk or whipping cream.

Could such a foam be used on a low-calorie diet?   *Yes.*

*

REVIEW

1. If a desert tribe lived solely on meat and milk products, would they be healthy?

   No, there are some that do and they lack some nutrients for good health such as ascorbic acid.

2. The substance in greatest quantity in milk is _____.

   Water.

3. *True or false.* Lactose, milk sugar, is quite soluble.

   False.

4. The U.S. Public Health Service's Milk Ordinance and Code is a *law/recommendation.*

   Recommendation; it may become a law or ordinance if passed by a state or local authority.

5. *True or false.* Grade A milk has no bacteria in it.

   False.

6. Fresh milk is *acid/almost neutral/alkaline.*

   Almost neutral.

7. When a sufficient quantity of acid is added to milk, it _____.

   Sours.

8. When milk sours, it forms a clabber which can be separated into _____ and whey.

   Curds.

9. When you combine a milk product with an acidic substance that may sour or curdle the milk, mix the *milk/acidic substance* into the *milk/acidic substance,* using good agitation to blend well.

   Acidic substance, milk.

10. *True or false.* Yogurt has more milk solids in it than does regular milk.

    True.; it is concentrated some before clabbering.

11. *True or false.* Pasteurized process cheese is less likely to become rubbery than is natural cheese when heated.

    True.

12. Riboflavin in milk is destroyed by *acid/heat/sunlight.*

    Sunlight.

187

13. *True or false.* The best grade of butter is AA or 93 score.   *True.*

14. Vanilla ice cream must be _____ per cent milk fat or it cannot be called ice cream.   *Ten.*

15. The maximum quantity of overrun allowed in ice cream is *100 per cent, 95 per cent, 90 per cent, 80 per cent.*   *100 per cent.*

16. Stabilizers in ice cream retard *overrun/formation of large crystals.*   *Formation of large crystals.*

17. *True or false.* Cheese is cooked before it is set with rennin.   *False.*

18. Which will have the highest quantity of milk fat: (a) cottage cheese, (b) creamed cottage cheese, (c) cream cheese, or (d) cheddar cheese?   *Cheddar cheese.*

19. Rennin comes from the _____ of calves.   *Stomachs.*

20. The surface surrounding the holes in Swiss cheese should be *shiny/dull* inside.   *Shiny.*

21. Ricotta cheese is made from _____.   *Whey.*

22. When you serve cheese with fruit as a dessert, serve it *cold/at room temperature.*   *At room temperature.*

23. To whip evaporated milk, _____ it thoroughly and then whip it.   *Chill.*

7

Soups

Soups are nutritious and appetizing foods that can be used as appetizers, snacks, or the main course. There are hot soups, cold soups, light or heavy, thick or thin soups. Time and skill are required to make a good soup from the beginning. Nevertheless, there are times when you may wish to make your own soup. At other times, with the numerous canned, frozen, and dehydrated soup mixes available today, you can enjoy excellent soups with very little labor.

The nutrients in a stock are very limited. Why do we say that soups are nutritious?

A broth and many thin soups are not very nourishing in themselves, but they may have nourishing foods added to the stock and they also whet the appetite. Soups with a milk base are distinctly nutritious.

*

White, brown, and clear stocks are the bases used for making many soups. The preparation of these stocks is discussed in Chapter 9. Clear stock is sometimes called a broth and served as a light soup to introduce a meal. If you clarify such a broth and add vegetables, seasonings, and additional meat, you have made a bouillon or a consomme. Bouillon comes from the word "boil," but we usually mean that it is a clear soup with a fairly definite meaty flavor. Consomme comes from the word "consummate" which means the highest. A consomme is a strongly flavored soup made from beef, veal, or chicken stock or a blend of these. It, too, is clarified, and the flavor is heightened by additional meat, vegetables, and seasonings.

Define broth, consomme, and bouillon.

Broth: a clear stock used as a soup. Consomme: strongly flavored clarified soup made of a chicken, beef, or veal stock or a blend of these. Bouillon: clarified soup with a meaty flavor.

189

CONSOMME AND BOUILLON

To make a consomme or bouillon, take rich beef, veal, or chicken stock and add raw lean ground beef, *mirepoix* (finely diced carrots and celery), seasonings, some well-browned onions, and egg whites. Use a recipe for exact proportion and measure accurately so you will consistently make a good soup. Add a bundle of herbs (*bouquet garni*) and seasonings in a cloth bag. Mix everything well when cold. Then, heat at a low simmer; do not boil. The protein in the eggs and meat coagulates and binds in most of the clouding materials. This coagulated mass rises slowly to the top and forms what is called a raft.

Would the expense and work involved indicate that purchase of the canned product is more expedient and less costly?

Yes.

*

If you boil a stock you are clarifying with egg protein, it will cloud. This mixture should be simmered for about four hours and subsequently poured carefully through several thicknesses of cheese cloth, taking care not to let any bits of material flow over. The sediment in the bottom is not used in making the soup. This process of carefully pouring off the clear upper portion of the soup and leaving the sediment or cloudy portion in the pan is known as decanting.

What should you do if your recipe says to decant the soup?

Carefully pour off the clear portion of the soup.

*

Consommes and bouillons can be varied to make many, many different kinds of light soups. *Consomme printemps* (springtime consomme) is made by adding cooked fresh young vegetables to clear soup. *Consomme royale* is made by adding a cooked custard of rich milk and eggs to it. French onion soup is made by adding cooked onion slices to a rich bouillon. These clear soups are not always served hot. You might like to try a chilled consomme or bouillon. These soups, which are of a jelly-like consistency when chilled, can be varied with the addition of a bit of lemon juice or other distinctive flavor. The adventuresome diner will enjoy an avocado bouillon prepared by chilling a mixture of mashed or diced avocado and bouillon. Serve with a dollop of sour cream to add a gourmet touch.

Would you call consomme and bouillon basic soups?

Yes, because they can be modified in many ways to produce a wide variety of soups.

*

Usually, a rich broth will have enough gelatin to set it lightly. If it does not, you can add a slight bit of pure gelatin to make it set. A soup that contains two per cent gelatin will set rather firmly when chilled. Since most bouillon and consommes contain some gelatin, as little as one tablespoon per pint of soup may be enough to give the desired tender gel.

A jellied soup can be made if the stock contains nearly _____ per cent gelatin.

Two.

*

CREAM SOUPS

Cream soups are thin cream sauces containing pureed cooked vegetables and appropriate seasonings. To make a cream soup, prepare a thin white sauce, add the pureed vegetable, heat quickly to the desired temperature, and serve. Delightfully subtle flavors and colors can be obtained from thoughtful use of vegetables. Cream-of-celery soup has a delicate yet pleasing flavor. The taste treat obtained in a cream-of-spinach soup may be a pleasing surprise to you.

Are there other vegetables that can be used to make cream soups?

Yes. Try mushrooms, asparagus, tomatoes, onions, potatoes, or almost any vegetable.

*

Strictly speaking, a cream soup contains only the pureed vegetable, but many times it is desirable to use the diced vegetable in the soup to add an interesting appearance and texture. If pieces of the vegetable are used, cut them in attractive shapes, yet be sure that they are small enough to be easily managed when eating the soup.

Is a soup containing sliced mushrooms and a white sauce a cream soup?

In the strictest sense, no, but such a soup is a popular soup and is actually titled cream-of-mushroom soup.

191

Cream soups contain milk, of course, and milk curdles rather easily because of its protein content. A curdled soup is definitely unattractive and is to be avoided. First of all, use fresh milk. As milk ages, it becomes increasingly acidic and curdles more readily. Do not oversalt the soup because salt also may cause curdling. However, do adequately season the soup before serving. Soup is less likely to curdle if it is served as soon as it is heated to serving temperature. Long holding encourages precipitation of the protein.

Which of the following factors discourage curdling in a cream soup?
A. *Diced rather than pureed vegetables.*
B. *Use of fresh milk.*
C. *Use of older but not sour milk.*
D. *A minimum amount of salt compatible with good flavor.*
E. *Serve as soon as heated to serving temperature.*

B, D, and E.

*

Considerable care must be taken if an acidic food is added to a cream soup. The classic example of the effect of acid is the preparation of home-made cream-of-tomato soup. The acidic tomatoes can quickly curdle milk if milk is added to the tomatoes. Minimize the problem by reversing this procedure and adding the acidic tomatoes to the white sauce. If the milk is fresh, this procedure will give you a beautiful smooth soup with a delicate pink color and a delightful flavor.

To make cream-of-tomato soup, add the _____ _____ to the _____ _____ _____.

Pureed tomatoes, hot white sauce.

*

Sometimes a cream soup is flavored with shellfish; this is called a *bisque*. It is not technically correct to call a cream-of-tomato soup a bisque, but you do find this term being applied to vegetable cream soups on occasion.

If you made a thin cream sauce and added pieces of shrimp to it, what would it be called?

Shrimp bisque.

*

A *bechamel* sauce, which is usually made with cream, gives a slightly more stable cream soup than one made with milk, but cream soup made by the *velouté* method is quite stable. To make a cream soup by the *velouté* method, make a *roux* by heating flour and butter together. Then add some stock and the pureed vegetable to make a moderately heavy *velouté* base. Blend in hot milk or hot cream slowly with good agitation to produce a smooth soup. If you thicken the hot milk mixture just slightly with a *roux* and the *velouté* just a bit less, you get a soup that is not likely to curdle.

Match the sauce in column A with the description in column B.

A	B	
1. Roux	A. Rich white sauce containing cream	1. B.
2. Velouté	B. A cooked mixture of butter and flour	2. C.
3. Bechamel	C. White sauce made using a thickened roux	3. A.

*

A cream soup can be finished with a liaison. One effective liaison in cream soups is a mixture of three parts cream and one part egg yolks. These ingredients help to thicken the soup, so it is necessary to reduce the amount of flour or other thickener used. This makes an extremely delicate and smooth-flavored soup. With the addition of the egg yolk, it is still more imperative to keep the soup below boiling and keep the holding time very short. The proteins in egg yolks will give a curdled appearance to the soup if they are overheated, just as they do in an overheated stirred custard.

A liaison **thickens/thins** *a soup a little when it is added.*   Thickens.

*

POTATO AND LEGUME SOUPS

Hearty pureed soups are made from starchy vegetables such as navy beans, split peas, Lima beans, or potatoes. These vegetables are cooked until they are soft enough to be rubbed through a sieve or put through a food chopper. The puree is then blended with a stock or a very thin white sauce. Flavor interest is gained by adding chopped meats or vegetables.

True or false. Pureed soups made from starchy products have enough naturally occurring thickening in them to make them as thick as desired.

Partly true; see the next item.

*

Cream soups do not always have to be thickened by using a dry starch in a sauce. Purees made from substances that lack enough starch to thicken the soup to a desired consistency can be thickened by adding a mixture of mashed potatoes. The mashed potatoes contain a large amount of starch, a practical and useful thickening material.

If a puree is too thin, you can thicken it with a starch thickener or even use _____ _____.

Mashed potatoes.

*

Purees may be quite light and be served to introduce a meal, or they may be substantial and form a meal or a major part of it. Pureed soups made of legumes supply a liberal quantity of protein and other nutrients as well as energy. With the addition of a bit of meat or cheese, a legume soup provides a good source of protein and can be served as the main course.

A puree   can/cannot   be light enough to begin a meal.

Can.

*

CHOWDERS AND OTHER SOUPS

A chowder is a special type of heavy soup originated by the early French explorers in New Foundland. Eventually it traveled to New England. The word "chowder" comes from the French word *chaudiere* which means "hot pot." A chowder contains bits of bacon or salt pork, diced potatoes, and onions and some other main seasoning such as fish, clams, lobster, corn, Lima beans, abalone, or mushrooms.

A typical chowder has a main seasoning ingredient and pieces of bacon or salt pork plus _____ _____ and _____.

Diced potatoes, onions.

*

There are two kinds of chowder. One has a milk base and is called New England or Boston chowder; the other is made from a clear fish stock and contains tomatoes. It has no

milk in it. This is called New York, Coney Island, or Philadelphia chowder.

*New England chowder has _____ in it; New York chowder
is unique because it contains _____.*

Milk, tomatoes.

*

A gumbo is a regional dish that originated in the South.
Contents of a gumbo include okra, rice, onions, and other
vegetables; *fillé*, a powder made from young sassafras leaves,
is the seasoning. It was a dish loved by the Cajuns, the
French Arcadians who fled from Canada and settled in
Louisiana. It is thought that they learned to make the dish
from African slaves who brought okra with them when they
were brought to the United States.

*Gumbo is distinguished because it contains okra, rice, and
a seasoning called _____, which is powdered young sassafras
leaves.*

Fillé.

*

A heavy puree, chowder, or gumbo can be served as a meal
in itself. If these soups are to be a main course, they should
contain a large amount of meat to provide needed protein.
A mutton or lamb stock filled with meat, vegetables, and
rice is called Scotch mutton broth and is heavy enough to
make a meal. A *petite marmite* soup, a peasant soup loved
by Henry IV of France, is a chicken soup containing many
vegetables and generous pieces of chicken. Soups high in
noodles and other starches are filling but not nearly as
nourishing as when meat is added. *Borsch* or *borscht* is a
popular low-protein soup containing beets.

*Would you say that we should observe the rule that light
or thin soups are used to introduce a rather substantial
meal, medium soups are used to start a light meal or to supplement a meal, and heavy soups make a complete meal or
almost a complete meal?*

*Yes, this is a rule
followed in menu
planning.*

*

Cold soups are a particularly bright note in menus for hot
weather. Jellied bouillon or consomme topped with a dash
of lime juice is very refreshing. Another favorite is *vichysoisse*, a cream-of-potato soup made from rich chicken stock,
mashed potatoes, and cream, and topped with a sprightly
garnish of finely chopped chives. Borscht may be served

cold also. Scandinavians are fond of a cold (or hot) fruit soup. This is thickened with tapioca or sago and may be seasoned with just a bit of red wine. A berry soup made of blueberries or huckleberries is similar to a fruit soup. A novelty soup called frosted sherry soup is made of fresh raspberries and seasoned with just a bit of sherry.

Have you ever tried a cold soup?

If not, you're due for a pleasant surprise.

*

Most soups should be accompanied by some garnish. Accompaniments such as crackers, bread sticks, or other crisp foods go well with soups. Purees are excellent with croutons (small toasted pieces of bread) . Cream-of-corn soup is delightful with crisp popped corn on top. A large, toasted, cheese crouton is usually put on top of French onion soup. Browned onion rings, crisp bacon bits, crumbled potato chips, breakfast cereals, seasoned whipped or sour cream, salted nuts, and crumbled blue or other cheese may be used to give distinction and flair to a soup.

Besides a garnish, it is desirable to accompany the soup with _____ or some other crisp food.

Crackers.

*

REVIEW

1. The base of a soup is a _____.

Stock.

2. A consomme is made from _____, _____, or _____ stock, or a blend of these.

Beef, veal, chicken.

3. A raft in making a clarified soup is:
   A. The coagulated portion that rises to the top.
   B. The sediment found on the bottom after cooking.
   C. The well-browned onions added to it for flavor and color.
   D. None of the above.

A.

4. *True or false.* You should not boil a stock when it is being clarified with lean ground meat and egg whites.

True.

5. It takes approximately a *one/two/three/four* per cent gelatin solution to set a soup stock.

Two.

6. A *bisque* is a _____ soup flavored with *fish/shellfish*.

Cream, shellfish.

196

7. The most stable cream soup is one made from: (a) white sauce, (b) cream sauce, (c) a *velouté* base, or (d) bechamel sauce.

A velouté base.

8. A liaison *can/cannot* be used to finish a cream soup.

Can.

9. *True or false.* The thickener of a pureed soup is always some starchy vegetable such as beans, mashed potatoes, or peas.

False; starch thickeners are used also.

10. A chowder is made either with a _____ base or a _____ _____ base which has tomato added to it.

Milk, fish stock.

11. A gumbo is seasoned by _____.

Fillé.

12. *Vichysoisse* is usually a *cold/hot* soup.

Cold.

8

Meat Cookery

Sometimes the word "meat" is used in a broad sense to mean the flesh of poultry and all types of fish as well as that of animals. It is often used in a more restricted sense to mean only red meats, that is, pork, beef, lamb, mutton, and any other animal flesh. In this chapter, the term "meat" usually will be used in the broad sense to include poultry and fish in addition to the various types of animal flesh. Since all flesh has characteristics in common when considered from the viewpoint of food preparation, this inclusive use of the term "meat" is convenient. When it is used in reference only to animal flesh, this will be clearly indicated in the context of the sentence.

As generally used in this book, the word "meat" means:
A.   Flesh of animals.
B.   Flesh of poultry.
C.   Flesh of fish and shellfish.
D.   All of these.

D. All of these.

*

All meats are excellent sources of protein, the B vitamins (thiamine, riboflavin, and niacin), phosphorus, copper, and iron. The proteins in meats are important nutritionally because they are complete proteins. This means that they contain all the essential amino acids needed by humans for growth and maintenance of body tissues. When meat is included in a meal, it gives you a satisfied feeling and delays hunger. You can say that meat has high satiety value. This satiety value is due to the protein in meat and also to the fat. Americans often have a greater amount of meat in their diets than they actually need for nutritional demands because meat is a favorite food despite its high cost.

Meat is an excellent source of **complete/incomplete** *protein. It is also important because it contains* **iron/lead** *and* **vitamin D/the B vitamins.**

Complete, iron, the B vitamins.

*

The composition of meat varies from one cut to another and from one type to another. The protein level ranges from 9 to 19 per cent. The fat content varies widely between cuts as they are served; trimming and preparation greatly influence the amount of fat in a serving. Between 50 and 75 per cent of meat is water. You will also find that meat contains a small amount of carbohydrate in the form of glycogen.

The largest single substance present in meat is:
A. *Vitamins and minerals.*
B. *Fat.*
C. *Water.*
D. *Protein.*
E. *All are equal.*

C. Water.

*

BUYING MEAT

If you tour a grocery store to check prices on many of the basic food items in your diet, you will quickly discover that meat is the most expensive food item you buy. It is also a very important food in a meal from the standpoint of nutritional value and pleasure of diners. If you are to prepare a flavorful, tender, and juicy cut of meat, it is necessary to first learn how to select meat. Selection can be based on the suitability of a cut for the type of preparation planned, the quality of cut needed for the specific purpose, and the actual cost of the cut.

Meat is suitably selected on the basis of:
A. *Quality of the cut.*
B. *Cost.*
C. *Suitability of cut for the recipe planned.*

A, B, and C.

*

When you shop in most stores, you will notice that there are two types of stamps on many of the carcasses hanging

199

in the cold storage locker. One stamp is placed on the carcass by a federal meat inspector. This circular stamp, which is imprinted on each wholesale cut of the animal, says Inspected and Passed. Any meat that will cross state lines and enter into interstate trade is required by federal law to undergo this federal inspection. Very little meat today is not inspected and passed for wholesomeness because most states now regulate inspection of meat marketed within their borders. If meat has been inspected by federal inspectors, it will bear the round federal stamp signifying that the meat animal was inspected for wholesomeness before slaughter. This stamp is also visible proof that the conditions during slaughter and in subsequent cold storage were sanitary. In other words, the meat is safe for your consumption when it leaves the packer if it bears the inspection stamp.

True or false. The circular federal inspection stamp tells you that the meat is tender and juicy.

False. The inspection stamp testifies to the wholesomeness or healthfulness of the meat but does not indicate general palatability of the carcass.

*

Lamb, mutton, beef, veal, and pork are all stamped on the fat side of the carcass with the inspection stamp. A purple-colored edible dye is used for stamping these meats. However, this purple dye does not stain the skin of poultry legibly. This problem was solved by attaching to poultry a cardboard or metal tag imprinted with the inspection stamp. Although fish are not legally required to be inspected, some fish plants today have voluntary federal inspection.

Poultry that has been inspected will be marked with **a circular purple stamp/a tag imprinted with the inspection stamp.**

A tag imprinted with the inspection stamp.

*

As an aid to consumers in selecting their meat, grading is often done. However, this is an optional service provided by the meat packers and is not required by law even when meat enters interstate commerce. Grading may be done according to federal grade specifications. If federal grades are used, the grade marked on the carcass must correspond to the quality of the meat. The characteristics required for each grade classification are very specifically spelled out to serve as a clearcut guide for the federal graders. Packers also have the option of using their own grade designations, or they may elect to not grade their meat at all. If packers use their own designations, they use their own selected names for grades. This is frequently done in the marketing of ham and other cured pork products.

True or false. Grading of meat is required by federal law.

False. Grading is an optional service which meat packers may elect to provide for consumers.

*

Federal grading standards are commonly used to evaluate beef and poultry. Pork, veal, lamb, and fish also are sometimes graded by federal standards and marketed with the federal grade designation. The federal grade stamp is a shield-shaped emblem bearing the appropriate U.S. grade designation such as U.S. Choice or U.S. Prime. The grade stamp is prominently displayed on most retail cuts of meat because it is applied with a repeating roller stamp that is rolled the length of the carcass. In contrast, the inspection stamp often is not visible on individual retail cuts. It is only required to be readily apparent on the large wholesale cuts. However, you can be certain that any meat bearing a federal grading stamp also has been inspected by federal inspectors.

When you see a shield-shaped stamp on a meat cut, this stamp will tell you **the meat is wholesome and safe to eat/ the quality or palatability of the meat.**

The quality or palatability of the meat.

*

The different types of flesh foods have different federal grade designations. The top quality in poultry and fish (when graded) is marked U.S. Grade A. The other two grade designations are U.S. Grade B and Grade C. Pork is graded alive from best to poorest quality, U.S. No. 1, U.S. No. 2, U.S. No. 3, Medium, and Cull. From highest to lowest, lamb is marked U.S. Prime, Choice, Good, Utility, and Cull. Mutton is very similar; that is, U.S. Choice is the top grade followed in sequence by U.S. Good, Utility, and Cull. The first four grade designations for beef, veal, and calf are U.S. Prime, U.S. Choice, Good, and Standard (in descending order). The last two grades in beef are Commercial and Utility and in veal and calf are Utility and Cull.

Name the U.S. grades for pork, lamb, and beef, beginning with the top grade and proceeding in sequence.

Pork *(live grades): U.S. No. 1, U.S. No. 2, U.S. No. 3, Medium, and Cull.* **Lamb:** *Prime, Choice, Good, Utility, and Cull.* **Beef:** *Prime, Choice, Good, Standard, Commercial, and Utility.*

Beef Chart

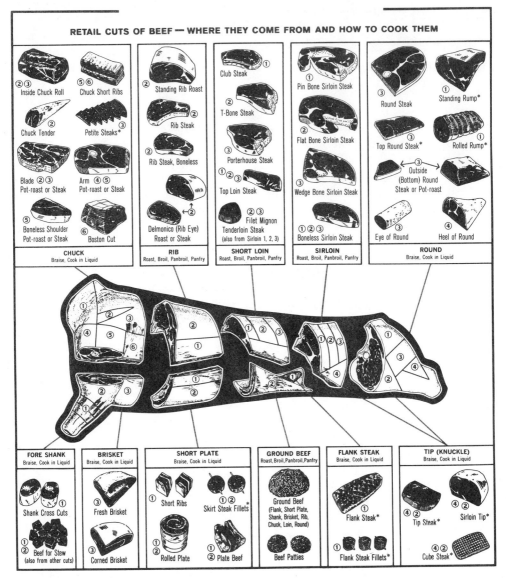

RETAIL CUTS OF BEEF — WHERE THEY COME FROM AND HOW TO COOK THEM

*May be roasted, pan broiled, or pan fried from high quality beef.
Source: National Live Stock and Meat Board

Beef is graded on the basis of two criteria: cutability and palatability. Cutability is determined by judging the amount of salable meat in relation to bone and fat. This is important to the wholesaler as well as to the retailer and the consumer, because a large amount of waste represents

Veal Chart

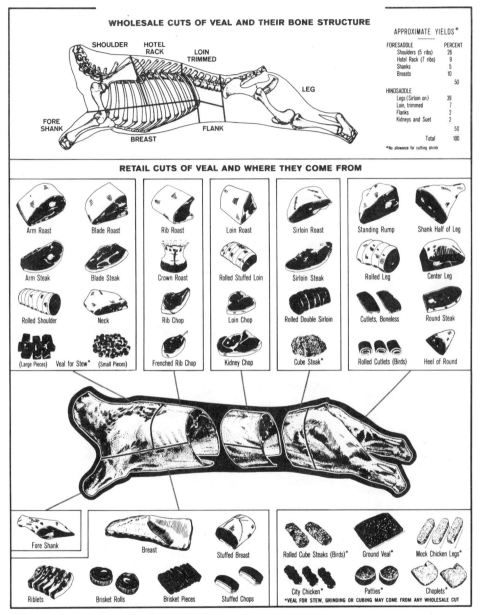

WHOLESALE CUTS OF VEAL AND THEIR BONE STRUCTURE

SHOULDER HOTEL RACK LOIN TRIMMED LEG

FORE SHANK BREAST FLANK

APPROXIMATE YIELDS*

FORESADDLE	PERCENT
Shoulders (5 ribs)	26
Hotel Rack (7 ribs)	9
Shanks	5
Breasts	10
	50
HINDSADDLE	
Legs (Sirloin on)	39
Loin, trimmed	7
Flanks	2
Kidneys and Suet	2
	50
Total	100

*No allowance for cutting shrink

RETAIL CUTS OF VEAL AND WHERE THEY COME FROM

Arm Roast   Blade Roast

Arm Steak   Blade Steak

Rolled Shoulder   Neck

(Large Pieces) Veal for Stew* (Small Pieces)

Rib Roast

Crown Roast

Rib Chop

Frenched Rib Chop

Loin Roast

Rolled Stuffed Loin

Loin Chop

Kidney Chop

Sirloin Roast

Sirloin Steak

Rolled Double Sirloin

Cube Steak*

Standing Rump   Shank Half of Leg

Rolled Leg   Center Leg

Cutlets, Boneless   Round Steak

Rolled Cutlets (Birds)   Heel of Round

Fore Shank

Breast   Stuffed Breast

Riblets   Brisket Rolls   Brisket Pieces   Stuffed Chops

Rolled Cube Steaks (Birds)*   Ground Veal*   Mock Chicken Legs*

City Chicken*   Patties*   Choplets*

*VEAL FOR STEW, GRINDING OR CUBING MAY COME FROM ANY WHOLESALE CUT

Source: National Live Stock and Meat Board

less profit to the merchandiser and greater expense to the consumer. Palatability must be judged when assigning a grade. To judge palatability, the grader examines the flesh to be sure that it is firm and has a fine texture and good

203

Pork Chart

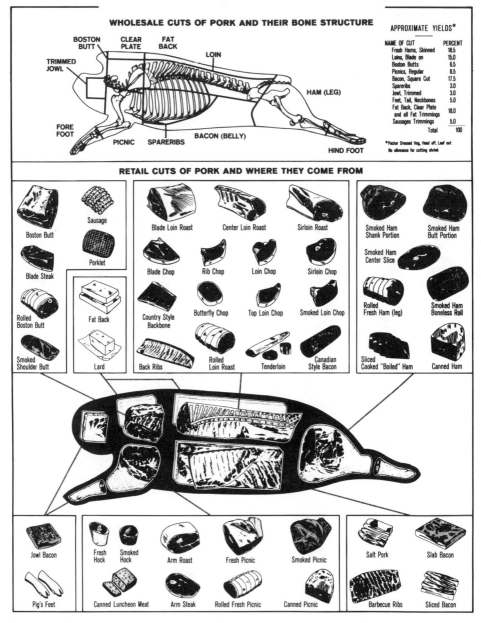

Source: National Live Stock and Meat Board

color characteristic of the type of meat. With beef, he also checks to see if there is fat (marbling) deposited within the muscle as well as around it. Pork that is highly marbled will be rated lower and beef that is highly marbled will be

Lamb Chart

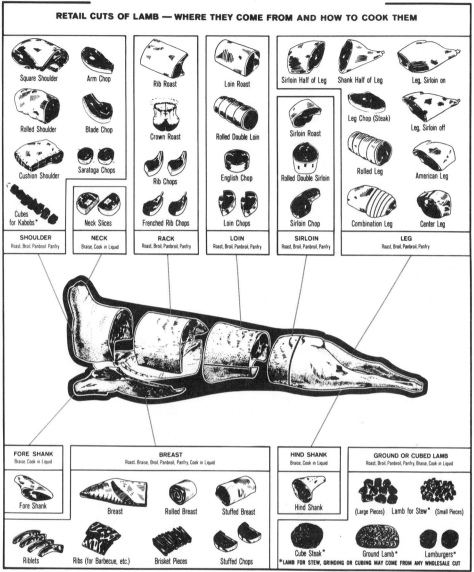

RETAIL CUTS OF LAMB — WHERE THEY COME FROM AND HOW TO COOK THEM

SHOULDER	NECK	RACK	LOIN	SIRLOIN	LEG

Square Shoulder · Arm Chop · Rolled Shoulder · Blade Chop · Cushion Shoulder · Saratoga Chops · Cubes for Kabobs* · Neck Slices · Rib Roast · Crown Roast · Rib Chops · Frenched Rib Chops · Loin Roast · Rolled Double Loin · English Chop · Loin Chops · Sirloin Half of Leg · Shank Half of Leg · Leg, Sirloin on · Leg Chop (Steak) · Leg, Sirloin off · Sirloin Roast · Rolled Leg · American Leg · Rolled Double Sirloin · Sirloin Chop · Combination Leg · Center Leg

SHOULDER — Roast, Broil, Panbroil, Panfry
NECK — Braise, Cook in Liquid
RACK — Roast, Broil, Panbroil, Panfry
LOIN — Roast, Broil, Panbroil, Panfry
SIRLOIN — Roast, Broil, Panbroil, Panfry
LEG — Roast, Broil, Panbroil, Panfry

FORE SHANK — Braise, Cook in Liquid
BREAST — Roast, Braise, Broil, Panbroil, Panfry, Cook in Liquid
HIND SHANK — Braise, Cook in Liquid
GROUND OR CUBED LAMB — Roast, Broil, Panbroil, Panfry, Braise, Cook in Liquid

Fore Shank · Breast · Rolled Breast · Stuffed Breast · Hind Shank · (Large Pieces) Lamb for Stew* (Small Pieces) · Riblets · Ribs (for Barbecue, etc.) · Brisket Pieces · Stuffed Chops · Cube Steak* · Ground Lamb* · Lamburgers*

*LAMB FOR STEW, GRINDING OR CUBING MAY COME FROM ANY WHOLESALE CUT

Source: National Live Stock and Meat Board

rated higher. This difference is due to the fact that pork is naturally high in fat and may become too greasy for greatest palatability if well marbled. Beef, however, will be more juicy and flavorful if well marbled.

Meat is graded according to its ____ and ____ when it is graded according to government standards.

Cutability, palatability.

Not all beef grades are available in retail markets. Some retail markets may carry U.S. Prime meat, but the majority carry U.S. Choice as their top grade. Only a small quantity of beef is graded as U.S. Prime and this grade commands a premium price. Much of the beef of this grade is purchased for use in restaurants or is sold in specialty meat markets where customers are willing to pay for the privilege of purchasing this quality. U.S. Choice is available in almost all grocery stores across the country; U.S. Good is sold in somewhat fewer grocery stores but is still readily available to the consumer. The lower grades of meat are not commonly carried in meat markets, but they are used largely in canned meats or sausage products. These lower grades are less expensive because they are less tender. However, this meat is desirable in such products because of the pleasing flavor.

True or false. Only U.S. Prime, Choice, and Good are considered to be useful for human consumption.

False. The lower grades have an excellent flavor and may be used in ground meat processed products very satisfactorily despite the reduced tenderness found in the unground meat from these grades.

*

You can make better selections at the meat counter if you look at the features that the graders consider when they judge carcasses. Check to see if there is a general blocky shape to the cut with a high ratio of meat to bone. This aspect of palatability is called conformation. You will also want to notice the finish of the cut. To judge finish, notice the amount of fat and the firmness of the fat surrounding the muscle; also note the marbling within the muscle. The quality is determined by the fineness of the muscle fibers, the color and general appearance of the muscle itself, and the maturity of the animal. Maturity is most easily judged by looking at the bone structure, but the amount of fat deposited around and within the muscle also is influenced by the age of the animal. A mature animal will have well-calcified bones compared to the softer, more porous bones of a young animal.

To judge meat at the meat counter, look for good _____, _____, and _____.

Conformation, finish, quality.

*

When you look at the different portions of a cut available at the meat counter, you will notice distinct differences between the various packages. There will be some packages from carcasses that rated very high within the grade designation and others that were so low that they almost fell into the next lower grade category. Despite these differences, you can expect to pay the same price per pound for any portion of the same cut. The wise consumer can get the best possible quality for her money by careful selection.

Why is it helpful to be able to recognize excellence in meat cuts?

You can get better cuts of meat for your money if you are able to make the best possible selection among the meats available at a certain price per pound.

*

Carefully defined terms are applied to meat animals of various ages. Young cattle that are marketed between the ages of approximately 3 and 14 weeks are called veal. Since the animals are so young, there is little time for fat to be deposited. Veal characteristically is not marbled and has very little fat surrounding the muscle. The cuts are smaller than the corresponding cuts of beef because the animal is not full grown. You will also notice that veal is lighter in color than is mature beef. Cattle marketed up to the age of a year, yet older than those used for veal, will be marketed as calf. This meat is more flavorful and darker than veal but lacks the marbling and fullness of flavor characteristic of beef.

Cattle marketed between the ages of _____ and _____ weeks are sold as _____; between the ages of 14 and 52 weeks, the meat is sold as calf; after that it is termed _____.

Three, 14, veal, beef.

*

You can use the fat as another key to determining the quality and age of beef. Ideally, beef fat is rather hard and creamy white. Yellow color in beef fat is an indication of an older animal with less desirable meat in many instances. Animals fed on grass rather than grain are likely to have a yellow color to the fat. Some breeds of cattle have a fat that is yellower than others.

*For the best beef, select cuts that have a **soft/hard** fat that is **white/yellow**.*

Hard, white.

*

Sheep are also labeled according to age at the time of slaughter. Animals less than a year old are marketed as lamb; those that are older are sold as mutton. Lamb is a relatively dark red and becomes progressively darker as the

207

animal becomes older. Mutton is a deep red. Since lamb is from a younger animal, it is not surprising that lamb has less fat covering the muscles than is found surrounding cuts of mutton. The fat of lamb is very hard and white.

Would the size of the muscle in a rib chop be a good indicator of whether a chop was lamb or mutton?

Yes. Mutton would be larger because the animal would have been full grown, whereas lamb would be from an animal slaughtered before full growth had occurred.

*

Pork is usually from animals slaughtered between the age of five months and a year. As soon as the young pigs reach a marketable weight, they are shipped to market before the fat in the carcass gets excessive. This practice gives better, less fatty pork and also reduces the cost of feeding the meat animals. Breeding stock is usually sold for use in sausage meats. The breeding stock are older, of course, and have a rosy-pink flesh compared with the light-pink color of the young meat animals.

Match the color of the flesh with the type of animal.

A	B	
1. Veal	A. Bright, deep red	1. C.
2. Calf	B. Bright, light red	2. B.
3. Beef	C. Tannish pink	3. A.
4. Pork	D. Light pink	4. D.
5. Lamb	E. Rosy pink	5. A.
6. Mutton	F. Bright, very deep red	6. F.
7. Pork breeding stock		7. E.

*

Poultry is commonly marketed at a rather young age. Rock Cornish hens are usually 5 to 7 weeks old. Chickens marketed as broilers or fryers are between 9 and 12 weeks old; a roaster is 3 to 5 months old. Some male chickens are de-sexed before they are a month old and are marketed as capons before they are 8 months old. Ducklings for roasting are also relatively young when marketed. Roaster ducklings are customarily under 16 weeks old. The fryer-roaster turkeys presently available are also less than 16 weeks old. Turkey hens and toms are usually between 5 and 7 months old, although they may occasionally be held until almost 15 months old before being marketed.

True or false. Turkeys are usually older than chickens at the time of marketing.

True.

*

Age affects quality of all types of flesh. More mature animals are heavier, their flesh is more coarse in texture, and they have more fat. Breed also influences quality. Breeds of chickens selected for fine egg-laying ability are not as meaty or as flavorful as breeds raised for their flesh. Dairy cattle and beef cattle show a similar comparison. Dairy cattle yield less meat per pound and have a poorer flavor to their meat than do beef cattle.

Is there a difference in the conformation (shape and build) of egg-laying chickens and those bred for their flesh? Between dairy and beef cattle?

Yes, in both types of animals. Egg-laying chickens and dairy cattle are rather angular, whereas the meaty breeds have a blocky, square conformation which grades high.

*

A yellowish flesh shows that their is some fat beneath the skin of poultry and is thus an indication of quality. This is in contrast to the bluish color apparent when there is a lack of fat. Another key to quality in poultry is the flexibility of the breast bone. The desired young poultry has a pliable cartilagenous breast bone. A fine-textured smooth skin is another indication of good quality in poultry.

*If a turkey has a plump breast, a yellowish skin with smooth texture, and a pliable breast bone, this turkey will probably be graded U.S. Grade **A/C**.*

A.

*

A high-quality bird is free of deformities and bruises. It has a good layer of fat along the back and some fat deposited in the cavity as well. When purchasing, note particularly the condition of the skin. Poultry that has been carefully dressed will have an unbroken skin which is cleaned of virtually all feathers and hairs. The cavity will have been cleaned well; the giblets and neck will be wrapped and packed in the cavity.

*An unbroken skin in poultry is important so that **drip loss during cooking will be minimized/the stuffing will not dry out during cooking**.*

Drip loss during cooking will be minimized.

209

If you are selecting a bird that is frozen, inspect the entire carcass carefully for freezer burn. This is a drying of the flesh caused by storing the bird in a bag with a small break in it. The result of such storage is a drying out or dehydration which causes a change in the color of the skin and produces a dry, powdery character in the meat. After roasting, the freezer burn will appear as a darkened area.

True or false.   Freezer burn makes a bird less attractive, but has little effect on the quality of the meat itself.

False. The area that is burned will be rather dry and flavorless.

*

The amount of bone in proportion to meat in all poultry is relatively high. Consequently, you must allow between 10 and 12 ounces per serving. For instance, a two and one-half pound broiler chicken can readily be split down the back and breast bone and then the breast portion divided from the leg region. This yields four portions: two portions consisting of the wing and half a breast each and two portions with the leg and thigh. A duck contains considerably more fat than does chicken and, consequently, it takes a three and one-half pound duck to provide four adequate portions. The duck would be split in the same way as the broiler after it was roasted. Turkeys contain a higher proportion of meat to the bone. Of the poultry, turkeys may be expected to yield the largest number of servings per pound. Turkeys over 18 pounds provide more meat per pound than do smaller turkeys. Roasting chickens provide the best ratio of meat to bone if they are four pounds or larger.

Rank the following in order of greatest yield per pound purchased: ducks, turkeys, and chickens.

Turkeys provide the most meat per pound, chickens are intermediate, and ducks provide the least. Usually, turkeys yield about 50 per cent or more meat per dressed pound, chickens about 35 to 40 per cent, and ducks and geese only approximately 25 per cent.

*

Poultry is marketed in several ways: as the whole bird, halved, quartered, or cut up. Fowl marketed whole, sometimes referred to as ready-to-cook, is eviscerated and the heart, liver, and gizzard are wrapped and packed in the body cavity along with the neck. Quartered, halved, or cut-up poultry are ready-to-cook birds that have been divided as indicated. Chickens are frequently purchased in one of these forms; occasionally turkeys are split also.

Would you expect cut-up poultry to be more expensive than poultry marketed as the whole bird?

Yes. The labor required to cut it up is expensive. Also the elimination of the neck and gizzard from the package of cut-up chicken makes it necessary to increase the price.

*

Fish are often available frozen at the meat counter, but you may wish to purchase the fresh item when possible. When selecting fresh fish, notice the appearance of the eyes, the gills, and the scales. The eyes should be bright, the gills should be red rather than dark colored, and the scales should not be easily removed. It is important that the flesh be firm and not flabby. Fish that is fresh will have virtually no odor; there should not be a strong fishy odor or a suggestion of ammonia. Skinned fish should show no trace of sliminess, but some fish will naturally have some slime on their outer surface.

Which of the following is an indication of fish that is spoiling?
A. Bright eyes.
B. Firm flesh.
C. Red gills.
D. A distinct aroma of ammonia.

D. A distinct aroma of ammonia.

*

Fresh clams, oysters, crabs, and lobsters in the shell should be purchased alive. You can tell that clams and oysters are alive by checking their shells to be sure they are tightly closed. Open shells indicate that they have died. Crabs, crayfish, lobsters, and other shellfish should show eye and antenna movement and the legs and tails should show tension and vigorous movement when picked up. Weak, apparently dormant shellfish are called "weaks" and may be a good buy. In some markets you can buy uncooked frozen shellfish such as soft shell crabs, shrimp, and lobster. These uncooked shellfish are said to be "green," although they are actually a rather dull color. Sometimes crabs, lobsters, and shrimp are cooked before being sold either chilled or frozen. The heat during cooking changes their color to a bright red or splotchy red.

True or false. When shellfish are cooked, they turn green and, consequently, are called "green."

False. They turn red when cooked; the term "green" means uncooked.

*

STORING MEAT

As soon as meat is skinned after slaughter, contamination by microorganisms begins. Meat is an excellent medium for their growth. Therefore, it is important to keep meat chilled at all times and to be extremely careful in handling meat. Any cutting or handling of meat increases the contamination. Avoid stacking meat, fish or poultry in deep layers where air cannot reach each piece readily. Air helps to keep meat fresh longer. You can tell when meat begins to spoil by noting the strong smell of ammonia and the moist, rather slimy and slippery surface.

If meat is sterile at the time of slaughter, why does it spoil?

Bacteria and other microorganisms are in the air and quickly invade a carcass. In addition, the intestinal tract is a rich source of bacteria which can move into other parts of the carcass if the tract is not quickly and neatly removed. This region within the cavity of the body is a particular problem in poultry. Sometimes antibiotics are added to the water used to clean the cavity of a chicken to help reduce the spoilage originating in that region.

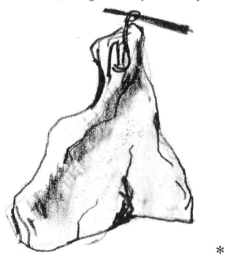

*

Various means can be used to retard bacterial growth in meats during storage. Occasionally you may wish to use a marinade in the preparation of a cut of meat. The acid in a marinade retards growth of microorganisms. Freezing is very effective in slowing the deterioration of meat. However, fat will slowly oxidize even during frozen storage, resulting in gradual development of a rancid flavor. This problem is worse if salt is added to the meat prior to frozen storage. The actual period for storage that is feasible varies with storage temperature and the type of meat. A maximum storage temperature for long-term storage is 0°F, and colder temperatures are still better. If meat is held above 0°F, the deteriorative changes will take place more quickly and safe storage time will be limited. Fluctuations in temperature during storage should be avoided if you wish to maintain the best possible condition of the meat throughout storage. The fat in all types of poultry and in pork is the factor limiting frozen storage. Six months is usually suggested as the longest feasible storage period, and quality is better if storage is limited to three months. Beef, veal, lamb, and mutton can be stored up to a year satisfactorily. It is important to wrap all meats securely in air-tight wraps such as polyethylene bags to prevent freezer burn and undesirable oxidative changes.

What would be the maximum time you would hold these meats in frozen storage (0°F or less): pork chops, chuck roast, turkey, and lamb?

Pork chops—three to six months. Chuck roast—up to a year. Turkey— three to six months. Lamb—up to a year.

*

Curing or salting is an effective way of delaying deteriorative changes in meat. This method of meat preservation was widely used long before the days of the locker plant and home freezer. A brine solution is used to impregnate meats such as pork being prepared as ham. The high salt content discourages the growth of microorganisms and enables you to store the cured meats satisfactorily in a cold room. Meats also may be smoked to slow microbiological growth. Smoking and curing are popular means of preserving fish as well as pork products. These two preserving methods change the flavor of the meat.

*Curing and smoking are means of _____ meats. These processes **do/do not** change the flavor.*

Preserving, do.

*

Meat should be stored in the refrigerator at temperatures just above freezing (up to 36°F), but wipe all surfaces with a damp paper towel first. Be particularly careful to wash the inner cavity of poultry since this area spoils very readily.

Store meat in the refrigerator at temperatures just **above/below** *freezing.*

Above.

*

COLOR IN MEAT

The red color in meat is caused by a protein pigment called myoglobin. A bit of red color is also contributed by hemoglobin, but most blood is drained from the animal after slaughter. This reduces redness that would otherwise be contributed by hemoglobin. Red meats such as mutton and beef contain more myoglobin than do lamb, veal, and pork. There is less myoglobin in chicken breasts than in the legs and thighs. Meat that has only a small quantity of myoglobin is called "light" or "white" meat; "dark" or "red" meat is significantly higher in this substance.

The redness in meat is caused mainly by **myoglobin/hemoglobin.**

Myoglobin.

*

Have you ever noticed that beef cuts sometimes look rather dark after they have been exposed to air for a time? This color change from the pleasing red color of myoglobin to the brown color of metmyoglobin takes place as the iron in myoglobin is oxidized to form the new pigment. This color change is a problem in the marketing of meat. A change also occurs in the pigment when meats are cured. Sodium nitrate, salt, and heat during the curing process will undergo a change to form the nitrite salt, which in turn combines with myoglobin to produce the familiar red color of cured meats. This compound, nitrosomyoglobin, is stable in heat, and the meat retains the same red color even after cooking is completed.

Match the pigment in column B with its description in column A.

A	B	
1. *Changes color when heated*	A. *Nitrosomyglobin*	1. *B.*
2. *Remains red when heated*	B. *Myoglobin*	2. *A.*
3. *Brown pigment formed when meat is exposed to air*	C. *Metmyoglobin*	3. *C.*
4. *Gray pigment in cooked meat*	D. *Hematin*	4. *D.*

214

FLAVOR IN MEAT

Much of the distinctive flavor of meat is carried in the fat or fatty substances. It is possible to modify some animal fats by controlling the diet. The feed will influence the flavor of the fat. Peanuts are fed to some hogs to produce an animal that will yield hams of superior flavor. Wild ducks that feed on wild rice in the northern marshes take on the flavor of the wild rice they eat; a duck that feeds along the coastal areas of the sea has a very fishy flavor. Corn-fed animals have a different flavor than those raised on pressed cottonseed cake.

True or false.   The flavor of meat is influenced by the animal's diet.   *True.*

*

The muscle contributes the meaty flavor. As an animal ages, the flavor in the meat becomes more intense. Many substances contribute to the flavor bouquet of meat. Some of the flavor components are nitrogen-containing compounds such as uric acid, creatinine, and ammonia. The acidity of a meat cut also influences flavor. You have noticed, no doubt, that the flavor differs somewhat from one muscle to another in the various parts of the carcass. A clear example of this difference can be noted when you compare the flavor of the leg and the breast of chicken.

*Which statement is **not** true?*
A. *Some of the flavor of meat comes from nitrogen-containing substances.*
B. *Muscles that get the most exercise have the least flavor.*
C. *Flavor is different in various parts of the carcass.*
D. *Meat of young animals has less flavor than that from older animals.*   *B.*

*

Flavor changes in meat during cooking. Some very volatile substances are lost and other new compounds form. The flavor of meat is modified in many recipes as the result of adding a variety of ingredients. The variation in flavor that can be achieved by adding herbs, spices, wines, and other ingredients is almost endless.

*Cooking   **changes**/fixes   the flavor of meat.*   *Changes.*

215

The idea of ripening or aging foods is not unique to beef. Cheese has long been aged, and flour customarily is aged (by the addition of bleaching agents) . The Chinese like to age their eggs until they develop a strong flavor. Preference for these various aged items varies widely, but many gourmets feel that aged meat has a superior flavor compared with the unripened meats.

Would you think you could safely state that most people in our country prefer aged beef?

Probably not. Enjoyment of the flavor of aged beef is a cultivated taste for many people, and numerous others have never had the opportunity to taste it.

*

Meat starts to age or ripen rapidly for the first two weeks and continues to age slowly after the initial aging period. Most aged beef is hung at least two to three weeks to achieve the characteristics desired by most people who prefer aged beef. During aging, several changes occur: flavor becomes more intense, muscles become more tender, and the meat turns a darker color.

Meat ages rapidly up to the end of the first   14/21   days of ripening and then the aging slows down.

14.

*

Only beef of the very highest quality can be ripened satisfactorily. Pork does not ripen in an agreeable manner because of the difference in its fat. Beef must have a heavy fat cover on the outside of the carcass if it is to age well. This fat cover protects the meat itself from molds and bacteria that would otherwise penetrate the meat and cause spoilage during the aging period. Aging conditions must be controlled to reduce growth of microorganisms during the long storage period. Since microorganisms do not grow as well in an acid medium, aging meat is sometimes wiped down with a vinegar-soaked cloth. Even with these precautions, some mold will grow on the fat surface and form "whiskers." At the end of the aging period, some of the fat cover is trimmed to remove these whiskers.

Why is only top quality beef aged?

Only top quality beef has a sufficiently heavy layer of fat to prevent microorganisms from entering the meat and to allow trimming of part of the fat layer at the end of the storage period when the whiskers must be removed.

*

Meat loses approximately one per cent of its moisture each day during ripening. This moisture loss increases the price for aged meat because of the small weight loss. The expense of the cold storage for long periods of time is also a significant factor in determining the price of aged beef. During aging the meat becomes acidic, which causes myoglobin to change to hematin at a lower temperature than it ordinarily does during cooking. If you want steaks to be rare or medium, they must be cooked a shorter time if they are aged than if they have not been aged.

Would meat that is soaked in an acid marinade tend to change color more quickly that meat that is not soaked in acid?

Yes. The meat changes color at a lower temperature because of the acid in the marinade.

*

Meat can be aged in a polyethylene bag which is shrunk tightly around the carcass to eliminate air. This bag prevents the loss of moisture and development of whiskers that are typical when meat is ripened in the open air. The odor of the aged meat is strong when the bag is first removed but it quickly disappears.

Meat aged in a polyethylene bag will be **more/less** *moist than meat aged in the open cold storage room.*

More.

*

JUICINESS IN MEAT

Meat naturally contains between 50 and 75 per cent water, although there is some variation from cut to cut and from carcass to carcass. The amount of water in meat is of great interest to the consumer because this water is necessary if the cooked meat is to be juicy. The ability of raw meat to hold or retain water is called its water-holding capacity. Meat with a low water-holding capacity will show a large drip loss during storage and will be rather dry when cooked. You will find that meat with a greater water-holding capacity is definitely more palatable. The ability of meat to hold water is influenced by the acidity of the meat. More acid meats cannot hold as much water as can less acidic ones. Pork holds water better than beef.

The ability of meat to trap or retain water in the tissues is called its _____ _____.

Water-holding capacity.

*

Frozen meats will lose some of their water from the tissues upon thawing because the ice crystals formed in freezing

217

rupture some cell walls and allow the juices to run out as the meat thaws. This drip loss resulting from damage to the cell walls is influenced by the rate of freezing. The loss is greatest when meat is frozen very slowly; slow freezing forms large crystals that break through cell walls. Very rapid freezing results in many smaller crystals and less damage is done to the structure of the meat.

*The rate of freezing **does/does not** influence the amount of drip loss when frozen meat is thawed.*

Does.

*

Moisture is lost from meat during cooking. Proper procedures will keep this loss to a minimum so that the meat itself will be as juicy as possible. During cooking, there will be drip losses that collect in the bottom of the pan and there also will be evaporative losses. The losses resulting from cooking of meat are referred to as shrinkage. Some meats may lose as much as 35 per cent or more of their original weight during cooking.

The moisture loss (plus fat loss) of meat in cooking is called ____.

Shrinkage.

*

Meat cookery procedures can be controlled to keep shrinkage to a practical minimum. Very high cooking temperatures increase moisture and fat losses. For moist meat cookery, moderate oven temperatures of 300°F to 325°F are preferred. It is also wise to avoid cooking meat too long, because this causes excessive moisture loss and the meat will be dry. You can demonstrate this loss of juiciness in meat that results from overcooking by roasting a turkey to the desired final temperature (185°F in the thigh) and then comparing it with turkey that has been roasted to 190°F or more. You will notice a significant drying of the meat during the extended roasting period.

*Loss of moisture from meat during cooking can be minimized by using a **low/high** cooking temperature and by avoiding **undercooking/overcooking**.*

Low, overcooking.

*

Fat also contributes to the juiciness of a cut of meat. When selecting meat, check the amount of fat within the muscle itself. Some fat is important if you are to prepare a meat

218

that seems juicy after it is cooked. However, a large quantity of fat in meat will give the meat a greasy taste and will have poor mouth feel. This may be a particular problem in pork and in lower-priced ground beef containing added fat.

*Fat **does/does not** contribute to the apparent juiciness of a piece of meat.*

Does.

*

TENDERNESS IN MEAT

The tenderness of a piece of meat depends upon the quantity of connective tissue it contains. Connective tissue is the protein that holds the protein fibers into the muscles we call meat. This connective tissue is present in greater quantities in some parts of the animal than in others. Muscles that receive a considerable amount of exercise are higher in connective tissue than are those in more protected areas of the body. Muscles along the back and loin of the animal are thus more tender than are those in the neck, shoulder, shanks, or legs.

The protein that holds muscles together is called _____ tissue.

Connective.

*

There are two kinds of connective tissue in meat. The white, abundant connective tissue is called collagen. Collagen is converted from tough connective tissue to tender gelatin in some meat cookery methods. This change will occur slowly when meat is cooked in moist heat. The addition of acids such as tomato juice, vinegar, sour cream, or lemon juice speeds the change of the connective tissue from collagen to gelatin. As you would expect, meat that is cooked long enough in moist heat will become very tender because of the change in the collagen. The other connective tissue is yellow and is aptly named elastin. This type of connective tissue occurs in bundles that are distinctly yellow and rubber-like. Cooking procedures will not alter this protein at all. The only way to tenderize elastin is to break it up into smaller pieces mechanically by pounding, cubing, dicing, or grinding. These procedures do not cause a chemical change in the composition of the elastin. Meat is more

219

palatable if elastin is removed rather than being added to ground meats.

Match the term in column A with the factors that pertain to it in column B.

A	B	
1. Collagen	*A. Tenderized by grinding*	
2. Elastin	*B. Tenderized by moist heat*	
	C. White connective tissue	
	D. Yellow connective tissue	
	E. Converted to gelatin more quickly by acid	*1. A, B, C, and E.*
		2. A and D.

*

The tenderness of meat is influenced by the conditions under which an animal is raised. Animals raised on the range where they have to forage for their food are less tender than are animals carefully pampered in a feedlot where they have nothing to do but eat. The breed of the animal and its age are also important factors in determining tenderness of meat. Stocky animals are now being bred that have a good ratio of meat to bone and these are carefully raised to produce as tender an animal as possible. In general, younger animals are more tender than older animals.

Which chicken do you think would have the most tender meat:
A. One who lives in a barnyard or one kept in a cage and fed a special mash to make it grow rapidly?
B. An older chicken or a young chicken?
C. An egg-laying breed or a meat breed?

A. The one kept in a cage and fed a special mash.
B. A young chicken.
C. A meat breed.

*

Meat can be tenderized by adding an enzyme to it. The enzyme tenderizes the meat by converting collagen to smaller protein molecules. Papain, either in powdered form or in solution, is the enzyme available for home use. This is prepared from papaya. Wherever the enzyme touches the meat, the conective tissue will be tenderized. The actual amount of change is determined by the length of contact time before cooking. Papain can cause so much breakdown of the collagen that the meat surface will actually become powdery. The chief disadvantage in tenderizers is that it is almost impossible to spread the tenderizer uniformly all over the meat and, consequently, some areas will be too powdery while other places may still be distinctly chewy. As a response to this problem, one large meat packer developed the procedure of injecting the enzyme preparation into the live animal shortly before slaughter. The enzyme then is distributed via the blood to all areas of the body to give uniform tenderization. This enzyme is inactive until meat is heated in the early stage of meat cookery. Then the enzyme reacts quite quickly throughout the meat until the temperature increases to the point where the enzyme is inactivated.

Papain is:
A. *A meat tenderizer.*
B. *An enzyme.*
C. *A substance from papaya.*
D. *A, B, and C are all correct.*

D. *A, B, and C are all correct.*

*

MEAT COOKERY METHODS

Meat can be classified into tender and less tender cuts. On the basis of this classification, you will wish to use either a dry heat cookery method or one using moist heat. Meats that are tender cuts are low in connective tissue. If they also contain a modest amount of fat, they may be suitably prepared by dry heat cookery methods. Cuts that are high in connective tissue and rather low in fat are more palatable when prepared by a moist heat method. You may wish to cook tender cuts of meat that are low in fat by a moist heat method so they they will be more juicy.

Tender cuts of meat are usually cooked by **dry/moist** *heat methods, and less tender cuts are prepared by a* **dry/ moist** *heat method.*

Dry, moist.

*

Dry Heat Cookery

Broiling, pan broiling, pan frying, roasting, and deep-fat frying are all classified as dry heat methods because the meat is cooked uncovered and evaporating moisture is allowed to escape. In dry heat meat cookery, the heat coagulates the muscle proteins to make them slightly less tender than they were before cooking was started. Almost no change occurs in collagen in dry heat meat cookery, with the total result that meats prepared by dry heat methods are actually slightly less tender than they were when you began to prepare them. There also is some drip loss, but this loss of moisture is offset by the apparent juiciness resulting from the melting of fat during cooking.

The basic dry heat cookery methods are: _____, _____, _____, _____, *and* _____.

Broiling, pan broiling, pan frying, deep-fat frying, roasting.

*

Broiling.   Broiling is done by exposing the meat directly to radiant heat. The source of heat can be a glowing element, hot coals, or a live flame. The temperature is regulated for broiling by moving the meat either toward the source of the heat or away from it. The heating source does not cycle on and off to regulate the broiling temperature. This cycling would change the heat from radiant heat to heating by convection. Very thick cuts of meat need to be placed farther from the heat source than do thinner cuts. This provides a lower temperature on the very thick cuts so that they are heated more slowly. With a slower rate of heating, the temperature will be spread in the meat more uniformly and there will be adequate opportunity to cook the center of the meat before the outer portion becomes too done. This precaution is less necessary with thinner cuts of meat because the heat will penetrate rather quickly to the center of the meat.

Broiled meats are cooked by **radiant heat/convection.**

Radiant heat.

*

Meat cuts suitable for broiling need to be a minimum of one inch thick. Thinner cuts become too dry when they are broiled. Of course, meats for broiling should be tender cuts and should be marbled. Prime and Choice T-bone, porterhouse, club, sirloin, and rib steaks of beef are excellent choices for broiling if they are cut at least one inch thick. Fat fish such as salmon also broil well. Bacon, ham, and other cured pork products may be broiled, but fresh pork is less suitable for broiling because it is difficult to heat the interior of the cut to the necessary internal temperature to kill any trichinae that might be present. Trichinae are the parasites that are occasionally found in pork. These are not detected by inspection and, consequently, pork needs to be cooked to at least 170°F to be certain that the organisms are killed. There is no hazard from them if they are heated until they die. Most hogs are not infected with trichinae, but it is still wise to be cautious when preparing pork. Veal is not broiled because it is very low in fat and is relatively high in connective tissue. Lamb chops are suitable for broiling if they are thick. Young chickens can be split and broiled.

*Meat to be broiled should be **tender/less tender** and should be cut **at least/less than** one inch thick.*

Tender, at least.

*

All steaks and chops to be broiled need to be scored through the fat and connective tissue at one-inch intervals around the edge of the cut. This is done to keep the meat flat while it is being broiled. If this step is omitted, the connective tissue will shrink during the broiling period and the meat will curl, causing some portions to be nearer the heat than others. It is impossible to broil meat uniformly unless it is flat and a uniform thickness throughout.

True or false.   Meat needs to be scored at one-inch intervals through the fat and connective tissue to keep the cut from curling.

True.

*

Broiled meats have a pleasing flavor if they are cooked carefully to give a pleasingly browned surface. Additional flavor can be added to the meat by broiling over charcoal. Charcoal from hardwoods rather than soft resinous woods is preferred. The charcoal should be lighted and burned until it is a glowing bed of coals before the meat is placed in position. Active flame causes considerable smoking of the meat and results in a less palatable and less healthful product. If fat drips onto the hot charcoal and ignites, the resulting sooty flame may burn the meat or deposit tars from the smoke. Open flame during charcoal broiling can be controlled by avoiding too hot a fire or by sprinkling water on the fire as flames erupt.

*To avoid charred, sooty meat when charcoal broiling, control the charcoal so that you have **an open flame/a bed of hot charcoal.***

A bed of hot charcoal.

*

When broiling in a broiler compartment of a range, it is important to use the broiler pan designed by the manu-

223

facturer of the range. Place the meat on the cold broiler pan. Do not cover the grid of the broiler pan with aluminum foil because this traps the melting fat on the grid and keeps the fat close to the source of the heat. The fat quickly overheats and ignites, causing a broiler fire. Broiler pans are designed to let the fat drain away to the bottom of the pan where the grid protects the fat from the intense heat of the broiler unit. By removing the fat from the broiling meat, you not only eliminate the fire hazard but you also avoid frying the meat in the drippings. As a result, broiled meats are less greasy and lower in Calories than either pan-fried or pan-broiled meats.

Meats to be broiled should be placed on a **cold/hot** *broiler pan; the grid* **should/should not** *be lined with aluminum foil.*

Cold, should not.

*

Meats should be broiled unsalted until the first side is a tempting brown and the meat is approximately half done. If a cut of meat is thick enough, you can insert a thermometer to indicate accurately the doneness of the meat. If a thermometer is used, it should be inserted from the side of the cut in a position half way between the upper and lower surface. You will need to push this in as straight as you possibly can so that the bulb of the thermometer is in exactly the center of the largest muscle in the cut. The thermometer should be in the muscle and should not be touching either bone or fat. When the meat is about half done, salt the cooked surface and then turn the cut with a pair of tongs. Tongs are used because they allow you to turn the meat easily without puncturing and allowing some of the juices to drain from the meat. Continue broiling the second side until the cut is done to the desired degree.

It is recommended that meat be turned **once/twice** *while being broiled and that the salt be added* **at the beginning/just before the meat is turned.**

Once, just before the meat is turned.

*

If you do not have a thermometer, you will find it necessary to use appearance and time to tell you when broiled meats should be turned and when they are done. Sometimes it is impractical to try to insert a thermometer in cuts that are just barely thick enough to broil. When you broil cuts

224

Table 8.1. Timetable for Broiling*†

Cut	Weight (Pounds)	Approximate Total Cooking Time (Minutes)	
		Rare	Medium
BEEF:			
Chuck steak (high quality) —1 in	1½–2½	24	30
1½ in	2–4	40	45
Rib steak—1 in	1–1½	15	20
1½ in	1½–2	25	30
2 in	2–2½	35	45
Rib eye steak—1 in	8 oz	15	20
1½ in	12 oz	25	30
2 in	16 oz	35	45
Club steak—1 in	1–1½	15	20
1½ in	1½–2	25	30
2 in	2–2½	35	45
Sirloin steak—1 in	1½–3	20	25
1½ in	2¼–4	30	35
2 in	3–5	40	45
Porterhouse steak—1 in	1¼–2	20	25
1½ in	2–3	30	35
2 in	2½–3½	40	45
Ground beef patties—1 in by 3 in	4 oz	15	25
PORK—SMOKED:			
Ham slice, tendered—½ in	¾–1	Ham is	10–12
1 in	1½–2	always	16–20
Canadian style bacon—¼ in slices		cooked	6–8
½ in slices		well done	8–10
Bacon			4–5
PORK—FRESH:			
Rib or loin chops ¾–1 in			20–25
Shoulder steaks ½–¾ in			20–22
LAMB:			
Shoulder chops—1 in	5–8 oz	Lamb chops	12
1½ in	8–10 oz	are not	18
2 in	10–16 oz	usually	22
Rib chops—1 in	3–5 oz	served	12
1½ in	4–7 oz	rare	18
2 in	6–10 oz		22
Loin chops—1 in	4–7 oz		12
1½ in	6–10 oz		18
2 in	8–14 oz		22
Ground lamb patties—1 in by 3 in	4 oz		18

*This timetable is based on broiling at a moderate temperature (350°F). Rare steaks are broiled to an internal temperature of 140°F; medium to 160°F; well done to 170°F. Lamb chops are broiled from 170°F to 175°F. Ham is cooked to 160°F. The time for broiling bacon is influenced by personal preference as to crispness.
†By permission of the National Live Stock and Meat Board.

without a thermometer, be sure to broil them on the first side until they have a pleasing appearance. This should take approximately half the time indicated in Table 8.1. The remainder of the recommended cooking time is used to complete the broiling of the second side. With considerable experience, you may be able to press on broiled meat with a blunt instrument and tell by touch if the meat is done. Raw meat yields far more readily than well-done meat, which is firm to the pressure of a blunt instrument. It may be more practical for you to make a very small incision right next to the bone when you think that the broiled meat is done to your satisfaction. Check to see if the meat next to the bone is the correct color to suit you. Beef may be broiled to rare, medium, well done, or any stage in between. Rare beef will be warm in the center and red, medium still will be pink in the center, and well done will have lost all traces of pink but still will be juicy. Most persons prefer lamb that has no pink color remaining. Ham, of course, cannot be judged by its color since no change occurs.

True or false. To determine if broiled meat is done to your satisfaction, it may be necessary to cut the meat right next to the bone to check the color of the meat.

True.

*

A thermometer takes the guess work out of broiling. If you wish to have beef steaks rare, broil the meat until the themometer indicates 140°F. For medium steaks, the thermometer should read 160°F; for well done, continue broiling to 170°F. Ham slices are done when the thermometer registers 160°F. Lamb is usually broiled from 170°F to 175°F.

The final internal temperature for broiled beef steaks may range from _____°F to _____°F; ham is broiled to _____°F, and lamb is cooked to _____°F to _____°F.

140, 170, 160, 170, 175.

*

Broiled meats may be varied in flavor by marinating them in an acidic mixture such as a sharp salad dressing for an hour or more before broiling. This helps to tenderize the meat as well as to vary the flavor. You may wish to try broiling meat with a sauce brushed on it. Usually these sauces are brushed on at frequent intervals during the broiling process. This practice of basting meats with a sauce is a variation of broiling and is referred to as barbecuing.

Barbecuing/marinating is done to modify the flavor of meat before broiling; *barbecuing/marinating* is a variation of broiling in which the meat is basted frequently with a sauce.

Marinating, barbecuing.

*

Pan broiling. Another quick, dry heat method of preparing meats is pan broiling. This process is done in a frying pan on top of the range rather than in the broiler compartment. Any meat that can be broiled can be pan broiled. In addition, thinner cuts of meat are pan broiled satisfactorily although they would be too dry if broiled in the broiler compartment. It is also possible to pan broil thin pork cuts satisfactorily. The range of meats that can be pan broiled is obviously large. Any tender beef steaks up to two inches thick will pan broil well. Ham slices pan broil very well, as do pork chops cut one-half inch thick. Since pork, to be safe, should be cooked to at least 170°F in the interior, it is unwise to pan broil pork chops an inch or more in thickness. Lamb chops pan broil very well. Bacon may be pan broiled, as can sliced liver, kidney, sweet breads, and brains.

True or false. Meats that are less than one inch thick may be pan broiled satisfactorily, but they will be too dry if broiled.

True.

*

To pan broil meats, score the fat and connective tissue around the edge of the cut and then place the meat in a cold frying pan. If preferred, the pan may be preheated slightly and the fat at the edge of the cut rubbed on the bottom of the pan to lightly grease it before the meat is put in. This helps to keep the meat from sticking. The meat is cooked at a moderate heat, and any fat that drains from the meat is removed as it accumulates. As soon as the meat is pleasingly browned on the first side, the meat is turned with tongs and the cooked side is salted. Thin pieces of meat usually will be done by the time the second side is browned adequately. However, thicker cuts may need to be turned at least one more time to complete cooking them to the desired end point before the surface is too brown. The appearance of the interior of the meat near the bone is the most practical means of telling when the meat is done.

The appearance can be checked by making a small cut in the meat next to the bone. Pork and well-done beef should not have any trace of pink showing, medium beef will look pink in the center, and rare beef will be pink to red in the center. Many people prefer lamb that has lost its pink color, although some like it pink.

When meats are pan broiled, the fat is **allowed to collect in the pan/drained off as it collects in the pan.**

Drained off as it collects in the pan.

*

Regulation of the temperature used to pan broil meats is important for producing a pleasing cut of meat. Thin cuts of meat may be pan broiled with a relatively intense heat to give a pleasing brown on the exterior because the heat penetrates through the thin piece of meat. Cuts of meat an inch or more in thickness need to be cooked at a more moderate temperature to allow time for the heat to penetrate to the center of the cut before the outside is too dark and dry. The desired degree of doneness of the meat also influences the rate at which the meat should be pan broiled. Rare pieces of meat need to be browned rather quickly on the outside so that the cooking will be completed before the interior becomes too well done. Well-done steaks need a lower temperature to permit more uniform heat distribution in the meat. Keep in mind that the heat always needs to be controlled in pan broiling to avoid overheating the fat and protein in the meat. When fat is overheated, it begins to smoke and quickly breaks down to an acrid, irritating chemical compound called acrolein. This is unpleasant for the cook as well as for the meat. The outer surface of meat quickly becomes burned, dry, and tough if the cooking temperature is too high. The heat should be regulated so that the fat never smokes or spatters badly, but if you should begin to overheat the meat, remove the pan from the heat and adjust the temperature control to a lower position.

Pan broiling should be done at **high/moderate/low** *temperatures; thick cuts of meat need to be pan broiled at* **higher/lower** *temperatures than thin cuts.*

Moderate, lower.

*

Pan frying. Pan frying or sautéing are two terms used to identify the dry heat meat cookery method in which meat is fried in shallow fat. Sometimes the terms grilling and griddle-frying are used to identify this process. The term "pan fry" is preferred to simply "frying," because frying could imply either shallow or deep-fat frying. Pan frying clearly indicates shallow fat. Pan frying is most successfully done in a heavy aluminum or cast iron skillet. These types of materials conduct heat uniformly to promote even browning of meats. Incidentally, these same materials are particularly well suited to pan broiling also. Stainless steel pans tend to spread the heat less evenly so meat may get too brown where it touches hot spots in the pan and be too light in other areas. When stainless steel is combined with another metal such as an iron core in a stainless steel skillet, the heating characteristics of the pan will be somewhat improved.

One requirement for successful pan frying is a **light/ heavy** *frying pan that distributes heat* **evenly/capriciously.**

Heavy, evenly.

*

The amount of fat used to start the pan frying process needs to be varied to suit the type of meat being fried. Pork chops or other meat cuts containing appreciable fat will pan fry very well if you place them in a frying pan containing only a light coating of fat. As these cuts are fried, additional fat will drain from the meat to help in cooking the cut. Meats that are low in fat are fried more successfully if approximately an eighth of an inch of fat is melted in the frying pan before the meat is added. If meats are lightly floured prior to pan frying, they may suitably be fried in about one-eighth inch of fat, but breaded or batter-dipped meats will absorb some of the fat in the pan and are best fried in approximately one-fourth inch of fat.

For pan frying, melt about _____ inch of fat for lightly floured foods and about _____ inch for breaded or batter-dipped items.

One-eighth, one-fourth.

*

It is just as important in pan frying as in broiling and pan broiling to have the meat remain flat throughout the cooking period rather than permitting it to curl. Therefore, the first step in preparation for pan frying is to score the fat and connective tissue surrounding the meat cut. The meat may be left uncoated or may be lightly floured, rolled in bread or cracker crumbs, or batter-dipped preparatory to

229

pan frying. Breading will stay on the meat better if the meat is first dipped in beaten egg or milk. Whether coating with flour, bread, or cracker crumbs, gently shake the coated meat before placing it in the frying pan. This removes loose material that would fall off and begin to burn in the fat during cooking. You might enjoy using prepared mustard as the adhesive to fasten breading securely to pork chops. If you let the breaded product stand 15 minutes before pan frying, the breading material will adhere well.

Before meat is pan fried, it should always be _____; you may also wish to dip the meat in _____, _____, or _____ crumbs, or in a batter.

Scored; flour, bread, cracker.

*

When the meat is placed in the frying pan, the fat should be hot enough to start frying the meat immediately. Control the temperature of the fat to avoid smoking of the fat and excessive spattering. Brown the meat to a golden brown on the first side before carefully turning it to brown the second side. If necessary, turn the meat more than once. The degrees of doneness for pan fried meats are the same as those for broiling or pan broiling; that is, beef may be rare, medium, or well done and pork is cooked to well done as are poultry and lamb. The primary distinction between pan broiling and pan frying is the amount of fat maintained in the pan. A minimum of fat is present at all times when meat is pan broiled. During pan frying, a layer of fat is present in the pan throughout the cooking period. Usually the fat is allowed to accumulate in the pan as it drains from the meat.

*In pan broiling, fat **is/is not** allowed to accumulate in the pan; in pan frying, fat **is/is not** allowed to remain in the pan as it drains from the meat.*

Is not, is.

*

Various fats or oils may be used for frying meats. Salad oils from cottonseed, corn, or other vegetable source are excellent choices for frying foods because they have a high smoke point. This means that they can be heated to a high temperature before the fat begins to break down and release acrolein. However, you may wish to use other fats and oils occasionally to add a particular flavor note to a meat. Butter adds a pleasing richness to the flavor of meats fried in it if the temperature is carefully controlled. Butter does decompose at a relatively low temperature and may give a burned flavor to meats if the butter is overheated during pan frying. Other animal fats with distinctive flavors that might be used for frying meats include lard and bacon fat. Olive oil and margarines are vegetable oil items that contribute distinctive flavors to meats if they are not overheated.

If you wanted a crisp, golden-brown surface on breaded pork tenderloin slices, should you use corn oil or butter for the frying fat?

Corn oil would be a better choice. Although butter would give a pleasing golden-brown surface and a good flavor to the meat, it requires considerably more care to use butter without burning than to use corn oil.

*

Pan frying is a suitable method for the tender cuts of beef that are cut into steaks. Thin pork chops and pork tenderloin, young poultry, and lamb chops are very satisfactory when pan fried. Lean fish (less than five per cent fat) are excellent when pan fried or deep-fat fried. Steaks or filets of halibut, perch, cod, sole, flounder, haddock, bass, and swordfish are excellent lean fish to pan fry. Some shellfish are well suited to pan frying. Sometimes the fat fish such as salmon, trout, shad, tuna, and mackerel may be pan fried, although they are often broiled or baked. Regardless of the cooking method selected for fish, avoid overcooking them. Fish and shellfish have very little connective tissue and are consequently tender before they are cooked. High temperatures and long cooking times only cause toughening of the muscle proteins. This explains why fish get less tender during cooking, while other types of meat high in connective tissue become more tender with extended cooking time.

True or false. Long cooking of fish and shellfish is not recommended because there is little connective tissue to tenderize and the long cooking only toughens the muscle protein present in the fish.

True.

*

Deep-fat frying. Proper selection and care of the fat used for deep-fat frying will improve the quality of deep-fat fried meats. Vegetable oils are definitely the best choice for deep-fat frying. Most shortenings presently available have emulsifiers added to them to improve their behavior in cakes and other baked products. However, fats containing emulsifiers

231

have a lower smoke point than do cottonseed, soybean, and corn oils; consequently the fats break down at a faster rate at the tempeartures required for deep-fat frying. Oils break down less rapidly if food particles do not fall off and remain in the hot oil. Loose breading materials and excess batter that drips from meat into the cooking oil speed deterioriation of the fat. Excess moisture clinging to the meat or salt crystals also limits the usefulness of the hot fat. Therefore, it is good practice to salt meats after they have been deep-fat fried and to wipe meat surfaces lightly with a paper towel to remove any excess moisture. Shake off loose crumbs before frying breaded meats.

Select the items that reduce the useful lifetime of an oil selected for deep-fat frying:
A. Salt particles.
B. Breading particles.
C. Excess drops of batter.
D. Water.

All of these reduce the length of time an oil may be used satisfactorily for deep-fat frying.

*

Deep-fat frying is most easily done in a thermostatically controlled deep-fat fryer, in a pan equipped with a thermostat, or in a pan placed on a thermostatically controlled heating unit. The oil should be preheated to 350°F before the meat is added. This temperature is excellent for frying meats. It allows the meat to cook slowly enough so that it can be done adequately in the center without burning the outside of the cut. It is sufficiently fast to cook the meat before it has an opportunity to become excessively greasy. Be sure to cook only enough meat to drop the temperature of the hot fat just briefly when the cold meat is added. Too much meat at one time drops the temperature of the fat excessively, and the recovery time to the desired frying temperature is so long that cooking time is noticeably extended. The result is greasy fried meats. Of course, meats that are fried at too high a temperature are either underdone in the center or too dark on the exterior. High temperatures reduce the useful frying lifetime of a fat as well as reducing the quality of the food fried in the fat.

_____ *control is essential if you are to deep-fat fry meats successfully.*

Temperature.

232

Deep-fat frying is a rapid method of cooking meat by dry heat cookery because the meat is in direct contact with the hot oil. Although the exact length of time required for deep-fat frying is determined by the size of the piece and the temperature of the meat before it is placed in the fryer, most meats can be deep-fat fried in three to six minutes when the fat is preheated to the correct frying temperature. If you do not have a thermometer, you can time the browning of a dry bread cube in the hot fat. For deep-fat frying, the fat should be hot enough to brown stale bread in 60 to 70 seconds. This check is obviously less accurate than using a thermometer.

The cooking time for deep-fat frying is **longer/shorter** *than the time required for broiling.*

Shorter.

*

Chicken and seafood such as oysters, scallops, fish fillets, and croquettes, are often deep-fat fried. The seafood and fish items are cooked just long enough to heat them through thoroughly. They toughen quickly in the intense heat used for deep-fat frying. Deep-fat fried chicken requires longer cooking to permit the heat to penetrate and cook the meat next to the bone. Check to be sure that the meat pulls free of the bone easily. If it does not, the chicken needs to be cooked longer.

If seafood that has been deep-fat fried is tough, it has been cooked too **little/much.**

Much.

*

One cause of kitchen fires is overheated fats used in deep-fat frying. Avoid splattering fat onto a hot element or into an open flame. Do not heat fats above 400°F and work carefully around hot fats. If fat should happen to ignite in the kitchen, smother the flames. If possible, put a lid quickly on the pan. Turn off the gas or electricity. Aluminum foil is effective in smothering a flame by keeping air away from the fire. Baking soda can be sprinkled on the flames. Do not pour water on a fat fire. A grease fire will spread if water is poured on it because the water causes the hot fat to spatter. If you should happen to receive a burn from hot fat or any other source while cooking, immediately apply ice to the burned area. Keep ice in direct contact with the burned area until medical help comes or until the ice pack has been applied for at least 20 minutes. Immediate

233

application of ice is important to reduce the damage inflicted by the burn.

*Put out grease fires by **pouring on water/smothering them with a lid or aluminum foil;*** *treat burns by applying* **cold butter/ice.**

Smothering them with a lid or aluminum foil; ice.

*

Certain precautions can be taken to minimize the possibility of fires and burns when deep-fat frying. Pans should always have the handles positioned over the counter rather than protruding out from the range. Never leave a pan of hot fat unattended. Remember that the fat will bubble up when moist foods are added to it; avoid putting so much fat in the container that the fat overflows when moist foods are added. Carefully add foods to hot fat to reduce the likelihood of splashing and spattering. Have the lid of the deep-fat fryer or a roll of aluminum foil within reach in case they should be needed. These precautions are intended to help you develop good safety habits in the kitchen. Deep-fat frying, like broiling and other cookery procedures, is only dangerous when you forget to consider kitchen safety. There is no good excuse for fires of any size in the kitchen. Think safety!

At all times in the kitchen, think _____.

Safety. Be safety conscious for yourself and others.

*

Fat can be used several times for deep-fat frying if it is properly handled. First of all, avoid excessive heating of fats when deep-fat frying. There is no reason to heat fats over 400°F. Keep preheating periods to a minimum and allow fats to begin to cool as soon as you are through deep-fat frying. The longer the fat is held at frying temperature, the more quickly it breaks down. Strain the fat after each use to remove food particles that may have settled to the bottom. If you are frying with oil, you can allow the oil to cool completely before straining it through several layers of cheese cloth to remove the foreign particles. If you are frying with a shortening that is solid at room temperature, it will be necessary to strain the fat while it is still sufficiently warm to be fluid. It is unwise to strain fat while it is still extremely warm because of the possibility of being burned by it. One other precaution may be observed to prolong the usefulness of solid shortenings. These should be warmed fairly slowly until they melt to avoid overheating the fat

234

in contact with the pan. Once the fat becomes fluid, you can increase the rate of heating without overheating a portion of the fat.

True or false. Straining fat or oil between use will increase its lifetime for satisfactory deep-fat frying.

True.

*

Deep-fat fried foods are at their best as soon as they are removed from the hot fat, dried on absorbent towels, and salted. They then have the desired crispness on the outer surface and a pleasing flavor reminiscent of the fat in which they were fried. However, fried foods very quickly begin to seem a bit soggy and greasy if they are allowed to stand. If fried foods must be held, they should be stored in a hot dry place such as a low-temperature oven set at 150 to 170°F. This temperature helps to retain the desired crispness without continuing to cook the meat.

When fried foods are done, they should be **drained on absorbent paper/dished directly onto a serving platter** *and* _____.

Drained on absorbent paper, salted.

*

Breading for deep-fat frying or pan frying gives the best product if you can make it adhere well to the meat throughout the frying process. You may wish to first dip the meat into seasoned flour before dipping it into a liquid and the breading crumbs. However, the flour is sometimes omitted. Suitable breading items include bread crumbs, cracker crumbs, cornmeal, and crushed cereals. For a breading that adheres well to the food, use very fine crumbs. Coarse crumbs give a looser breading that does not cling to the meat as well. In addition, coarse crumbs absorb more fat during frying. As a guide in knowing how much breading material will be needed, plan on breading to equal about one-third the weight of small meat items such as shrimp. For larger items like veal cutlets, you will need breading to equal about one-fourth the weight of the meat. For large pieces like chicken, the breading will need to be about 15 per cent of the weight of the meat.

If you want a meat product to hold its breading well and absorb a minimum of fat, bread it with **fine/coarse** *crumbs.*

Fine. You also will discover that breading stays on better if it is allowed to dry on the meat for about 15 minutes before the meat is fried.

235

Foods may be batter-dipped for deep-fat frying. Batters for deep-fat frying are made with flour, eggs, milk, and salt with various seasonings and baking powder used occasionally as optional ingredients. If you wish a thick crust on the meat, make a thick batter. For a thinner crust, make a thinner batter that will coat the meat completely but still flow fairly readily. Dip the meat in the batter and allow the excess to drip into the bowl of batter. Then, quickly slip the meat directly into the hot fat.

Why is the basket omitted when frying batter-dipped meats?

Try it once and you will quickly understand. The batter sticks to the basket and fries to it. Then you have to tear the food from the basket and much of the batter mixture pulls away from the food.

*

Occasionally, meats are wrapped in a dough and deep-fat fried. The dough used for this is a little richer than a biscuit dough but contains less fat than is used for pastry. If too much shortening is used in the dough, the finished product will be distinctly greasy. To prepare these fried pies, use about as much dough as filling. For example, you can make fried chicken pies by sealing two ounces of chicken filling in two ounces of rolled dough. This is called either a fried pie or a turnover.

Why should you be sure to seal the dough well around the filling when making a fried pie?

A tight seal keeps the dough from becoming so grease soaked and it also keeps the moist filling from coming in direct contact with the fat and speeding the breakdown of the fat.

*

Roasting. Roasting is the dry heat method of meat cookery done in the oven. Since roasting is to be dry heat, meat that is roasted must be cooked without a cover of any kind. If a cover is placed on the roasting pan or if aluminum foil is tightly wrapped around the meat, moisture is trapped around the meat as it evaporates from the surface and this results in a moist environment for the meat rather than the dry one recommended. Much confusion exists about roasting of meat, primarily because of the piece of kitchen equipment referred to as a roaster or roaster pan. This is the familiar large oval pan with a cover. Since the roaster has a cover, it suggests to many people that the cover should be used when meats are cooked in this pan. Of course, for true roasting the cover should not be placed over the turkey or other large meat cut that might be roasted in this pan.

When roasting a turkey in a roaster, ***do/do not*** *use the cover that comes with the pan.*

Do not.

236

Meats suitable for roasting include standing rib roast or rolled rib roast of beef, rack and leg of lamb, pork loin roast, crown roast of pork or lamb, ham, poultry, whole fish, and large cuts of fatty fish. Meats with a layer of fat surrounding the roast are particularly well suited to this method of preparation, but lean cuts that are tender can be roasted if they are basted with butter or a sauce periodically throughout the roasting period. Bacon strips traditionally are arranged on the upper surface of a roast tenderloin of beef to provide the fat necessary to prevent drying of this lean cut. Poultry is usually basted with melted butter to keep the skin from drying out when it is roasted. Fish steaks may be roasted in a butter sauce.

Meats that will roast satisfactorily are **tender/less tender** *cuts. If the meat is lean, tender cuts need to be _____ with melted fat when they are roasted.*

Tender, basted.

*

You may have wondered why ham and fish were listed among the meats that are suitable for roasting, because the usual expression is baked ham or baked fish. Actually, roasting and baking are synonyms for the dry heat cookery of meat in an oven. This type of meat cookery is done at moderate oven temperatures that permit the heat to penetrate slowly through the meat to cook the center of the meat before the outside of larger cuts becomes dry and burned. Smaller roasts are prepared at 325°F and larger ones are roasted at 300°F. These low temperatures tell you that roasting is a much slower process than the other dry heat methods of meat cookery.

True or false. *Baking of meats is done at 300°F and roast-is done at 325°F.*

False. Baking and roasting are two words that designate the same process. The temperature is determined by the size of the cut to be prepared.

*

Contrary to other dry heat cookery procedures, the meat is not scored for roasting. Roasts with the rib bones retained can be prepared for roasting by simply placing the rib roast upright, with the bones serving as a rack to hold the meat out of the drippings. A shallow pan is used to hold rib roasts as well as any other roast. Other roasts are placed on a rack or trivet to hold the meat out of the drippings throughout the roasting process. In all cases, the fat side of the roast should be at the top so that the melting fat in the cut will baste the meat as it drains toward the roasting pan.

The meat thermometer is carefully inserted so that the bulb is in the center of the largest muscle and is not resting in fat or touching bone. Then the meat is ready to be placed in the oven and the temperature set. Meats do not need to be watched carefully during roasting, because the heat is low enough that the meat is very unlikely to burn. Occasional checking is all that is needed. The low sides on the roasting pan permit good circulation of air for uniform cooking of the meat.

Rib roasts **do/do not** require a rack for roasting; other cuts **should/should not** be placed on a rack.

Do not, should. The rib and chine bones of the rib roast provide a built-in rack for this cut.

*

Roasting is most accurately and satisfactorily done using a meat thermometer. With the aid of a thermometer, you can tell just how done the interior of the meat is without having to guess or try to see what the interior really looks like. Timetables for roasting (Tables 8.2 and 8.3) are merely

Table 8.2. Timetable for Roasting*

Cut	Approximate Weight (Pound)	Oven Temperature Constant	Interior Temperature When Removed from Oven	Approximate Cooking Time (Minutes per Pound)
BEEF:				
Standing rib†	6–8	300–325°F	140°F (rare)	23–25
			160°F (medium)	27–30
			170°F (well)	32–35
	4–6	300–325°F	140°F (rare)	26–32
			160°F (medium)	34–38
			170°F (well)	40–42
Rolled rib	5–7	300–325°F	140°F (rare)	32
			160°F (medium)	38
			170°F (well)	48
Delmonico (rib eye)	4–6	350°F	140°F (rare)	18–20
			160°F (medium)	20–22
			170°F (well)	22–24
Tenderloin, whole	4–6	425°F	140°F (rare)	45–60 (total)
Tenderloin, half	2–3	425°F	140°F (rare)	45–50 (total)
Rolled rump (high quality)	4–6	300–325°F	150–170°F	25–30
Sirloin tip (high quality)	3½–4	300–325°F	150–170°F	35–40
VEAL:				
Leg	5–8	300–325°F	170°F	25–35
Loin	4–6	300–325°F	170°F	30–35
Rib (rack)	3–5	300–325°F	170°F	35–40
Rolled shoulder	4–6	300–325°F	170°F	40–45

Table 8.2 Timetable for Roasting (*Continued*)

Cut	Approximate Weight (Pound)	Oven Temperature Constant	Interior Temperature When Removed from Oven	Approximate Cooking Time (Minutes per Pound)
PORK, FRESH:				
Loin:				
Center	3–5	325–350°F	170°F	30–35
Half	5–7	325–350°F	170°F	35–40
Blade loin or sirloin	3–4	325–350°F	170°F	40–45
Roll	3–5	325–350°F	170°F	30–35
Picnic shoulder:	5–8	325–350°F	185°F	30–35
Rolled	3–5	325–350°F	185°F	40–45
Cushion style	3–5	325–350°F	185°F	35–40
Boston shoulder	4–6	325–350°F	185°F	45–50
Leg (fresh ham) :				
Whole (bone in)	10–14	325–350°F	185°F	25–30
Whole (boneless)	7–10	325–350°F	185°F	35–40
Half (bone in)	5–7	325–350°F	185°F	40–45
PORK, SMOKED:				
Ham (cook before eating) :				
Whole	10–14	300–325°F	160°F	18–20
Half	5–7	300–325°F	160°F	22–25
Shank or butt portion	3–4	300–325°F	160°F	35–40
Ham (fully cooked) :‡				
Half	5–7	325°F	130°F	18–24
Picnic shoulder	5–8	300–325°F	170°F	35
Shoulder roll	2–3	300–325°F	170°F	35–40
Canadian style bacon	2–4	300–325°F	160°F	35–40
LAMB:				
Leg	5–8	300–325°F	175–180°F	30–35
Shoulder:	4–6	300–325°F	175–180°F	30–35
Rolled	3–5	300–325°F	175–180°F	40–45
Cushion	3–5	300–325°F	175–180°F	30–35

*By permission of the National Live Stock and Meat Board.
†Ribs which measure 6-7 inches from chine bone to tip of rib.
‡Allow approximately 15 minutes per pound for heating whole ham to serve hot.

intended as guides to help you plan approximately what the total roasting time will be so you will know how soon to begin roasting a cut for a meal. As you will discover through experience, the heat penetration of roasts will vary from one cut to another. Time alone is not an accurate gauge of how hot a cut of meat may be in the interior. Only a thermometer can tell you this. The final interior temperature for roasting of various cuts is the same as has been indicated for other types of dry heat meat cookery. Rare beef roasts are done when the interior reaches 140°F, medium roasts are heated to 160°F, and well-done roasts to 170°F. Pork

Table 8.3. Timetable for Roasting Turkey

Oven Weight (Pounds)	Oven Temperature	Cooking Time (Minutes per Pound)
8–10	325°F	25–20
10–14	325°F	20–18
14–18	300°F	18–15
18	300°F	15–13
20	300°F	15–13

loin roasts are heated to 170°F, other fresh pork cuts to 185°F, and hams to 160°F. Picnics are most palatable when heated to 170°F. Lamb is roasted to 175°–180°F. Stuffed poultry should be roasted until the temperature in the middle of the dressing reaches a minimum of 165°F. Some people prefer to continue heating poultry until the temperature in the dressing reaches 170°F. This higher temperature is a safety measure to be certain that any microorganisms (staphylococci, streptococci, and salmonellae) that might be in the stuffing or in the cavity of the fowl are killed.

*Timetables/thermometers are useful guides in the roasting of meats, but the greatest accuracy for determining the endpoint of roasting is obtained with a **timetable/thermometer.***

Timetables, thermometer.

*

The temperature of the meat cut at the beginning of the roasting period influences the total time required to reach the desired end point. Ordinarily, meat is removed from refrigerator storage at a temperature of about 34°F and is quickly prepared for roasting. Frozen meats may be thawed before roasting is begun or they may be roasted from the frozen state. Large frozen roasts require about one and one-half times longer roasting time than thawed roasts. For obvious reasons, the thermometer must be inserted in the middle of the roasting period when frozen meats are roasted.

Allow about _____ times longer to roast large meat cuts from the frozen state than you would allow for refrigerated roasts.

One and one-half.

You will notice that nothing has been said about salting roasts before roasting them. You may salt roasts or rub them with other seasonings before roasting them. However, the flavors penetrate only a fraction of an inch into the meat and little flavor change actually is accomplished. Usually, meat should be salted as each piece is carved and served. Sometimes you may wish to insert minced onion or garlic into openings in the meat. These flavors are held well by the fat they contact in a roast. You may add chopped vegetables such as onions, carrots, celery, or parsley to the roasting pan. These items have little influence on the meat flavor, but they do enhance the flavor of the drippings. The French term for these chopped mixed vegetables is "*mirepoix*."

Salt is usually placed on the **whole roast/individual slices** *to enhance the flavor of the meat.*

Individual slices. Salt penetrates poorly into meat during the roasting period.

*

There is one variation of oven roasting that is sometimes used because of its convenience when cooking large quantities of chicken, chops, or steaks. This variation is "ovenizing" and the final product is very similar to pan fried meats. For this procedure the meat is lightly greased by dipping it in oil or melted butter before arranging it in a shallow baking pan. This allows the meat to brown nicely without basting when it is roasted in a 350°F oven. Ovenized meats are thinner cuts of meat and should be salted before being placed in the oven.

True or false. Ovenizing is a variation of a dry heat cookery method.

True.

*

There is residual heat in roasted meat that will continue to cook the meat a short time after it is removed from the oven. The actual heat rise that will occur is largely determined by the size of the roast being prepared. An increase of only a degree or two occurs in small roasts, but in the large roasts used in quantity cookery the temperature may rise as much as 15°F in rare roasts and 25°F in well-done roasts. These large roasts need to be removed from the oven before the thermometer reaches the desired final temperature to allow for the anticipated residual heat rise. They are then allowed to stand half an hour at room temperature to firm the meat. This makes the meat easier to carve and the individual servings look more attractive. Family-size roasts should stand about 20 minutes before being carved.

True or false. To give better appearance to the individual portions, roasts should be held at room temperature 20 to 30 minutes before carving.

True.

*

To give variety to a meal, fish and poultry are frequently roasted with a dressing. These dressings are selected to enhance the natural flavor of the meat with which they are prepared. They may be made with a variety of ingredients for flavor and texture variations, but usually they are made with a cereal product as the base. Bread stuffings made from white or whole wheat bread cubes or corn bread are popular. Rice and wild rice are also frequently used as the base of stuffings. The seasonings may be widely varied and are usually selected from sage, saffron, thyme, marjoram, curry powder, and poultry seasoning. Onions, celery, green pepper, raisins, apples, oysters, water chestnuts, chestnuts, and other nuts are just some of the other items added for texture and flavor notes.

Some foods commonly used as the base for dressings are _____, _____, _____, and _____.

White bread, whole wheat bread, rice (sometimes wild rice), corn bread.

*

Dressings can be either moist or dry. If only a small amount of liquid is added to the dressing, the roasted dressing still will be light and easily separated on a plate, even though a small amount of the meat drippings will have been absorbed into the dressing. Moist dressings are made with more liquid and, as a result, they are heavier and more compact. Dry dressings are usually served with a gravy or sauce; moist dressings do not need a gravy to increase their palatability.

*If you did not plan to make gravy from the drippings of a roasted turkey, you would probably stuff the turkey with a **moist/dry** stuffing.*

Moist.

*

Dressing expands a bit when it is baking because of the drippings it absorbs from the meat. Allow for this expansion when stuffing the fish or poultry by lightly packing the dressing into the cavity rather than pressing it in firmly. When stuffing poultry, be sure to stuff the neck opening as well as the body cavity. This gives a pleasingly plump contour to the fowl when the neck flap is secured over the dressing. Dressing increases the roasting time compared to the same meat prepared without a stuffing. You can estimate the total time required by weighing the stuffed item before roasting. If you do not have a scale, estimate the weight of the dressing and add that weight to the weight of the meat itself. The dressing required to fill a bird approximates the weight of the dressed fowl including the neck and the bag of giblets. This weight is usually marked on the poultry as purchased and can be used as a guide to the total weight to use in figuring the approximate roasting time of the stuffed fowl.

*When meat is roasted with a dressing, it is necessary to **add/subtract** the weight of the dressing **to/from** the weight of the meat.*

Add, to.

*

Moist Heat Cookery

Moist heat cookery methods are designed to soften the connective tissue in less tender cuts of meat as well as to coagulate the muscle proteins. Dry heat methods cause almost no change in connective tissue and, consequently, are only suited to tender cuts of meat. All of a beef carcass except the loin and rib area is classified as less tender. This means that many beef cuts should be prepared by moist heat cookery. All cuts of veal are suitably prepared by moist heat cookery because of the mild flavor of the meat, the abundance of connective tissue, and the lack of marbling. Thick pork chops are prepared by moist heat cookery to be certain that they are cooked sufficiently in the center before they get too dark on the surface. The breast and shanks of lamb are appropriately cooked with moist heat. Although most poultry is prepared by dry heat methods, stewing hens are old and less tender and should be cooked with moist heat.

Moist heat cookery tenderizes meat by slowly changing collagen, the white connective tissue in meat, into gelatin. Gelatin, of course, is very tender, especially when hot. By converting the collagen to gelatin, you have greatly tenderized the meat. With proper technique, less tender cuts of meat will become more tender than the tender cuts of meat. In fact, they should become so tender that they can

243

be cut with a fork easily. This change in tenderness occurs rather slowly. Therefore, it is necessary to allow usually a minimum of two hours, and often three hours, to prepare less tender cuts of meat successfully.

Moist heat cookery methods will convert **collagen/elastin** *to* ____.

Collagen, gelatin.

*

The change from collagen to gelatin takes place more rapidly if some acid is added to the meat during the cooking period. Two acidic liquids that are used to help tenderize meat are tomato juice and lemon juice. By soaking meat in an acidic marinade, you can also speed the tenderizing of less tender cuts of meat. The German recipe, sauerbraten, makes use of acid in a marinade to shorten the actual cooking time. Sauerbraten is made by marinating rump roast in a vinegar marinade for a day or longer before the meat is cooked. During the refrigerated storage of the marinating meat, some change begins to take place in the collagen, and the actual cooking period required to make the rump roast fork tender is greatly decreased.

Which would you expect to cook to fork tenderness more quickly: rump roast cooked with water or rump roast cooked with lemon juice and water?

Rump roast cooked with lemon juice and water.

*

Moist heat cookery provides an excellent opportunity to introduce a wide variety of flavors to meats. As previously suggested, acidic liquids may be added. Such juices as lemon, pineapple, tomato, lime, and orange are popular additions to various types of meats. Sour cream and assorted marinades also are flavorful liquids for moist heat meat cookery. Sometimes soups and soup concentrates also are used to add flavor to meats.

True or false. Moist heat cookery provides more opportunity for flavor variations than does dry heat meat cookery.

True.

*

There are two main methods of moist heat meat cookery: braising and stewing. There are other terms applied to moist heat cookery, but these are actually variations of these two main methods. Swissing is a term used to indicate a braising method using tomato juice, tomato sauce, or tomato paste. Pot roasting is a term for braising beef roasts cut from the chuck or shoulder area. Steaming is also a variation of braising in which only a small amount of liquid is used and the simmering period is quite short. This is a popular method of cooking fish. Fricasseeing is another braising method commonly used for preparing poultry and veal. Rabbit and lamb may also be fricasseed. Fricasseed meats are cooked in liquid without preliminary browning of the meat. Sometimes meats are braised in a casserole; this method has led to the term casseroling or "jugging" (an English term for cooking rabbit, chicken, or other meat in a casserole).

Some terms for moist heat cookery that are variations of the two main methods are: _____, _____, _____, _____, and _____.

Swissing, pot roasting, steaming, fricasseeing, casseroling (or jugging).

*

Braising. Braising is perhaps the more commonly used method of moist heat meat cookery because it is widely adaptable to the preparation of many types of meats. Preliminary preparation for braising may include removing excessive fat, cutting connective tissue, and pounding to break up some connective tissue. Pounding is a mechanical means of helping to tenderize less tender cuts of meat. This technique may be used in the preparation of meats to be cooked as Swiss steak as well as for such items as veal cutlets. Usually meat that is to be tenderized in this manner will be pounded until it is about half of its original thickness. Flour may be pounded into the meat during the mechanical tenderization process. Flour helps to absorb the juices during the cooking period and also contributes to a uniform color during the browning of the meat.

Pounding is done to **tenderize/toughen** *a meat cut by* **physically/chemically** *breaking up some collagen.*

Tenderize, physically.

*

Frequently, browning is the first step in braising meat. The meat is browned in a small amount of fat to attain a uniformly pleasing brown on all surfaces. The heat should be controlled at a moderately high temperature during browning to achieve the desired color reasonably quickly but without the smoking and spattering that occur when fat is overheated and begins to break down. The meat should be browned until it achieves the color you wish to have when braising is completed. This browning period is important because of the change in the color of the drippings as well as the improved appearance of the meat. If browning has been done well, the drippings will have the brown color necessary for making gravy that is a pleasing color.

Browning improves the color of the _____ and the _____.   *Meat, drippings.*

*

After meat is browned (if you have decided to have it browned), a small amount of liquid is added and the pan is covered. At this point you need to decide whether to complete the braising process on top of the range or in the oven. If you elect to complete the tenderizing of the meat on top of the range, reduce the heat to maintain a simmer throughout the remainder of the cooking period. At no time should the liquid be hot enough to boil, because boiling makes the muscle proteins less tender than they would be if the temperature is held just below the boiling point. One word of caution is particularly necessary if this second phase of braising is done on top of the range. Be careful to check regularly to be certain that there is still liquid present in the pan and that there is liquid underneath the meat. It will probably be necessary to replenish the liquid one or more times during the long, slow period required to tenderize the collagen. Even with a well-designed frying pan or Dutch oven, there is some evaporation and this liquid needs to be replaced. If all the liquid evaporates, the meat once again begins to brown in the drippings and may quickly pass beyond the point of a pleasing brown to an undesirable burned state. Do not be fooled by simply seeing fluid in the pan, because hot fat is fluid too. If you can hear a sizzling sound, you can be certain that the meat is once again being cooked without water, and liquid should be added immediately. If you are uncertain whether the liquid is fat or water, drop a small quantity of water in the pan. It will spatter if there is only hot fat present.

During the simmering period, you occasionally should check that there is still some fat/water present in the pan.

*

Meat is less likely to go dry in the oven than it is over direct heat, but you will still have some evaporation from the pan. Therefore, it is prudent to note the moisture present in the covered pan in the oven. It is possible to lose all the added liquid even in the oven; if this happens, it must be replaced with whatever liquid has been selected for use in the recipe. If you neglect to add more liquid before it all evaporates, add the necessary moisture and lift the meat to be certain that it is released from the pan and surrounded by a thin layer of liquid. This precaution prevents further burning or scorching where meat may be adhering to the pan. The oven should not be set hotter than 325°F for the second phase of braising, and 300°F is often preferred. Higher temperatures increase the rate of evaporation of liquid from the pan and are detrimental to the tenderness of the meat.

Outline the procedure for braising meat.

1. *After preliminary trimming and pounding (if desired), the meat may be browned as the first step. Browning is optional, however, and is not always done in braising.*
2. *A small amount of liquid is added to provide the moisture needed to convert collagen slowly to gelatin. The pan is covered and the meat is simmered until done.*

*

Meat that is braised is always cooked until it is well done. Braised meats are never rare. There is no particular reason to use a thermometer when braising meats. Neither can cooking time be used as the criterion for doneness. The simplest and best test is to insert a cooking fork into the meat to judge the tenderness. If the fork can be inserted and withdrawn very easily, the meat is done. However, if some force is needed to push the fork into the cut, further simmering is necessary to increase the tenderness. The actual time required for braised meats to be done varies considerably and is influenced by the type of meat, the size of cut, the grade of meat, and the use of acid liquids. Approximate times for braising are given in Table 8.4.

The best test for doneness of meat that has been braised is:

A. *The thermometer measures the desired final temperature.*

B. *The pink color has just disappeared from the meat next to the bone.*

C. *The correct amount of time has elapsed.*

D. *A fork can be inserted and withdrawn from the meat easily.*

E. *None of these.*

D.

*

Table 8.4. Timetable for Braising*

Cut	Average Weight or Thickness	Approximate Total Cooking Time
BEEF:		
Pot roast	3–5 lb	3–4 hr
Swiss steak	1½–2½ in	2–3 hr
Fricassee	2-in cubes	1½–2½ hr
Beef birds	½-in (× 2 in × 4 in)	1½–2½ hr
Short ribs	Pieces (2 in × 2 in × 4 in)	1½–2½ hr
Round steak	¾-in	1–1½ hr
Stuffed steak	½–¾ in	1½ hr
PORK:		
Chops	¾–1½ in	45–60 min
Spareribs	2–3 lb	1½ hr
Tenderloin:		
Whole	¾–1 lb	45–60 min
Fillets	½-in	30 min
Shoulder steaks	¾-in	45–60 min
LAMB:		
Breast, stuffed	2–3 lb	1½–2 hr
Breast, rolled	1½–2 lb	1½–2 hr
Neck slices	¾-in	1 hr
Shanks	¾–1 lb each	1–1½ hr
Shoulder chops	¾–1 in	45–60 min
VEAL:		
Breast, stuffed	3–4 lb	1½–2½ hr
Breast, rolled	2–3 lb	1½–2½ hr
Veal birds	½-in (× 2 in × 4 in)	45–60 min
Chops	½–¾ in	45–60 min
Steaks or cutlets	½–¾ in	45–60 min
Shoulder chops	½–¾ in	45–60 min
Shoulder cubes	1–2 in	45–60 min

*By permission of the National Live Stock and Meat Board.

A wide variety of cuts can be braised satisfactorily. Any veal cut is well suited to braising. Rump roast, round steak, flank steak, chuck roast, arm roast, and blade pot roasts are beef cuts that are customarily braised. Pork chops an inch or more thick, particularly those with stuffing, are prepared by braising to be certain that they are cooked sufficiently for safety. Lamb breast or lamb shanks also may be braised. Sometimes, large frying chickens are braised for a relatively short time to complete cooking the meat through to the bone.

Pork chops one inch thick or thicker should be cooked by **dry/moist** *heat cookery for safety's sake.*

Moist.

*

Sometimes a variation of braising is done by putting the meat in parchment or brown paper and wrapping it tightly to seal the meat in the paper. The meat is then cooked in the paper, a process called *"en papillote."* Pompano, a very fine Gulf sole, is often prepared *"en papillote"* with seasonings and chopped sea foods. It is considered one of the finest delicacies.

Would you classify pompano en papillote as a fish prepared by moist heat or by dry heat cookery?

Moist heat cookery.

*

Fish to be braised should be firm enough to not break up easily when cooked. It can be skinned, cut into fairly large pieces, and gently simmered in a casserole or pot. Usually, lean fish are braised in a rich sauce and fatty fish in a less rich sauce. When a thin liquid is used to braise fish, the liquid is often poured from the fish after braising is completed and then is thickened into a sauce suitable for serving over the fish.

What do you think is likely to be a major problem when braising fish?

The problem is to keep the fish from breaking up. Fish have so little connective tissue that they tend to break apart very easily.

*

When braising fish, keep the total cooking time short. Allow time to heat the fish through and to blend the flavors of the items used to season it. Avoid prolonged cooking—this tends to dry out the fish and break it up unnecessarily. In addition, the proteins in the fish muscle will become less tender with long exposure to high temperatures. Since fish

249

is so low in connective tissue, the main purposes of braising are to warm the meat and to blend the flavors. Tenderizing connective tissue is not an objective of fish cookery.

The two main purposes in cooking fish by braising are to _____ _____ _____ and _____ _____ _____.

Warm the meat, blend the flavors.

*

Stewing. Stewing and cooking in liquid are synonymous terms for the other method of moist heat meat cookery. As in braising, meats for stewing may be browned first if desired. However, this step is omitted when stewing poultry or cooking fish in liquid. Then, liquid is added to cover the meat. The liquid most commonly used is water. Throughout the stewing of meat, the liquid should be maintained at simmering rather than at boiling to promote tenderness of the muscle proteins.

How does stewing differ from braising?

The main difference is in the amount of liquid used. Braising is done with just enough liquid to cover the bottom of the utensil, and stewing is done with sufficient water to cover the meat. Several liquids are commonly used in braising, but water is usually chosen for stewing.

*

In both braising and stewing, flavors are carried quite satisfactorily into the meat. Therefore, it is practical to season the meats when the liquid is added. Add sufficient salt and other seasonings to give a pleasing flavor to the finished product. More salt will be required in stewing than in braising because of the larger quantity of water in contact with the meat. Usually the meat will have a pleasing flavor if approximately one teaspoon salt is added per quart of water.

*When preparing meats by moist heat cookery methods, **do/do not** add salt when the liquid is added.*

Do.

*

Frequently when stewing meat, you may wish to cook vegetables along with it. This is most satisfactory when the vegetables are added near the end of the cooking period. They should be timed to complete cooking just when the meat is finished. If the vegetables are added too soon, they will become mushy and unidentifiable before the meat is done. The actual length of time required for the vegetables is influenced by the type of vegetable and the size of the pieces. Add the largest pieces first and then add diced vegetables and frozen ones later. Large cubes of potatoes require approximately 20 minutes, sliced carrots and diced celery cook in approximately 10 to 15 minutes, and frozen vegetables will be done in about 8 minutes.

If you were making a beef stew, in what sequence would you add the following ingredients: frozen peas, potatoes, beef stew meat, boiling onions, and carrots?

Beef stew meat first followed in order by potatoes, boiling onions, carrots, and frozen peas.

*

Fish are sometimes prepared by a specific type of cooking in liquid known as poaching. Poaching is done by gently simmering the fish in just enough liquid to cover it until the meat is heated through thoroughly. This is a very short time compared with the usual long stewing period (average of approximately three hours) required to tenderize beef, and other types of meat that may be stewed. If fish is poached, it is usually placed on a rack so that it can be lifted in and out of the liquid easily without breaking it up. Salmon and oysters poach well, as do lean fish, brains, and small pieces of veal. Milk is sometimes chosen as the cooking liquid, or fish might be poached in a rich fish stock called a "court bouillon."

Why do you think milk or fish stock might be selected in place of water for poaching some fish?

These liquids give added flavor and they also give a whiter product, a characteristic that is desirable when preparing white-fleshed fish. If the poaching liquid is rich enough, it can be served with the item being poached. This is done when oysters are poached, and the product is known as oyster stew.

*

Some fat fish are excellent when poached, chilled, and served cold with mayonnaise. Such fish are most easily prepared by poaching them unskinned on a greased rack. After poaching, remove the skin from the top side and cover the fish with a damp cloth during the chilling period that follows. The chilled fish then can be turned over and the skin removed from the second side.

True or false. Sometimes poached fish are served chilled.

True.

251

Fish stews are popular in several countries under various names. In France, a *matelote* is a fish stew served with a slightly thickened sauce. *Chioppino*, an Italian dish, is very similar. *Bouillabaise* is a fish soup that is becoming increasingly popular in the United States and is perhaps the best known of the fish stews.

Fish stews should be cooked a **shorter/longer** *time than beef or other meat stews.*

Shorter. A long time is required to soften the connective tissue in beef, but fish is very low in collagen content.

*

VARIETY MEATS

Variety meats include the internal organs such as heart, kidney, liver, brains, tripe (stomach of cattle), and sweet breads (thymus glands). Chitterlings, the small intestines of young pigs, are sometimes available. In some regions, the testicles of young lambs or calves may be eaten. Tongue also is considered a variety meat. A great delicacy of the Near East is the eye of the young roasted goat. It is always saved for the guest at the feast!

Variety meats are important in the diet because many of them are high in nutrients. But why do some population groups who do not realize their nutritive value, consider them to be such delicacies?

They learn to like them and prize them because there is such limited availability of variety meats.

*

Kidneys from young animals may be broiled or pan fried (Table 8.5). Heart and kidney from older animals require moist heat cookery methods, however, because they are less tender cuts of meat. If heart is partially cooked by moist heat, it can be stuffed and then baked to complete its preparation. Tongue is often cured and smoked, but regardless of its treatment prior to purchase, it needs to be prepared by moist heat cookery. Liver from young animals becomes more acceptable when it is braised to aid in tenderizing it.

Would you say that the cooking of variety meats is based on the amount of connective tissue in the item?

Yes. This is generally true.

Table 8.5.   Timetable for Cooking Variety Meats*

Type	Broiled Minutes	Braised† Minutes	Braised† Hours	Cooked in Liquid Hours	Cooked in Liquid Minutes
LIVER:					
Beef:					
3–4 lb piece			2–2½		
Sliced		20–25			
Veal (calf):					
Sliced	8–10				
Pork:					
Whole (3–3½ lb)			1½–2		
Sliced		20–25			
Lamb:					
Sliced	8–10				
KIDNEY:					
Beef				1–1½	
Veal (calf)	10–12			¾–1	
Pork	10–12			¾–1	
Lamb	10–12			¾–1	
HEART:					
Beef:					
Whole			3–4	3–4	
Sliced			1½–2		
Veal (calf):					
Whole			2½–3	2½–3	
Pork			2½–3	2½–3	
Lamb			2½–3	2½–3	
TONGUE:					
Beef				3–4	
Veal (calf)				2–3	
Pork ⎱ Usually sold					
Lamb ⎰ ready-to-serve					
TRIPE:					
Beef	10–15‡			1–1½	
SWEETBREADS:					
Beef	10–15‡	20–25			15–20
Veal (calf)	10–15‡	20–25			15–20
Lamb	10–15‡	20–25			15–20
BRAINS:					
Beef	10–15‡	20–25			15–20
Veal (calf)	10–15‡	20–25			15–20
Pork	10–15‡	20–25			15–20
Lamb	10–15‡	20–25			15–20

*By permission of the National Live Stock and Meat Board.

†On top of range or in a 300°F oven.

‡Time required after precooking in water.

Some variety meats have strong flavors which may be subdued with special treatment for wider appeal. Kidneys may be partially cooked in water which is discarded and then replaced with fresh water. This procedure minimizes the strong flavor of the kidney. Occasionally, liver and kidneys from young animals may be soaked in milk before they are sautéed or broiled to reduce their flavor a bit. Another technique to reduce the flavor of variety meats is to soak them in salted water or vinegar prior to cooking in fresh water.

Special cooking methods for variety meats are designed to reduce **cooking time/flavor.**

Flavor.

*

Blanching is a process used to treat some variety meats preliminary to their actual final preparation. For instance, sliced beef liver may be briefly immersed in boiling water and the stringy, connective tissue removed. Sweet breads and brains may be briefly heated in water and vinegar to alter the flavor and to bleach the white variety meats before they are freed of their membranes. Other meats may be dipped in boiling water to give them more firmness before they are cooked. Thus, blanching is really a means of pretreating meats to prepare them for their final cooking method. The meats are partially cooked or parboiled by this process.

When meats are blanched they are **fully cooked/parboiled.**

Parboiled.

*

DECORATING MEATS

For special occasions you may wish to decorate a ham, tongue, turkey, or fish. The cooked meat is chilled and skinned in preparation for the mayonnaise or white sauce that has been blended with gelatin. The mayonnaise or white sauce is carefully poured over the meat, which is then refrigerated to set the coating. When the first layer has set, a second coating is applied and chilled. This process is continued until the coating is between one-eighth and three-sixteenths of an inch thick. A design is then set in, using decorative pieces of carrot, olives, green pepper, pimiento, or other items. Bouillon or consomme chilled almost to the point of setting is poured carefully over the designed meat. The meat is again refrigerated until the glaze is firm.

Why is the last clear gelatin-containing coating applied?

To seal in the decoration and give the meat an extra sheen.

*

Such decorative work is called *"chaud-froid"* (show-fraw). This French expression literally means hot-cold, which indicates the food's history of being chilled following cooking. These meats are served cold, of course.

Would you plan to prepare such an item the day you will be serving it?

No. Allow two days and preferably three: a day to cook and chill the meat, a day to decorate it, and a day to rest up!

*

Some meats are glazed to improve their appearance. Hams may be cooked, scored and decorated with fruits or whole cloves, and then covered with a fruit topping or brown sugar glaze. Roast duckling may be served with an orange glaze. Care should be taken to make the glaze thick enough to cling to the meat when hot. Thin glazes drain badly from the hot meat and do not give the desired shiny appearance.

True or false.   Glazes are sometimes served over meats to improve their appearance and modify their flavor.

True.

*

Sauces are another means of garnishing meats to make them more attractive and flavorful. Poached fish may be enhanced by serving it with a delicately flavored Hollandaise sauce. A sirloin steak, carefully broiled and served with *maitre d'hotel* butter (melted butter seasoned with just a bit of lemon juice and cayenne), is truly a gourmet item. These are but two of many sauces that may be used to brighten the entree. As you work more with foods, you will become acquainted with many sauces and their uses.

Is it worthwhile spending time to decorate food items occasionally?

Yes, for we eat to satisfy social and psychological as well as nutritional needs. Attractive food adds satisfaction to life by helping us to enjoy our food more.

*

CARVING MEAT

Most meats are carved by cutting across the grain to give an attractive slice that is easily cut and tender to chew. One exception is a large broiled steak that is carved. In this case, the steak is sliced with the grain to give each person a slice of each muscle in the steak. This is most easily done by first removing the bone and then carving the muscles. In Porterhouse steak, each serving consists of a slice of the loin and one of the tenderloin. A sharp knife is important to carve any meat cut neatly and easily.

How many slices of meat are contained in a serving of Porterhouse steak?

Two, a slice of the tenderloin and of the loin.

*

To simplify carving of poultry, first remove the wing by cutting through the wing joint at the breast; remove the leg and thigh where the thigh joins the carcass. This exposes the large breast muscles of the fowl for uniform slicing across the grain. It is desirable to have a hot platter to receive the wing, leg, and thigh as they are carved from the carcass. The dark meat can then be sliced easily from the leg and thigh to be served along with the white breast slices. The carved slices of breast may be placed directly on the dinner plates or be arranged on the hot platter. When one side has been carved completely, turn the platter around and repeat the same carving procedure on the second side of the carcass.

To carve a turkey, first remove the _____ and then the _____ and _____ to make it easier to carve even slices of the _____.

Wing, thigh, leg, breast.

*

To carve a leg of lamb, ham, or pork, cut several slices from the side of the leg containing the knee cap. Then turn the leg over and rest the meat on the cut surface. The meat is now in position for you to make perpendicular slices to the bone. These slices can be freed from the bone by holding the knife parallel to the bone and carefully cutting to release the slices.

Place the following in proper sequence for carving a ham.
A. Cut thin slices to the bone, starting at the rear of the leg.
B. Cut several slices from the thin muscle side of the leg.
C. Cut the slices free from the bone area.

B, A, and C.

When all the meat is sliced on the big muscle side, turn the leg around and slice the small muscle. If the leg is small, it may be necessary to cut slanting or on a bias to get good slices.

True or false. A leg has two muscle areas that can be sliced against the grain.

True.

*

The direction you move when carving a leg is different in the large muscle than it is when the small one is carved. The shank of the leg preferably will be at the carver's right. If the meat is placed in this position, the knee cap will be toward the right and the aitch bone will be toward the left of the carver. Carving of the large muscle is easiest if you start on the right above the knee cap and proceed toward the aitch bone on the left. To carve the smaller muscle, cut on a slant or bias beginning at the aitch bone and progressing toward the knee cap.

The large muscle is carved **across the grain/on a bias,** *and the small muscle is carved* **across the grain/on a bias.**

Across the grain, on a bias.

*

For persons who enjoy the art of carving and are adept in this skill, it may be a pleasure to carve the meat at the table when guests are present. In other cases, it may be wiser to carve in the kitchen. This is particularly true if a duck or chicken is to be carved into four parts, because this job requires a heavy knife and a cutting board for best results. Meat carved in the kitchen can be arranged and garnished attractively on a platter before bringing it to the table.

True or false. It is definitely preferable to carve meat at the table.

False. Some meats are difficult to carve gracefully and some carvers may be rather graceless in front of guests. At such times the kitchen is a suitable place for the performance.

*

To carve a duck or chicken into four portions, place the bird breast side down on a cutting board. Place the tip of a heavy knife at the neck and cut straight along the back all the way through the tail. This splits the entire back of the bird so that the carcass can be wedged open by using the knife as a lever at the neck end of the back. Now you can insert the knife to cut open the breast side of the bird lengthwise. This is done by cutting with the knife from the body cavity outward through the skin covering the breast.

257

Finally, each half is divided between the breast and the thigh to give four servings of the fowl. This operation can be done very satisfactorily with poultry shears rather than with a carving knife.

Would you carve a turkey in the manner just described?

No. The servings would be much too large if a turkey were carved into quarters in the preceding manner. In addition, the bones are not as soft and easily cut as they are in a young chicken or a duck.

*

REVIEW

1. The meat that is usually highest in fat is: (a) beef, (b) lamb, (c) pork, (d) poultry, or (e) fish.

   (c) Pork.

2. *True or false.* All fish and shellfish are low in fat.

   False.

3. Juiciness in meat is contributed by _____ in the muscle and also by _____.

   Water, fat.

4. When meat is cooked, the muscle proteins are _____.

   Denatured (or coagulated).

5. Shrinkage in meat is due to loss of _____ and _____.

   Fat, water.

6. A circular stamp on a piece of meat indicates *wholesomeness/quality or grade;* a shield-shaped stamp indicates *wholesomeness/quality or grade.*

   Wholesomeness, quality or grade.

7. The top grade for poultry is: U.S. Prime, U.S. First, U.S. Premium, or U.S. Grade A.

   U.S. Grade A.

8. The two types of connective tissue in meat are _____ and _____.

   Elastin, collagen.

9. U.S. Prime is the top grade for beef. What is the next grade?

   U.S. Choice.

10. Meat is graded on the basis of _____ and _____.

    Cutability, palatability (includes conformation, finish, and quality).

11. Softness and porosity in bone are indications that the animal carcass is that of a *young/old* animal.

    Young.

12. Which is usually firmer: beef fat or pork fat?

    Beef fat is firmer, but pork fat should not be soft and greasy.

13. In each pair, which is older?
    A. Calf or veal.
    B. Lamb or mutton.
    C. Baby beef or calf.
    D. Broiler chicken or roaster chicken.
    E. Capon or hen.

A. Calf.
B. Mutton.
C. Baby beef.
D. Roaster chicken.
E. Hen.

14. *True or false.* Most pork on the market is less than a year old.

True.

15. You can distinguish between the fat from pork and from lamb by noting the _____ and the _____.

Color, firmness. (Pork fat is softer and has a pinkish tint, while lamb fat is hard and white).

16. *True or false.* A turkey weighing over 18 pounds will have more meat on its carcass per pound than one under 18 pounds.

True.

17. If oyster shells are open, what can you conclude about the oyster?

The oysters have perished and may not be a good buy because they may have started to spoil.

18. Which of the following pigments is important for the red color in fresh meat?
    A. Myoglobin.
    B. Hematin.
    C. Nitrosomyglobin.
    D. All are responsible.

A. Myoglobin.

19. *True or false.* The fat in meat makes an important contribution to the flavor of meat.

True.

20. Ripening of some prime beef is done to increase the _____ and alter the _____ somewhat.

21. Why is veal never aged or ripened?

Tenderness, flavor.
Veal lacks the heavy fat cover needed to protect the meat during the weeks of ripening.

22. Meat becomes *more/less* tender when it is cooked a long time uncovered in a hot oven.

Less.

23. The connective tissue that is not changed by cooking is _____.

24. Why does some meat remain red after it is cooked (ham, for example?)

Elastin.
The nitrite salts in cured meat are combined with hemoglobin to form a stable, red pigment that does not change in heat.

25. Collagen can be changed to _____ by *short/long* cooking in *dry/moist* heat.

Gelatin, long, moist.

26. Fish has *much/little* connective tissue in its flesh.

Little.

27. The enzyme most often used to tenderize meat is _____.

Papain.

28. Beef is called rare when it has reached _____°F, medium at _____°F, and well done at _____°F.

140, 160, 170.

29. Broiling of meat is done when heat is transmitted by *conduction/convection/radiation*.

Radiation.

30. The conversion of collagen to gelatin is aided by the presence of _____.

Acids.

31. In broiling, the thicker the piece of meat the *closer/farther* it should be placed from the heat source.

Farther.

32. When pan broiling, the fat is *removed from the pan as it accumulates/allowed to accumulate in the pan*.

Removed from the pan as it accumulates.

33. The dry heat method in which meat is cooked in an oven is called _____ or _____.

Roasting, baking.

34. Roasting is usually done at _____°F or _____°F, the higher temperature being recommended for *large/small* cuts of meat.

300, 325, small.

35. When roasting meat, the roasting pan *should/should not* be covered.

Should not.

36. Meats for broiling should be a minimum of _____ thick.

One inch.

37. The base of many dressings is *bread/sauteed vegetables/seasonings*.

Bread.

38. In pan frying, fat is *allowed to accumulate in/removed from* the pan.

Allowed to accumulate in.

39. The presence of water or food particles *lengthens/shortens* the length of time a fat can be satisfactorily used for deep-fat frying.

Shortens.

40. Deep-fat frying is a *dry/moist* heat cookery method.

Dry.

41. *True or false.* Pure water at atmospheric pressure will not get hotter than 212°F, but fat can be heated above 400°F.

True.

42. *True or false.* A breading that clings well to meat is made using fine crumbs and egg.

True.

43. Braising is a *moist/dry* heat method of cooking.

Moist.

44. In braising, a *small/large* amount of liquid is added.

Small.

45. Which of the following is not a special case of braising?
    A. Swissing.
    B. Sautéing.
    C. Fricasseeing.
    D. All are braising.

B. Sautéing is a dry heat method.

46. Vegetables are added to stews *at the beginning of the cooking period/later so they will be done at the same time as the meat.*

Later so they will be done at the same time as the meat.

47. In moist heat cookery, the liquid should be *boiling/simmering.*

Simmering.

48. Fish stews *can/cannot* be made.

Can. Matelote or chioppino are fish stews of national repute.

49. Which can be held in frozen storage longer: beef or pork?

Beef. The fat in pork limits its frozen storage to approximately three months.

50. If you wish to poach fish, select a *fat/lean* fish.

Fat.

51. When meat is soaked in a marinade, the flavor is *changed/unchanged* and the cooking time is *lengthened/shortened.*

Changed, shortened.

52. Which cut is not considered a variety meat: (a) liver, (b) kidney, (c) brains, (d) sweetbreads, or (e) chop?

(e) Chop.

53. When carving meats, it is usually better to cut *across/with* the grain of the meat.

Across.

54. Hamburger has a *shorter/longer* safe storage life than does a steak.

Shorter.

55. Fish spoil *more/less* easily than do other types of meat.

More.

9

Gravies and Sauces

In meat cookery it is common to use a gravy or sauce as a meat accompaniment to enhance the flavor of the meat. A pork, veal, or pot roast served with a rich flavorful gravy has more appeal to many people than does the meat itself. Gravies are prepared from the meat drippings and are, therefore, somewhat limited in variety. However, the list of sauces you can serve is almost endless, because sauces are not directly derived from the meat but may be made from substances not related to the meat item served.

The variety of gravies that can be served is   less/greater than the number of sauces that are possible.

Less.

*

Gravies and sauces are different items. A gravy is made from the drippings and carries the definite flavor of the meat from which it came. A sauce is a subtle blend of flavors and has no definite meat flavor predominating. Although sauces may be made from drippings or a stock, they may often be prepared without using any liquid or fat from the meat. Familiar examples of nonmeat sauces are raisin or cherry sauces for ham.

A gravy is flavored by the meat _____ while a sauce carries no definite _____ flavor but is a _____ of flavors.

Drippings, meat, blend.

*

[1]In some food preparation classes, parts of this chapter that are designed for classes oriented toward quantity foods work may be omitted.

Gravies and some sauces are thickened sometimes and served unthickened at other times. The "red-eye" gravy popular in the South with ham and *au jus* served with roast beef are examples. Seasonings are perhaps more important to sauces than gravies since these are the substances that help to make a blend of flavors. It is important to taste gravies and sauces before serving to be certain that you have achieved exactly the shading of flavor desired.

To make a gravy or a sauce, we need a liquid and _____; a thickener such as starch may be added if desired.

Seasonings.

*

When meat is cooked by moist heat methods, there is usually a sauce or broth that can be used to make a gravy. In dry heat meat cookery, there is often only a limited quantity of drippings from which to make gravy. Therefore, you may wish to make a sauce for meats cooked by dry heat; for meats cooked by moist heat, you might elect to make a gravy, but sauces may be used effectively at times.

It is more customary to serve a gravy with meat cooked by **dry/moist** *heat.*

Moist.

*

Many people think sauce cookery can only be done by an experienced chef. When you think of some of your favorite foods, you will realize that sauces are often prepared at home. Sauces are prepared for spaghetti and meatballs, for macaroni and cheese, for souffles, and for creamed vegetables. Mayonnaise is a pleasing sauce with boiled cauliflower. A Spanish sauce is excellent with an omelet. Sauces can be as simple as sautéing mushrooms and serving them over a steak. When you brush broiled lamb chops with melted butter, you are using a sauce called *maitre d'hotel* butter. As Americans have learned to eat a greater variety of cosmopolitan foods and have widened their experience with food, the use of sauces in food preparation has been increasing. The mixes that are readily available in the stores today are timesavers that encourage cooks to try more sauces. Even canned soups can be used to make sauces that will add a special touch to foods.

Are there any words of caution you should heed when planning sauces?

Be discriminating in your use of sauces. Plan them to enhance a product, not to disguise a poorly prepared food.

263

The use of sauce in a meal is important. Serve enough sauce to moisten and garnish a food, but avoid having the food swimming in the sauce. Usually one sauce for a main course is sufficient. Too many sauces in one meal will be monotonous. When you consider sauces appropriate for a meal, visualize their total impact—their flavor, texture, and general appearance—in relation to the course with which the sauce is to be served. Select the sauce that will make the greatest total contribution.

Is it correct to say that selection and service of a sauce are as important as preparation?

Yes. Even a perfectly prepared sauce will not seem appetizing if it does not combine well with the other foods in the meal, or if it is served in such a large quantity that it dominates the plate.

*

Sauces and gravies should be selected for the food they accompany. Traditionally, some foods and sauces go together. *Au jus* gravy goes naturally with roast beef; a horseradish sauce made of whipped cream and horseradish is another good combination with roast beef. Hollandaise sauce over a poached egg which rests on a grilled slice of ham and toast makes Eggs Benedict. Deep-fried oysters and tartar sauce are a natural combination. Brown gravy and pot roast are an inseparable pair. Lean fish served with a rich Newburg sauce takes on new meaning in the taste area, and a Cumberland sauce with its tartness and tanginess goes well with the flavor of wild game. A delicate chicken breast should be served with a mellow Supreme sauce. Experience and sense of taste will help you select fine combinations of sauces or gravies with other foods.

A gravy is intended for a particular meat item; a sauce is selected as a delicate complement to the flavor of the meat served.

Why is a gravy easier to select than a sauce?

Which food in column B would go well with the sauce in column A?

A	B	
1. Sour cream sauce	A. Poached cod, a lean white-fleshed fish	
2. Tomato sauce		1. C.
3. Hollandaise sauce	B. Roast turkey	2. F.
4. Sweet, tangy brown sauce	C. Baked potato	3. A.
5. Orange sauce	D. Venison, lean and strong flavored	4. D.
6. Cranberry sauce	E. Duck, fat	5. E.
	F. Spaghetti, bland	6. B.

Sauces often are used to give contrasting flavors to meats and the other foods served in a meal. A tart mint sauce can be served to modify the flavor of lamb and to remove any mutton flavor. The sensation of fattiness in a pork roast can be reduced by serving a tart applesauce. Hollandaise sauce is slightly acidic and is effective when served with poached salmon.

Why would a meal consisting of fruit cup, baked ham, raisin sauce, candied sweet potatoes, green peas, apple and date salad with marshmallow bits, cornbread, honey, and pineapple sherbet not give a good taste sensation?

It is much too sweet.

*

LIQUIDS FOR SAUCES AND GRAVIES

The liquid used for a gravy is frequently water but milk, stock, wine, tomato sauce, beer, cream, vegetable juice, or any other liquid may be used.

True or false. The liquid used for making gravy or a sauce is always water.

False; a wide variety of liquids is used.

*

Formerly, when the making of sauces was more or less reserved for the professional chef, a stock pot to serve as a base for sauces was an essential component of the kitchen organization. Today, we can purchase canned or dried stocks which duplicate in almost every way those that were so laboriously prepared by the chef.

True or false. To make a good sauce, one must have stock made the old-fashioned way.

False.

*

If you wish to make a stock, simmer meat and bones with vegetables and seasonings. There are basically two kinds of stock, white and brown. A white stock is made from unbrowned meat and bones; a brown stock is made from browned meat and bones. Good white stock can be made of unbrowned beef, veal, chicken, or even fish and fish bones. Beef, veal, and chicken are also suitable for a brown stock after they have been browned in the oven or on top of the range.

The two basic stocks made in the kitchen are a _____ and a _____ stock.

White, brown.

*

Chopped vegetables, a *mirepoix,* are added to the meat and bones in the stock pot and liquid is added to cover. For quantity food preparation, it will take a pound of chopped vegetables, four pounds of meat and bones, and five quarts of water plus seasonings to make a good stock. Smaller quantities can be prepared in the home. A stock started in hot water will be clearer than one started in cold water, but it will have slightly less flavor.

To five quarts of water, add _____ pounds of meat and bones and a _____ of chopped vegetables; if you want a clearer stock, start it in **hot/cold** *water, but if you want a more flavorful stock, start it in* **hot/cold** *water.*

Four, pound, hot, cold.

*

Salt the stock lightly at the start. This gives more flavor to the stock because many flavor compounds are soluble in salted water. Simmer just long enough to extract flavors but not so long that the stock begins to cloud. Fish will cloud most easily, chicken and veal next, and beef and lamb or mutton least easily.

True or false. You can simmer a fish stock longer than a beef stock.

False.

*

The stock is usually seasoned during cooking with bay leaves, pepper corns, cloves, marjoram, or other seasonings. When the stock is done, it is carefully drained through a strainer or cheese cloth into a container and quickly cooled. Stock may be kept in a refrigerator for an absolute maximum of seven days.

Do you think that stock is a perishable item and that care would have to be taken to see that it is chilled rapidly and kept under refrigeration afterwards?

Stock spoils very readily and must be maintained at a temperature below 40°F during storage. You can cool it quickly by putting the pan in the sink as soon as the stock is done and circulating cold, running water around it. Refrigerate it as soon as it cools.

THICKENING AGENTS

Gravies and sauces are usually thickened with flour, but cornstarch, waxy maize starch, or other starch thickeners may be used. Sometimes eggs or other protein materials may be used instead of starches. Eggs, flour, and bread crumbs make a rather cloudy or opaque gravy or sauce; rice starch, tapioca, potato starch, waxy maize, and cornstarch give a much more translucent mixture.

If you want a clear sauce or gravy, what should you use for thickening?

Select rice starch, cornstarch, tapioca, or potato starch for clarity.

*

Flour and other starch thickeners differ in their thickening power. Among the various flours available, cake flours contain the most starch and, consequently, give the most thickening. Pastry flours are next; all-purpose flours provide the least thickening. Bread flours contain a high amount of gluten which gives a rather stringy or tenacious quality to the sauce or gravy. This may be desirable if you are making a gravy for hot beef sandwiches and want a gravy that would cling to the sandwich and moisten it. However, if you are making a gravy to serve over turkey and dressing, a thinner, less clinging gravy is desirable. If you want to use flour for thickening this gravy, use a reduced amount of cake flour because the gluten in cake flour is not so tenacious. The desired thin viscosity is easily achieved by using a small quantity of the flour.

Cake flour is highest in starch, followed by pastry and then bread flour.

Which among these three flours, pastry, cake, or bread flour; contains the most starch and will give the thickest gravy if equal quantities of flour are used?

*

The clarity of the sauce or gravy will also be affected by the thickener used. All types of flour will give a cloudy mixture because of the protein in them. Cake flour will be the least cloudy of the flours. Sometimes you may want a clear mixture; at other times the clarity is rather unimportant. A fruit sauce over ham should be thickened with cornstarch because clarity is desired; regular gravy served with a pot roast can be thickened with all-purpose flour and be very satisfactory.

The clarity of the sauce or gravy **is/is not** *affected by the starch thickener used.*

Is.

*

Acids cause a chemical change in starches. This change reduces the thickening power of starches. If a sauce or gravy has a tart acid ingredient in it, you will need to use a larger quantity of starch than you might expect to get the desired thickening.

If you add wine or tomatoes to a sauce or gravy and then thicken it, will you need **more/less** *thickener than if you did not add them?*

More. Both of these foods are distinctly acidic and will reduce the thickening ability of the starch thickener.

*

Waxy maize or modified starches are particularly useful when you want to serve a clear hot sauce because they are as thick when hot as when cold. Another advantage they have is that they may be used very satisfactorily in products that are to be frozen and then thawed. Sauces or gravies thickened with most starches will separate and be unpalatable when they are frozen and thawed, but waxy maize can undergo freezing and thawing and still bind the liquid in the product.

What advantages do waxy maize or modified starches have?

They give a pleasingly clear paste, are as thick when hot as when cold, and do not release liquid when a frozen sauce or gravy is thawed.

*

Adding starch or flour to a hot liquid usually produces lumps. To avoid this, the starch thickener is mixed very well with the liquid to form a perfectly smooth slurry. This is then added slowly to the hot liquid while you stir thoroughly. You may have better results if you mix some of the hot liquid with the slurry and then pour this mixture slowly into the hot liquid. Be careful to stir well all the time you are heating the sauce or gravy if you wish to have a smooth product.

A slurry is a mixture of liquid and a _____.

Starch (or starch thickener).

*

To avoid lumps, always use good agitation and be sure that you do not have the liquid too hot when it is mixed with the slurry. This will also hold for the *roux* which is discussed next.

What mechanical process is the secret to avoid lumps when you add a slurry to thicken a sauce or a gravy?

Agitation, always good agitation.

*

A *roux* is a thickening agent made by blending fat and flour thoroughly and then cooking this paste a couple of minutes without browning the mixture. For quantity recipes, it may take as long as 10 minutes to cook the *roux*. The fat in the *roux* makes a richer, more flavorful sauce or gravy than is made by using a slurry. The fat is also valuable because it separates the starch granules and decreases the likelihood of lumps in a sauce made with the *roux*.

A roux is a thickener used for sauces and gravies and is made with _____ and _____.

Fat, flour.

*

A rule that is used in blending a *roux* into liquids is: blend a cold *roux* into a hot liquid and a hot *roux* into a cold liquid. This rule produces a mixture that is below the temperature at which starch swells and thus gives you ample opportunity to mix the *roux* into the liquid smoothly before the starch gets hot enough to swell. Slurries must always be cold when added. If they are heated, they will get so thick that it is almost impossible to blend them in smoothly.

A _____ can be added either hot or cold, but a _____ can only be added cold.

Roux, slurry.

*

Not all starches thicken within the same temperature range. Flour, cornstarch and other thickeners begin to swell around 150°F, but the final temperature for maximum thickening varies. Flour and cornstarch must be heated almost to boiling to achieve maximum thickening and eliminate the raw starch taste. Tapioca and some modified starches reach their maximum thickness approximately 20 degrees below boiling and begin to thin if overheated.

True or false. Starches achieve maximum viscosity at approximately 212°F.

False. Cornstarch and flour do, but some starches begin to thin at such a high temperature.

*

For maximum palatability, cook a gravy or sauce until the product is thickened and the raw starch flavor disappears.

269

In small quantities, this is accomplished by boiling a corn-starch or flour-thickened product for about three minutes. As you would expect, a long heating time is required to eliminate the raw starch flavor in a large quantity.

Why is cooking continued after a sauce comes to a boil?

The extended cooking time is necessary to complete gelatinization of the starch and remove the raw starch flavor.

*

Sauces and gravies usually are made over direct heat in a heavy saucepan, because it will transmit heat evenly yet slowly enough to reduce the hazard of scorching or lumping. If a heavy aluminum saucepan is not available, it might be necessary to use a double boiler. Of course, the sauce will take longer to heat in a double boiler, but it is possible to gelatinize the starch even though the sauce will not quite reach 212°F over boiling water.

True or false.   All gravies and sauces should be made in a double boiler.

False. A heavy saucepan is usually a better choice.

*

Sauces are "finished" by adding melted butter, wine, cream, or other liquids just before they are served. This gives the sauce the final delicate taste and texture desired. After a finishing ingredient is added to a sauce, the sauce should not be boiled.

When some ingredient is added to a sauce just before it is served and it is cooked only a bit after this addition, we say we have _____ the sauce.

Finished.

*

One of the finest finishing ingredients we use is a liaison. This is a mixture of three parts heavy cream and one part egg yolks. A part of the thickened hot sauce is added to the liaison and blended in well. This mixture is then added to the hot thickened sauce with good agitation. The temperature is raised to about 190°F to thicken the eggs, and the sauce is served. A liaison is used to thicken white, cream, *bechamel,* or *velouté* sauces. It is especially effective for thickening delicate cream soups or bisques.

*A liaison is a finishing agent that also thickens sauces and soups and is **one/three** part(s) heavy cream and **one/three** part(s) egg yolks.*

Three, one.

270

A good sauce or gravy should have a sheen to it that gives it a look almost like varnish. It should have a smooth, creamy consistency if thickened and a delicate flavor. It can do much to vary a food. A sauce can add color to some foods, flavor to others, and moisture to still others.

Hesseltine and Dow,[2] in their cookbook, say: "Anyone who values her reputation as a cook should look to her sauces." Do you agree?

Experience will tell you the answer.

*

GRAVIES

Gravies are made from meat drippings or meat juices which are usually diluted with water, stock, or other liquid. Before thickening the drippings, remove excess fat and use some of this fat to make a *roux*, if desired. Two tablespoons of fat are enough for each cup of liquid used.

True or false.   If a roux is used, some of the fat from the drippings will be used to blend with the flour; the rest of the fat will be removed before the drippings are thickened.

True.

*

The quantity of flour or cornstarch per cup of liquid to use to thicken a sauce or gravy is as follows:

Thickness Desired	Amount of Thickener Flour (or Cornstarch)		Amount of Fat
Thin	1 Tbsp	1½ tsp	1 Tbsp
Medium	2 Tbsp	1 Tbsp	2 Tbsp
Heavy	3 Tbsp	1½ Tbsp	3 Tbsp
Very heavy	4 Tbsp	2 Tbsp	4 Tbsp

Notice that you must use twice as much flour as cornstarch to get the desired amount of thickening. Most gravies are made by using the proportions for a medium sauce. Occasionally you might use a thin sauce. The other two sauces are not served with meat; they are used in souffles and croquettes.

If an acid ingredient is a part of the liquid, would you use more or less thickener?

More.

[2]M. Hesseltine and U. M. Dow. *The New Basic Cook Book* (Rev. Ed.). Houghton Mifflin, Boston, 1967.

Remember, not all gravies are thickened. An *au jus* from a roast or the drippings from pan-fried meat may be moistened with water, stock, or other liquid; then the drippings from the meat are worked up with a fork into the liquid. Heating during this process of extraction brings the drippings into solution more quickly. This liquid may then be spooned over the meat or over other foods for added flavor.

True or false. All gravies are thickened.   *False.*

*

SAUCES

If you can make a "mother" or basic sauce, you can then make many variations from the basic sauce. These variations are called secondary or "small" sauces by the chef. There are several basic sauces. The main ones are brown or *Espagnole, velouté, bechamel,* white, tomato, butter, and Hollandaise. Mayonnaise, French dressing, and a number of other cold sauces also may be varied to give many combinations to go with meat.

A basic sauce is sometimes called a _____ sauce; a sauce derived from the basic sauce is called a secondary or _____ sauce.

Mother, small.

*

Not all sauces are made from stocks or liquids that come from meat. When a sauce is made from a liquid such as tomato puree or milk and contains no meat essences, the sauce is called a neutral sauce. Some sauces, such as tomato sauce, may sometimes contain meat essences and thus be a meat sauce; at other times a tomato sauce may be a neutral sauce because it contains no liquid derived from meat.

Is catsup a meat or neutral sauce?

Catsup is a neutral sauce because meat essence is not used in its preparation.

*

One of the most common basic meat sauces is the brown or *Espagnole* sauce. It is made from a brown stock; frequently a browned *roux* is used. To make a browned *roux*, the flour is browned either in a pan over direct heat or in an oven. Since some thickening power is destroyed in the browning, additional flour must be used to thicken the sauce satisfactorily. Vegetables and seasonings are simmered in the thickened sauce for a long time to blend flavors. Usually this sauce is strained before serving.

*Browned flour has **more/less** thickening power than regular white flour.*

*

Another widely used basic sauce is the *velouté* or velvet sauce. It is made by taking a white stock, frequently made from chicken or veal, and thickening this with a white *roux*. It is then cooked with vegetables and seasonings and strained before use. This sauce is used with chicken, veal, fish, or other light meats. It resembles a fricassee gravy. The flavor should be rich and delicate.

A velouté sauce is made from a _____ stock.

White.

*

A *velouté* is not a white sauce. A white sauce is a neutral sauce with milk as the liquid. A *velouté* sauce is also not a *bechamel* sauce, although if it is made of rich veal or chicken stock, it can be made into a *bechamel* sauce. A *bechamel* is made exactly like a *velouté* except rich veal or chicken stock is used and the sauce is then finished with cream or rich milk. A *bechamel* sauce might be accurately described as a rich *velouté* sauce. *Bechamel*, like *velouté*, is a meat sauce and should not be confused with neutral white or cream sauces.

*A bechamel sauce is made from a **milk/stock** base and finished with _____ _____ or _____.*

Stock (chicken or veal), rich milk, cream.

*

Actually, a *bechamel* sauce is not a basic sauce since it can be made from a *velouté*. But since the *bechamel* is used to make so many varieties of other sauces, it is often considered to be a true basic sauce.

*A bechamel sauce **is/is not** a basic sauce.*

Let's say it is; however, purists will argue with us that it is not.

*

You can make a *velouté* sauce variation by using a fish stock as the liquid. When you do this, you have what is called a white sauce or sauce *vin blanc*, because dry white wine is usually part of the liquid. This sauce is widely used for fish dishes.

273

A sauce vin blanc is made from _____ stock, but otherwise is a typical velouté.

Fish.

∗

A tomato sauce is made with a blend of a rich meat stock and tomato paste or puree. It is thickened with a *roux* and then simmered with vegetables and seasonings, strained, and served. As has been noted, sometimes a neutral tomato sauce is prepared by omitting the meat stock. The flavor difference between the meat and neutral tomato sauce is not great. Sometimes a tomato sauce is cooked down (reduced) until it is thick and the starch thickener is omitted. This long simmering procedure is usually followed when the sauce is to be served over spaghetti or other pasta.

*A tomato sauce   **is/is not**   always made with a meat stock and it   **is/is not**   always thickened.*

Is not, is not.

∗

A white sauce is seasoned milk that is thickened. A cream sauce is a white sauce to which cream or rich milk has been added. If a *roux* is used, the fat and flour are blended together in a double boiler. Cool or hot milk that has been seasoned is blended in and the mixture is stirred throughout the cooking of the sauce. Seasonings may be crushed parsley and peppercorns, grated onion, bay leaves, and other flavorful herbs.

A cream sauce differs from a white sauce in that rich milk or _____ is added to a white sauce.

Cream.

∗

Hollandaise is an emulsion made by slowly beating melted butter into egg yolks over hot water and then adding lemon juice or vinegar after the emulsion is made. Although this sauce is usually served warm, it can be served cold. If it is served cold, it usually is quite firm due to the solidified butter.

*Hollandaise sauce contains melted butter beaten slowly into   **hot water/egg yolks/vinegar/lemon juice.***

Egg yolks.

∗

Skill is required to make a good Hollandaise. The egg yolks are beaten to stage one in a pan. This pan is then set into quite warm, but *not hot,* water. Gradually, the melted butter is beaten into the yolks to form an emulsion.

The mixture is then removed from the hot water and beaten slowly for awhile. Then lemon juice or vinegar and seasonings such as salt and white pepper are added. The excellent Hollandaise mixes now available in stores should make this delicate sauce more widely used. The skill required to make this sauce with the raw materials is not required in using the mix. Good, easy-to-make Hollandaise sauce can readily be prepared in a blender.

Why do you suppose quite warm water is used rather than hot water to make a Hollandaise sauce?

Hot water would cook the egg yolks and cause them to lose their ability to emulsify. The purpose of the water is to warm the yolks enough to blend with the butter without causing the butter to solidify.

*

Sometimes a Hollandaise sauce will break, causing the butter to separate from the yolks. All that is required to reform the emulsion is to slowly beat the broken emulsion back into a fresh egg yolk. This is the same procedure that is followed when a mayonnaise emulsion breaks as it is being prepared.

A broken emulsion in a Hollandise sauce can be reformed **by beating egg yolk into the broken emulsion/by beating the broken emulsion into an egg yolk.**

By beating the broken emulsion into an egg yolk.

*

Mayonnaise is a cold sauce made by beating salad oil into eggs or egg yolks. If the egg yolks are warm or a tablespoon of hot water is first blended with the eggs, a more permanent emulsion is formed. Vinegar or lemon juice and seasonings such as salt, white pepper, mustard, and paprika are added to the emulsion. The making of mayonnaise is discussed at greater length under salad dressings.

Mayonnaise is an emulsion made by beating _____ _____ into eggs or egg yolks.

Salad oil.

*

By blending a bit of lemon juice and cayenne into softened butter, we can make a butter sauce. If this is melted, it becomes *maitre d'hotel* sauce. A lightly browned butter becomes a *meuniere* sauce and a darkly browned one becomes a *beurre noir* (black butter).

Match the sauce in column A with its description in column B:

A	B	
1. Butter sauce	A. Softened butter with a bit of lemon juice and cayenne	
2. Maitre d'hotel sauce	B. Melted seasoned butter	1. A.
3. Meuniere sauce	C. Lightly browned seasoned butter	2. B.
4. Beurre noir sauce	D. Darkly browned seasoned butter	3. C.
		4. D.

*

The basic softened butter may be combined with many minced items or pastes. For instance, anchovy paste added to this butter is used for anchovy canapés. Minced lobster meat added to this butter sauce makes delicious canapés. A garlic butter may be made and used over steaks.

Is softened butter seasoned with cayenne and lemon juice a basic sauce?

Yes, and it has many variations just as do the other basic sauces.

*

Just a few of the ways in which a basic sauce can be varied to make a secondary sauce are indicated below.

Basic Sauce	Ingredients Added	Name of Secondary Sauce
Brown sauce	Chopped mushrooms	Mushroom sauce
Brown sauce	Red dry wine, chopped green onions, and beef marrow	*Bordelaise* sauce
Brown sauce	Tomato sauce, mushrooms, and white wine	*Chasseur* or hunter sauce (for game)
Brown sauce	Orange and lemon peel and juice, ginger, cayenne, port wine, and dry mustard	Cumberland sauce (for game)
Velouté sauce	Cream	Supreme sauce
Velouté sauce	Egg yolks and cream	*Poulette* sauce
Tomato sauce	Onions, garlic and green peppers	Creole sauce
Bechamel sauce	Green peppers, mushrooms, and pimientos	*A la king* sauce
Bechamel sauce	Sherry wine (dry)	Newburg sauce
Bechamel sauce	Grated Parmesan cheese	*Mornay* sauce
Hollandaise sauce	Tarragon vinegar	*Bernaise* sauce
Hollandaise sauce	Whipped cream	*Mousseline* sauce
Hollandaise sauce	Dill pickles, anchovy, capers, and mustard	*Remoulade* sauce (for fish)
Mayonnaise	Dill pickles, onions, and parsley	Tartar sauce
Butter sauce	Bread crumbs to melted butter	*Polonaise* sauce
Butter sauce	Sliced almonds to lightly browned butter	*Almandine* sauce (for fish)

REVIEW

1. *True or false.* We have a greater variety of gravies than we have sauces.　　*False.*

2. *True or false.* All sauces and gravies must be made; they cannot be made from mixes or canned soups.　　*False.*

3. A gravy is made from drippings of meat and carries the flavor of the meat from which it came; a sauce *has/has no* special flavor predominating.　　*Has no.*

4. The only liquid that can be used to make a neutral sauce is one that has no ＿＿＿ essence in it.　　*Meat.*

5. There are two basic stocks, a ＿＿＿ and a ＿＿＿ stock.　　*White, brown.*

6. To make a rich stock in quantity, use five quarts of water, ＿＿＿ pound (s) of bones and meat, a pound of vegetables, and seasonings.　　*Four.*

7. Stocks *will/will not* keep in a refrigerator for a week.　　*Will.*

8. Cake flour has *more/less* thickening power than all-purpose flour and *more/less* thickening power than corn-starch.　　*More, less.*

9. Which gives the clearest paste after it is cooked: bread flour, cornstarch, or waxy maize starch?　　*Waxy maize starch.*

10. A slurry is a mixture of flour and ＿＿＿.　　*Liquid.*

11. A *roux* is made by cooking ＿＿＿ and ＿＿＿ together.　　*Fat, flour.*

12. To make a medium white sauce, use ＿＿＿Tbsp of flour or ＿＿＿Tbsp cornstarch for each cup of liquid.　　*2, 1.*

13. Blend a *cold/hot* roux into a hot liquid and a *cold/hot roux* into a cold liquid.　　*Cold, hot.*

14. *True or false.* All starches thicken at the same temperature.　　*False.*

15. *True or false.* To finish a sauce means to bring it to a boil.　　*False.*

16. A liaison is a mixture of ＿＿＿ part (s) cream to ＿＿＿ part (s) egg yolks.　　*Three, one.*

17. *True or false.* Some gravies and sauces are not thickened.　　*True.*

18. A basic sauce may be called a ＿＿＿ sauce and a secondary sauce may be called a ＿＿＿ sauce.　　*Mother, small.*

19. Another name for the basic brown sauce is ＿＿＿ sauce.　　*Espagnole.*

20. A *velouté* is made from a *brown/white* stock.　　*White.*

21. A *bechamel* sauce is made from a rich stock of chicken or veal and then thinned with ＿＿＿ ＿＿＿ or ＿＿＿.　　*Rich milk, cream.*

22. *True or false.* *Velouté* sauce, white sauce, and *bechamel* sauce are the same thing.　　*False.*

23. To make a cream sauce, take white sauce and add ＿＿＿ to it.　　*Cream (rich milk).*

24. *True or false.* Tomato sauces are always thickened.　　*False.*

277

25. To make a Hollandaise sauce, beat _____ _____ into _____ _____ _____; then add a bit of vinegar or lemon juice and seasonings.

Melted butter, egg yolks.

26. To make mayonnaise, beat salad oil into eggs or egg yolks and seasonings, and then add _____ _____ or _____ and seasonings.

Lemon juice, vinegar.

27. If you modify a basic butter in each of the following ways, what sauces have you made?
    A.   *Just melt it.*
    B.   *Add grated garlic to it.*
    C.   *Brown it lightly and add slivered almonds.*
    D.   *Brown it darkly.*

A. Maitre d'hotel sauce.
B. Garlic butter.
C. Almandine sauce.
D. Beurre noir.

10

Egg Cookery

One of our most nourishing and versatile foods is the egg. Occasionally, as in a Caesar salad or an eggnog, eggs are used without being treated; more commonly they are cooked either alone or with other ingredients to make a variety of food products. In this chapter, attention will be directed to the preparation of products in which eggs are cooked alone or are the main ingredient in the recipe. By studying this information and observing the actual use of eggs in the laboratory, you will gain a greater understanding of both the role of eggs in cookery and the chemical and physical behavior of protein.

Eggs are important in food preparation because they are _____ and can be prepared in **only a few/many ways.**

Nourishing, many.

*

Eggs are valuable in the diet as sources of several nutrients. The protein content of eggs is of interest from the standpoint of both nutrition and food preparation. The specific contributions of egg protein as a thickening agent and as a foam base are paramount in this chapter. The total protein in eggs is a complete protein. This animal protein food contains significant amounts of all the essential amino acids needed by man and is easily digested by most people. Nutritionists recommend that protein be supplied in each meal, and eggs are often the protein food in American breakfasts.

The protein in egg is **a complete/an incomplete** *protein; that is, it contains* **most/all** *the amino acids that man needs to have supplied in his diet.*

A complete, all.

279

Eggs also supply vitamins and minerals. One excellent source of iron is egg yolk. The high iron content of yolks helps you to maintain your hemoglobin count at approximately normal. The yolk is also a good source of vitamin A. You will find that egg yolks vary in color. Darker, more orange-colored yolks are higher in vitamin A than are the light-yellow yolks. Vitamin D is also present in eggs laid by chickens exposed to sunlight. Yolks are rather high in fat (about 33 per cent).

Yolks are particularly good sources of _____ and _____.   *Iron, vitamin A.*

*

Whites are good sources of protein and the B vitamins. They are not good sources of iron or fat. The slightly yellow-green color of uncooked egg white is evidence of the presence of riboflavin. A considerably larger percentage of water (87 per cent) is contained in the whites than in the yolks (only 49.5 per cent).

Does this large amount of water in the white mean that egg whites are not nourishing?   *No. Whites are excellent sources of protein and riboflavin.*

*

QUALITY OF EGGS

Eggs are graded as an aid to consumers. Candling is the method most often used for evaluating egg quality. This process utilizes a very simple piece of equipment which you could construct if you want to see just how eggs are graded in the shell. Simply cut a hole, slightly smaller than an egg, in the side of a large can. Place a light bulb in the inverted can and hold the egg over the illuminated hole. The bright light will dimly silhouette the egg yolk and the air space. As you quickly twirl the egg in front of the light, you will be able to see some movement of the yolk. You can also tell how large the air space is.

Candling is a method of grading eggs **in the shell/out of the shell.**   *In the shell.*

*

Eggs of high quality will have a large quantity of thick white and only a little thin white. The thick white prevents the yolk from moving freely in the shell; the yolk silhouette during candling will be indistinct and well centered in the egg. The air space will be small. Such an egg is considered to be top quality and would receive a rating of Grade AA.

What two important signs of egg quality can be seen during candling?

Appearance of the yolk and size of the air space.

*

Some eggs of Grade AA quality are marketed with the label of Fresh Fancy Quality. When you see this you know two things: (1) the eggs are of Grade AA quality and (2) grading was done within the past 10 days. This sign of quality is the most meaningful of the grades used in marketing of eggs. If the eggs are not purchased within 10 days of grading, the Fresh Fancy Quality sign must be removed from them. This grading is done under the Quality Control Program of the U.S. Department of Agriculture.

Are eggs that have just been graded as Grade AA comparable in quality to eggs marked Fresh Fancy Quality?

Yes.

*

With the exception of Fresh Fancy Quality eggs, the grade designation assigned when the egg is candled remains with that egg until it is sold in the store. These grades are Grade AA, Grade A, and Grade B, with Grade AA being the highest and Grade B the lowest. As you would expect, the price is slightly higher for Grade AA than for Grade A, which in turn is higher than for Grade B.

The highest egg grade that can be used on eggs without a limit on marketing time is **Grade AA/Grade A.**

Grade AA.

*

The marketing operation can make a big difference in the quality of an egg by the time it is actually prepared by you, the consumer. Eggs will maintain excellent quality during marketing if they are kept in a slightly humid, chilled area. However, they begin to deteriorate quickly if they are stored without refrigeration. The white will become thinner, the air space will get larger, and the yolk will break more easily when the egg is broken out of the shell. Obviously, if these changes occur, the egg actually will be of lower quality than the grade marked on the box.

281

If you see eggs displayed on an unrefrigerated counter in a store, would you expect eggs marked Grade AA to be of excellent quality?

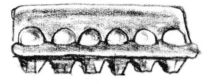

No. Eggs begin to lose quality quickly at room temperature, but the law does not require regrading of eggs to establish actual grade at time of purchase.

*

When you break eggs out of the shell, note their appearance. If your egg does not pile around the yolk well and if it tends to spread over a very large area, it is apparent that the egg has changed in quality from the time it was graded. Another point to check is the surface of the yolk. As eggs change to a lower quality, the yolk tends to flatten and spread rather than retaining its plump, round shape.

True or false. If you buy a carton of Grade AA eggs, you are assured that the eggs are the highest quality.

False. Eggs that are Grade AA quality when candled will be sold in the store under that grade regardless of the deterioration that may take place during the marketing operation.

*

You will notice that eggs are sold by size as well as by grade. The size of the egg does not influence quality at all. Jumbo is the largest size; a dozen eggs this size weighs not less than 30 ounces. The other sizes are, in decreasing order: extra large, 27 ounces minimum for one dozen; large, 24 ounces; medium, 21 ounces; small, 18 ounces; and peewee, 15 ounces. Note that there is a three-ounce difference between each size.

If medium eggs sell for 60 cents a dozen, large eggs would be a bargain if they sold for anything less than _____ cents per dozen.

69; the large eggs being 3 ounces or one-seventh heavier could sell for one-seventh more; one-seventh of 60 is 8⁴⁄₇.

*

If you are buying eggs to use in recipes, you will probably find that large or extra large eggs are the best sizes to buy. Jumbo size, although useful for fried eggs and other purposes where eggs are used individually, is larger than needed and may result in a less desirable product. Small or peewee eggs also present problems in preparing a recipe because they are so much smaller than the size expected in the recipe.

If a cake recipe requires two eggs, you should use **jumbo/ large/small** *eggs to get the best results.*

Large.

*

The color of the egg shells sometimes makes a difference in price. On the Boston market, many buyers prefer brown eggs and are willing to pay a cent more per dozen for them. On the New York market, buyers prefer white eggs and pay a cent more a dozen for them. The color of the shell has no effect on the nutrition or taste of the egg, however.

True or false. Brown eggs are more nutritious than white eggs.

False.

*

Most eggs on our markets are freshly laid, but sometimes we buy eggs that have been preserved by dipping them in oil and storing under refrigeration surrounded by inert gas. If stored properly, high-quality eggs may be a better quality after the six months of storage than low-quality eggs recently laid.

Why are eggs often dipped in oil before storage?

The oil covers the pores of the shell; this helps keep microorganisms from going into the shell and slows the loss of moisture from the egg.

*

Dried eggs can be purchased and used satisfactorily for many cooking purposes. The equivalents of dried eggs to fresh eggs are:

1 whole fresh egg	$= 2$ Tbsp dried egg plus $2\frac{1}{2}$ Tbsp water
1 fresh yolk	$= 1\frac{1}{2}$ Tbsp dried yolk plus 1 Tbsp water
1 fresh white	$= 1$ Tbsp dried white plus 2 Tbsp water

True or false. Dried egg yolks take more water per tablespoon than dried whole eggs.

False.

*

Frozen eggs are available either whole, as yolks, or as whites. These are not too frequently available for home use, but they are sold extensively for commercial use. All frozen or dried eggs that go across interstate boundaries must be pasteurized to destroy any bacteria that might be in them. The temperature of the eggs is brought up to 138°F and held long enough to destroy pathogenic bacteria. This makes it possible to use dried or frozen eggs for salad dressing or other dishes where the eggs are not cooked or are only lightly cooked.

Why don't the eggs thicken when they are heated this way?

Good question; you'll find out later.

＊

Eggs and egg products are excellent cultures for the bacteria *Salmonellae*. Fresh eggs may be contaminated with *Salmonellae* as they pass through the intestinal tract of the hen. *Salmonellae* invade the egg through small pores in the shell. Cooking can destroy these bacteria; but if the egg is not cooked or only lightly cooked and then allowed to stand in foods for four hours or more at room temperature, food poisoning can result.

Would you say there is possible danger of Salmonellae *poisoning in meringues, mayonnaise, salad dressing, Hollandaise sauce, or a stirred custard?*

Yes, in these the egg is used either raw or only lightly cooked.

＊

As an egg ages, it becomes more alkaline and develops hydrogen sulfide, a sulfur compound that smells like some hot springs. This is sometimes called a rotten egg odor. Long cooking time or high temperature encourages the development of this odor and flavor even in a freshly laid egg.

The strong odor and flavor that sometimes develop in eggs are caused by a compound called _____ _____.

Hydrogen sulfide.

＊

Sulfur is a component of the white but can easily combine with the iron in the yolk to form ferrous sulfide, a greyish-green, strong-tasting compound. You can see this as a layer around the yolk of hard-cooked eggs that have been overcooked. Occasionally, omelets may have a grey-green, custard-like layer on the bottom. This is also ferrous sulfide. Formation of this objectionable sulfur compound can be prevented if you use proper cooking procedures and make certain that your eggs are of good quality.

Ferrous sulfide forms when sulfur in the **yolk/white** *combines with iron in the* **yolk/white.**

White, yolk.

＊

Acid can be added to eggs to make them less alkaline. This change reduces the chance of the formation of ferrous sulfide when you cook an egg. The acids that are sometimes used in egg cookery include cream of tartar, lemon juice, vinegar, and tomato juice. You will notice that acids also stabilize foams and bleach the flavone pigment in the whites.

When you add cream of tartar to an egg mixture, the mixture becomes more **acidic/alkaline.**

Acidic. Cream of tartar is an acid so it makes eggs less alkaline.

*

Coagulation is the process of changing the characteristics of a protein by applying heat, mechanical agitation, or various other devices. One obvious change in egg protein is the change in solubility. Coagulated egg white, for example, will not flow because it has been changed into a less soluble form. The transition from a translucent, almost transparent material to the opaque, coagulated egg white can be observed readily. The accompanying change in tenderness is of particular concern in protein cookery, because cooking procedures influence the tenderness of the finished product.

The change that protein undergoes when it is heated sufficiently is called _____.

Coagulation.

*

Have you ever noticed that the whites of some fried eggs are less tender than the whites of others? Long cooking or too high a temperature will cause egg proteins to become rather tough. Controlled cooking will produce tender, palatable egg products.

True or false. If a hard-cooked egg had a tough white, you could conclude that it was either cooked too long or it was cooked at too high a temperature.

True.

*

Egg white becomes jelly-like when heated to approximately 140°F and firms at about 149°F. The yolk begins to coagulate at around 144°F and completes the process at approximately 158°F. A whole egg coagulates at around 156°F.

Will eggs pasteurized at 138°F thicken? Why?

At last you know. The pasteurization temperature reached (138°F) is too low to coagulate the egg proteins.

*

The addition of other foods to eggs changes the coagulation temperature of the egg proteins. Usually you must heat these mixtures to a higher temperature than those indicated above if you wish to coagulate the egg. For example, the coagulation temperature of an egg diluted with a cup of milk is 176°F. Sugar also raises the coagulation temperature. Regardless of what the egg is mixed with, 190°F is about the highest temperature it can reach without coagulating.

Match the coagulation temperature in column A with the condition or statement in column B.

A	B	
1. 149°F	A. The coagulation temperature of egg white	1. A.
2. 156°F	B. The coagulation temperature of whole egg	2. B.
3. 190°F	C. The highest temperature an egg mixture can go without coagulating the egg	3. C.
4. 176°F	D. The coagulation temperature of egg yolk	4. E.
5. 158°F	E. The coagulation temperature of a whole egg in a cup of milk	5. D.

*

FRIED EGGS

To fry eggs, select a frying pan sized to hold just the number of eggs you wish to prepare. You will have the best results if you use a skillet made of heavy aluminum or cast iron so that the heat will be uniform all around the pan. For the best fried eggs, it is necessary to use top-quality fresh eggs. Fresh Grade AA eggs will pile well around the spherical yolk rather than flowing freely around the entire pan.

Can you think of two other reasons why fresh eggs are so important for frying?

The yolk is less easily broken in a fresh egg and the flavor is fresher and more pleasing.

*

There are two ways to prepare fried eggs. One method uses an abundance of fat and the other method uses a minimum of fat and some steam. To use the first method, melt enough butter or other fat in a skillet to form a layer about one-eighth inch deep. When the butter begins to bubble, gently slide the eggs into the pan. Use a moderate to low heat to cook the eggs without toughening the protein. Controlled heat also avoids the hard crust that forms on the bottom and edge of the egg when a high heat is used. Usually, eggs fried in this way are basted to cook the upper surface and coagulate the white that covers the yolk. As soon as the white is coagulated completely and the yolk surface is veiled with coagulated white, the egg should be served.

How do you baste an egg?

Spoon the hot fat in the frying pan over the surface of the frying egg. Keep repeating this until the yolk appears veiled.

*

The second way of frying eggs is to use just enough melted butter to cover the skillet. Use controlled heat to fry the eggs. When the white begins to coagulate, add about one-half teaspoon water for each egg and quickly cover the pan tightly. The trapped steam will coagulate the upper surface of the eggs. As soon as the white coagulates to soften the bright color of the yolk, the eggs should be served. Eggs prepared in this manner or with the use of an abundance of fat are sometimes called "country-style" eggs.

Describe a well-prepared fried egg.

A well-prepared fried egg will have a white that is tender, yet firm and a yolk that is only slightly thickened. The white covering the surface of the yolk will be coagulated.

*

Some people prefer fried eggs that are a bit different from country-style eggs. One familiar way is to prepare eggs "sunny-side up" by not coagulating the white over the yolk. The term "over-easy" means to fry the egg without basting until the white is coagulated and then turn the egg over briefly before serving.

Match the cooking method in column A with the terminology in column B.

A	B
1. Basted with hot fat	*A. Sunny-side up*
2. Fried on both sides	*B. Country-style*
3. Fried without coagulating the white over the yolk	*C. Over-easy*
4. Adding a small amount of water and a cover to steam the surface	

1. B.
2. C.
3. A.
4. B.

*

A Teflon-lined pan makes it unnecessary to use fat in cooking eggs unless you want the flavor of butter or other fat. Regular pans also can be conditioned so that eggs will not stick when cooked in them. The pan is heated

287

with salt and fat and the hot fat rubbed into the pan vigorously with a pad. The salt acts as an abrasive, cleaning the bottom of the pan thoroughly so it can be lightly coated with fat and no other product. The pan is then cleaned thoroughly *without* washing. Now only a small quantity of fat is required to keep eggs from sticking in such a pan. If the pan is washed, the coating is destroyed and must be replaced by the same process, a process often called "conditioning." To keep a conditioned pan ready for use, simply wipe it with soft, absorbent paper toweling so no fat remains. This pan is set aside and used only for eggs.

Will bacteria or other undesirable microorganisms be a problem if such a pan is not thoroughly washed?

No, the pan is wiped clean and microorganisms will not develop readily in such a pan.

*

SCRAMBLED EGGS

To make good scrambled eggs, beat the eggs, along with milk or cream and salt to season, with a rotary beater until the whites and yolks are completely blended but not foamy. The liquid is added to make the finished product more tender. Pour this mixture into a skillet lightly coated with melted fat that is gently bubbling. As the egg mixture coagulates, slowly stir the cooked portions to permit the uncooked egg to run under and also be cooked. Slow stirring is recommended so that the egg will remain in large pieces rather than very small pieces. A moderate heat is used to help keep the eggs tender. You ordinarily should heat scrambled eggs until they are completely coagulated but still shiny.

Liquid is added to scrambled eggs to **lower the coagulation temperature/make the finished product more tender.**

Make the finished product more tender.

*

Scrambled eggs are judged by appearance, tenderness, and flavor. First, let us consider appearance. Well-prepared scrambled eggs are a uniform yellow color, with no flecks of unmixed white and no suggestion of browning on the surfaces. The surfaces are shiny, yet the eggs are coagulated. When you sample the eggs, they should be tender and flavorful.

288

Classify the following characteristics of scrambled eggs according to their desirability.

1. Uniform yellow color	A. Desirable	1. A.
2. Dull dry surface	B. Undesirable	2. B.
3. Flecks of white		3. B.
4. Browning evident on the surface		4. B.
5. Shiny surface		5. A.

*

Sometimes when scrambled eggs need to be held, it is advised that eggs be scrambled to the hard dry stage and then moistened with white sauce. This higher temperature in the eggs kills the bacteria. If eggs are eaten when they are prepared, you do not need to be concerned about the possibility of bacterial contamination because there is too little time for *Salmonellae* to multiply. The danger comes when the eggs are left to stand and the *Salmonellae*, being in an excellent culture medium, multiply rapidly. By scrambling eggs until dry, you dramatically reduce the viable bacterial count and extend the safe holding period.

The bacteria most commonly found in eggs are _____.   Salmonellae.

*

EGGS COOKED IN WATER

Eggs cooked in water may be prepared in the shell or out of the shell. Those cooked in the shell are hard cooked, medium, soft cooked, or coddled. Poached eggs are broken out of the shell before being coagulated in simmering water.

Eggs cooked out of the shell in simmering water are called _____ eggs.   Poached.

*

A medium egg soft cooks in simmering water at sea level in four to five minutes. If the egg is cold or larger, it will take slightly longer. A very cold egg may have such a rapid expansion of air inside when dropped into simmering water that it splits.

True or false. Drop a cold egg into boiling water and boil about three minutes to soft cook it.   *False. Warm the egg in tepid water for a minute first and then simmer it. Don't boil it if you are at sea level.*

Eggs in the shell placed directly into boiling water may crack from the expansion of air in the air space. To prevent this, the large end of the egg can be pierced with a pin to allow the air to escape easily as it expands. Or, as indicated, the eggs can be warmed so the air expansion is not as great.

Why is the large end rather than the small end of the egg pierced?

Because the air space is at the large end.

Would an old egg crack sooner than a fresh egg if dropped cold into boiling water? Why?

Yes. The old egg would have a larger air space than the fresh one.

*

When water is boiled at a high altitude, its temperature is lower than is water boiling at sea level. Therefore, in the mountains you can boil eggs rather than simmering them. At sea level, simmering temperatures range from 180° to 211°F; eggs are usually cooked at the upper end of that range. Yet at 6000 feet, the temperature of boiling water is only 200°F.

Will the same length of time be required to simmer a soft-cooked egg at sea level and to boil a soft-cooked egg at 6000 feet?

No. The egg boiled at 6000 feet will take slightly longer to coagulate because the actual temperature of the water is lower.

*

Temperatures just below 212°F are ideal for soft- or hard-cooking eggs because they coagulate the egg proteins in a reasonable time without toughening. However, even at this moderate temperature, overcooking will decrease the tenderness of the finished product. Four to five minutes in simmering water at sea level is the length of time required to coagulate completely the white and slightly thicken the yolk, as is desired for a soft-cooked egg. A simmering period of 20 minutes produces a firm, yet tender white and a coagulated yolk.

A tender egg product is prepared by carefully controlling the "two t's," _____ and _____.

Time, temperature.

*

Ferrous sulfide formation around the yolk of a hard-cooked egg can be a problem. Part of this problem can be curbed by using fresh eggs and avoiding too long a cooking time.

You also can reduce the chance for formation of ferrous sulfide if you plunge the egg into cold water as soon as it is cooked and keep the water cold until the egg is chilled. Although soft-cooked eggs do not develop ferrous sulfide, you may wish to plunge them briefly in cold water to stop the cooking at the correct stage of doneness.

If you boil eggs until they are completely hard cooked and then let them stand in the air to cool off, what is likely to develop around the yolk?

A dark ring from the formation of ferrous sulfide.

*

If you want to peel hard-cooked eggs, crack them well and start peeling at the large end. Since the egg peels more easily at the air sac area, you can get a good start on peeling the egg this way. Placing the egg under running water helps to remove the shell. The water gets under the shell and helps loosen it from the egg. Extremely fresh eggs sometimes are harder to peel than ones that are older.

To remove the shell of hard-cooked eggs, start at the **small/large** *end.*

Large.

*

Coddling is cooking eggs in the shell. The eggs are put into a pan and boiling water is poured over them—a pint of boiling water per egg. The pan is covered and set aside in a warm place. A coddled egg is soft cooked in about seven minutes.

A coddled egg is softer than a boiled or simmered egg. Why?

The higher the heat the more firm the cooked protein will be, and coddled eggs are heated in water that is not maintained at a simmer.

*

Poached eggs are usually heated in water, but they can be poached in milk or cream. Use a double boiler or very heavy pan to prevent scorching the cream or milk. The liquid should be maintained at a simmer throughout the poaching process to keep the protein tender and prevent the breaking up of the white. If water is used, you may wish to add two teaspoons of vinegar and one teaspoon of salt per pint of water. The salt adds flavor and the salt and vinegar together assist the egg in bunching together in cooking. Use of the vinegar is optional.

291

Why do you avoid boiling poached eggs?

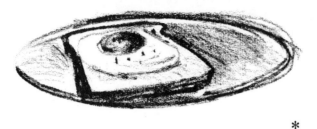

The eggs will toughen and an active boil might break up some of the egg white, especially if boiling occurs before the egg has coagulated on the outside.

*

To poach an egg, first break the egg into a saucer and inspect it. Heat two inches of water to simmering in a pan at least three inches deep. Then slide the egg gently toward the edge of the pan. If you drop the egg directly into the water, the force of the fall flattens the egg. Simmer the egg until the white coagulates and the yolk begins to thicken. Remove the egg from the water with a slotted spoon.

Will a Grade B egg poach satisfactorily?

No. The white will flow and spread too far in the water.

*

Eggs can be baked (or shirred) by breaking them gently into a greased baking dish and baking them in a 350°F oven until done. A better product is obtained if the greased baking dish is heated beforehand so that the eggs bake faster from the bottom than the top and the yolks remain somewhat soft. Sometimes eggs are baked by heating them gently until coagulation begins and then finishing them under the broiler, but the intense heat of the broiler may toughen the egg protein excessively. For variety, try pouring cream and diced cheddar cheese or other ingredients over the egg before baking. This makes an excellent breakfast dish or even is suitable for a luncheon. Bacon strips or ham can be partially cooked and then placed in the baking dish with eggs that are to be shirred. Other variations can be prepared by serving shirred eggs topped with creamed chicken, chicken livers in a sauce, a Spanish sauce, or some other product than blends well with the eggs. You might like to try heating corned beef hash in a greased baking dish until it is very hot and then indenting the hash at intervals to house the eggs to be shirred. Bake this casserole until the whites are firm and the yolks are just coated over the top. A bit of gravy can be served over each portion of egg and its nest of hash. By warming the hash before adding the eggs, you are cooking the eggs from the bottom as well as the top and they will cook through without overbaking.

A shirred egg is a _____ egg.

Baked.

*

EGG FOAMS

Whole eggs, yolks, or whites can be beaten to a foam for use in many food products. Egg white and egg yolk foams are used to give a light airy texture and increased volume. Eggs are beaten to four different stages, depending upon the use of the foam.

Stage 1. Whole eggs are beaten until well blended; they are usually called "well-beaten eggs" in recipes.

Stage 2. In this foamy stage, whole eggs develop a fairly fine yet fluid foam, egg whites have coarse bubbles and the mass flows, and yolks are somewhat lighter in color but are still fluid.

Stage 3. Whole eggs are quite stiff and light in color and peak slightly. Yolks are lemon colored and hold soft peaks. Whites do not flow and the peaks just bend over; but at the upper end of stage three, the peaks of whites stand up straight.

Stage 4. It is difficult to beat whole eggs or yolks to this stage. The whites are dry, stiff, and brittle and they flake off and can be cut into rigid parts. The whites are dull appearing; such whites have no use in cookery. They have expanded so much that they cannot stretch more during baking. When the air in them expands in the oven, the cell walls break and the structure collapses.

Egg whites beaten to stage 3 have peaks that _____ _____, while at the upper level of stage 3 they stand _____.

Bend over, upright.

*

Stage 1 eggs are used for French toast and omelets, for coating foods, for adding them to other ingredients in batters and doughs, and for scrambling.

*We ordinarily beat **whites/yolks/whole eggs** to stage 1.*

Whole eggs.

293

Stage 2 is the point at which cream of tartar or other acid is added to egg whites. This is also the time when you begin to add sugar when making meringues. Stage 3 is the desired end point for egg whites to be used in souffles, sponge cakes, angel cakes, and soft meringues. Upper stage 3 is correct for egg whites in chiffon cakes and hard meringues. Stage 4 eggs are beaten too much for satisfactory use in food products.

Match the product in column A with the correct stage in column B.

A	B	
1. Sponge cake	A. Stage 1	1. C.
2. Souffle	B. Stage 2	2. C.
3. Scrambled eggs	C. Stage 3	3. A.
4. Chiffon cake	D. Upper stage 3	4. D.
5. Begin adding sugar	E. Stage 4	5. B.
6. Add cream of tartar		6. B.
7. Soft meringue		7. C.
8. No use		8. E.

*

Older eggs beat to a larger volume of foam than do very fresh eggs, because the older eggs have a larger amount of thin white. Thin white is more extensible than is the viscous thick white. This effect of egg freshness on volume of a foam is most noticeable when a rotary hand beater or wire whip is used. The difference disappears to some extent when an electric mixer is used; the mechanical beater has enough power to break up the fresh egg and extend it. However, if fat is present, egg whites do not whip well regardless of the type of beater used.

Will whites with a small amount of yolk in them whip well? Why?

No. Egg yolks contain fat.

*

Volume of an egg white foam is very important because of its effect on the volume of the finished product in which it is used. You also will find that the stability of an egg white foam is essential to the preparation of foam-containing products. You can increase the stability of an egg white foam by adding a small amount of cream of tartar or lemon juice. These acids retard the gradual collapse that occurs in an uncooked foam.

OMELETS

There are two types of omelets, French and fluffy. The French omelet uses whole eggs beaten to stage 1. Beat until the eggs are completely blended and have no streaks of yolk or white; avoid foam development. Usually, some liquid (a tablespoon per egg) is added to make the omelet more tender. Milk, cream, or tomato juice are flavorful liquids to use in an omelet.

Why is it important to avoid building up a foam while beating the eggs?

If eggs are foamy, it is more difficult to cook the omelet and the finished product will be rather porous.

*

The frying pan should have its usual eighth of an inch of fat heated until the butter or margarine bubbles. Use a medium to medium-low heat to coagulate the omelet. Instead of stirring as you do when making scrambled eggs, use a narrow spatula to lift coagulated areas and allow the upper uncooked mixture to flow underneath. This procedure is used only as long as there is enough uncoagulated egg to flow under the lifted area. When finished, the omelet should be one continuous, circular mass. Check the bottom of the omelet to be sure that it is a golden brown. If it is too pale, turn the heat up briefly to brown it. Some people prefer to make a French omelet by stirring rapidly and shaking the pan as the egg mixture cooks, rather than using the procedure explained above.

True or false.   A plain omelet is made in exactly the same way as scrambled eggs.

False. Many things are similar but the cooking techniques differ significantly.

*

When the eggs are almost cooked at the top and the bottom is browned, fold the omelet in half. Before folding, you may wish to add creamed chicken, seafood Newberg, chicken livers and gravy, chopped cheese, bacon, ham, a Spanish tomato sauce, or even marmalade or jam.

Is it true that you can fill an omelet with jam or marmalade?

Yes. This can be served as a luncheon entree or as a dessert; if as a dessert, it may be flamed with rum or brandy.

*

An omelet filled with jam or marmalade can be "burned." To do this, sift powdered sugar lightly over the top of the cooked omelet. Then move a hot instrument such as a poker lightly across the top, burning the sugar in places.

This burning caramelizes the sugar and gives an excellent flavor to the omelet.

True or false.   Burning an omelet means to scorch it.

False. You merely burn some of the sugar that has been dusted over the top.

*

An omelet may have a sauce served over it. For example, if you fill it with creamed chicken, creamed mushrooms, or a seafood Newberg, thin the remaining sauce a bit and pour it over the omelet. This gives moistness and added flavor.

What would you think of using canned condensed soup such as cream of chicken, mushroom, or cheese for filling an omelet with the remainder as a sauce over the omelet?

This is an excellent time-saving idea.

*

Fluffy omelets are made by beating the separated yolks and then the whites to stage 3. The two foams are folded carefully to blend them. This is put into a frying pan containing one-eighth inch of bubbling hot butter or margarine. Use a moderate heat for one minute to start cooking and then quickly pop the omelet into a preheated 325°F oven to bake until done (about 15 to 20 minutes). Throughout the entire cooking period, do not disturb the omelet. When done, it is folded in half just the same as a plain omelet. Fluffy omelets are often filled with a sweet filling for a dessert, but they may be filled with creamed foods, bacon, cheese, meats, or other foods.

If you beat the whites and yolks to a stage 4 foam, would the fluffy omelet be successful?

No. The egg foams would not blend well and they would collapse when they expanded from the heat. The omelet would turn out heavy and soggy with very poor volume.

SOUFFLES

The word "souffle" means to blow up. A souffle is quite similar to a fluffy omelet. It is usually made from a starch-thickened base such as a thick white sauce. A small amount of this sauce is stirred quickly into beaten yolks and then returned, with thorough stirring, to the sauce. Chopped chicken, tuna, ham, cheese, or other foods may be added to this base. Then, egg whites beaten to stage 3 are carefully folded into the base until no streaks remain. This is baked in an oven preheated to 325°F. The souffle is done when a knife inserted in the center comes out clean. Underbaking and overbaking will cause a souffle to fall. Treat it with respect! A sauce usually is served over the souffle when it is served. A coconut souffle served with strawberry preserves or a sauce made from thickened fresh or frozen strawberries is a delightful dessert.

What do you think causes a layer on the bottom of a souffle?

Several things might cause this: (1) the sauce might have been too thin, (2) the egg whites were not beaten to the right stage, (3) the sauce and whites were not folded together adequately, and (4) the souffle may not have been baked immediately in a preheated oven.

*

CUSTARDS

Egg custards are sweetened milk and egg mixtures that are coagulated by heat. These mixtures can be baked to form a solid clabber or mass, or they can be stirred in a pan to make a slightly thickened sauce. Eggs can bind in more milk than water, fruit juice, or other liquids because the milk protein supplements the thickening accomplished by the egg protein. One large egg will set one cup of milk. Yolks make the firmest custard, whole eggs are next, and whites make the weakest clabber.

Increasing the proportion of egg to milk or other liquids **increases/decreases** *the firmness of the clabber formed in a custard.*

Increases. The increase in egg provides more protein for thickening.

*

If egg protein is overheated, it will squeeze out some of the liquid. This separation of the watery liquid from the coagulated custard is called syneresis. Either baking at too high an oven temperature or heating too long will cause syneresis.

The separation of liquid from a coagulated custard is called ____.

Syneresis.

297

If the custard is baked in the oven in a pan of hot water and if the custard is preheated to 140°F before it is baked in the oven, the chances for a smooth shiny clabber are enhanced. To prepare the custard, first mix together the eggs, sugar, and flavorings while bringing the milk almost to a boil. Pour the scalded milk into the egg mixture while stirring rapidly. Bake the custard in a 350°F oven. When a knife inserted halfway between the center and edge comes out clean, the custard is done.

True or false. Souffles and custards are done when a knife inserted in the center comes out clean.

False. This is the right test for a souffle, but a custard will be overbaked if this test is used. The residual heat in a custard will finish the coagulation process if you bake the product until it is set halfway between the center and edge.

*

If you use about one and one-half eggs per cup of liquid, a custard pie filling will be firm enough to cut and serve easily. Rice, bread, or other starchy substances, when added to a custard pudding, reduce the amount of eggs needed to set a custard. The starch in these products provide thickening so you can reduce the amount of egg to one egg per one and one-fourth cups of milk (four-fifths egg per cup of milk).

Match the ratio of egg to milk given in column A with the custards listed in column B.

A	B	
1. 1 egg to 1 cup milk	A. Rice custard pudding	1. B.
2. ⅘ egg to 1 cup milk	B. Regular baked custard	2. A.
3. 1½ egg to 1 cup milk	C. Custard pie	3. C.

*

Pumpkin pies, custard pies, and some other pies are basically custards; they are done when they test done halfway between the center and the edge. Avoid longer baking. These pies will have a soggy bottom crust if they are overbaked.

If a pecan pie has a soggy bottom crust, this is likely due to _____.

Overbaking.

*

Another variation of a baked custard is a timbale. This is a chopped or pureed vegetable mixture added to a small amount of unsweetened custard made using about one and

298

one-half eggs for every cup of milk. Bake this as a custard and then serve the product as a vegetable. Timbales may be varied by adding chopped chicken, ham, or other items in place of the vegetables. Some timbales are thickened with a very thick white sauce rather than with a baked custard. Just as with fondues, you discover that there are different kinds of timbales.

Why do you suppose the proportion of egg to milk is high in the making of these custard-type timbales?

Because the large proportion of chopped or pulped item to the custard weakens the custard, we must add more eggs for strength.

*

Another food product made from eggs is a fondue, which resembles a souffle but is slightly more compact and far less fragile. Bread from which the crusts are usually removed is placed either as a slice or in cubes in a baking pan. The flavoring mixture desired, which may be the same as used in a souffle (chopped chicken, ham, cheese, vegetable pulp, etc.) is added and then topped with a seasoned mixture of milk and eggs in the proportion of one whole egg to a cup of milk. The mixture is baked as a custard and usually is served with a sauce. The bread helps to give an openness to the product so that it is delicate and light.

Would you say that a fondue made in this manner would be between a custard and a souffle in texture?

Yes. The fondue discussed is just one type of fondue. This is often called a casserole fondue to distinguish it from the Swiss fondue, which is a thickened mixture of cheese and dry white wine flavored with kirsch. You may also be acquainted with fondue bourguignoune, the fondue made by quickly heating small pieces of beef in oil and dipping these pieces, held on a long fork, into various sauces.

Do you think it is possible to overbake a casserole fondue?

Yes. Syneresis occurs in this product when it is overbaked.

*

A stirred soft custard is made in the same proportions as a regular custard but it is cooked in a double boiler over simmering water and stirred constantly while being heated.

Care must be taken to ensure that all the custard in the container is stirred continually. This precaution assures more even heat distribution and a more uniform finished product.

To make a stirred custard, you should stir **occasionally/ constantly** *being careful to stir* **all areas/the sides** *of the pan.*

Constantly, all areas.

*

A stirred custard curdles or breaks more easily than a baked custard. Above 185°F a stirred custard will curdle rapidly. If you use a thermometer to judge doneness, cooking should stop when this temperature is reached. It is wise to continue to stir the custard after it is removed from the heat. This is a good precaution because some coagulation will take place due to residual heat.

If a stirred custard were cooked over boiling water, would you expect it to curdle?

Yes. The temperature of the custard rises so quickly that you cannot cool the custard before the egg protein gets too hot and curdles the custard. A slow rate of heating makes it much easier to avoid curdling the mixture.

*

There are two clues to tell when a stirred custard is done. If you have a thermometer, constantly note the temperature as it approaches 180°F. There should be a brief period when the temperature stands still or rises very slowly. As soon as you notice a faster temperature rise, immediately cool the custard. Even if you have been unable to notice the change in the temperature rise, it is wise to remove the custard from the heat when the thermometer indicates 185°F. Stirred custard quickly curdles above this temperature. The second test is to note how the stirred custard coats a silver spoon. When finished, the hot stirred custard will cover a spoon like cream does. If the custard curdles, beat it rapidly for a couple of minutes as it cools.

A stirred custard is done when the temperature reaches _____°F or when the mixture _____ _____ _____.

185, coats a silver spoon.

*

Stirred custards are used widely as sauces for desserts. The English trifle is basically a stirred custard. A custard combining sweet marsala wine and egg yolks served over lady fingers is the famous Italian dessert *zabaglione*. The French make many soft as well as baked custards. A floating island pudding is the popular French dessert *oeufs a la neige*. A Spanish cream is a stirred custard to which gelatin has been added; whipped cream is folded in when the mass is cool and is allowed to set.

Would you be surprised to see a soft or baked custard served in other countries?

No. Custards seem to be universal; most national cuisines include them, probably because the ingredients are usually available.

*

MERINGUES

Meringues are baked, sweetened, egg white foams. Sometimes variations are made by adding other ingredients and flavorings. The two basic types of meringues are soft and hard meringues. Their differences are due basically to the amount of sugar in each. A hard meringue contains twice as much sugar as a soft meringue.

The basic ingredients for a meringue are _____ _____ and _____.

Egg whites, sugar.

*

Soft meringues are used for toppings for pies, puddings, and floating islands. Hard meringues are served as desserts or can be a base for an elaborate dessert. With a pastry tube, you can shape the meringue mixture into rings which are filled with fruit, ice cream, or pudding mixtures after baking. A hard meringue is also the shell for angel pies, some tortes, and soda cracker pies. Layers of hard meringue also can be alternated with fruit, puddings, or other sweetened mixtures.

For a lemon meringue pie you would make a **soft/hard** *meringue, and for a lemon angel pie a* **soft/hard** *meringue.*

Soft, hard.

*

To make a soft meringue, allow the egg whites to warm to room temperature before starting to beat them. Beat them quickly to stage 2 (foamy stage). When you reach stage 2, add the cream of tartar and salt (if used). Gradually start to dust granulated or powdered sugar while continuing to beat to stage 3. Use a ratio of two tablespoons of sugar to

301

each egg white. Gently spread the finished meringue on the pie, being careful to seal it well to the crust and to avoid high peaks that will burn during baking. Bake the meringue-topped pie in a 350°F oven for 15 minutes or until it is a pleasing brown.

For a soft meringue, use _____ tablespoon(s) of sugar per egg white; bake in a _____°F oven for about _____ minutes.

Two, 350, 15.

*

Occasionally, you may have noticed beads of syrup on a meringue. This problem can be eliminated by adding the sugar gradually and being careful to add all of it by the early phase of stage 3. This technique allows adequate time for the sugar to dissolve completely in the whites while they are still sufficiently moist. If the sugar is added too late, the whites will be a little too dry to dissolve all the sugar, and beads will form on the baked meringue.

Beads on the surface of a soft meringue are caused by _____ _____.

Undissolved sugar.

*

Regular granulated sugar can be used for meringues, but there are two other sugar products you might wish to consider using. Superfine or dessert sugar is also a pure granulated sugar. It differs from granulated sugar only in the size of the crystals. If you compare dessert sugar with granulated sugar, you will readily see how very tiny the dessert sugar crystals are. These small crystals dissolve much more readily than do the granulated sugar crystals. The other sugar you may wish to use is powdered sugar, which is ground granulated sugar mixed with 15 per cent cornstarch. The added cornstarch helps to hold the moisture in a meringue.

The three types of sugar that may be used in a meringue are: _____, _____, and _____.

Granulated, dessert (or superfine), powdered.

*

Sometimes meringues tend to weep; that is, moisture drains from the meringue and collects between the meringue and the filling. This liquid also soaks into the edge of the crust. To avoid this problem, be careful to beat the meringue until the peaks just barely bend over and be sure to bake the meringue to a golden brown. Underbaked meringue always weeps. This tendency to weep can also be reduced by adding one-half teaspoon of cornstarch for each egg white. Sift the cornstarch lightly over the meringue just before beating is completed.

Weeping in a meringue can be reduced by using _____ teaspoon of cornstarch per egg white, beating to stage _____, and baking to a _____.

*

A soft meringue may slip on a pie filling. This can be caused by moisture draining from the meringue or by failure to seal the meringue carefully to the edge of the crust. Some ways of reducing weeping have just been mentioned. To this list, you should add the idea that meringues weep less if they are placed on a hot pie filling than they do on a cold one. The hot filling helps to bake the lower part of the meringue which receives only limited oven heat during the brief time a meringue bakes. For a banana cream pie, it is necessary to cool the filling before assembling and topping with meringue to avoid cooking the bananas, but other meringue pies can be assembled and the meringue baked while the filling is still quite hot. The other problem leading to slippage can be easily remedied by using a rubber spatula to gently push the meringue into each of the small indentations at the edge of the crust. The meringue will then be locked to the sides.

Why is weeping reduced if you put a meringue on a hot filling?

The heat from the filling helps to bake the egg white quickly at the bottom of the meringue before the foam can collapse and let liquid drain out.

*

A hard meringue contains twice as much sugar as a soft meringue. Of course, this means that you need to dissolve four tablespoons of sugar in each egg white, a not inconsiderable job! Begin by warming the whites to room temperature and then beat to stage 2. At this point add either lemon juice, vinegar, or cream of tartar to help stabilize the foam. You may prefer to use lemon juice or vinegar because their liquid helps to dissolve the sugar. Very slowly begin adding the sugar and continue beating until the peaks stand up absolutely straight. This is a lengthy process which is greatly aided by the use of an electric mixer. If the meringue seems gritty when done, beat in a small amount of water (not more than one tablespoon per white) to dissolve the sugar.

303

If you were telling a friend the secret of preparing a hard meringue, what should you tell her?

Have the eggs at room temperature. Add cream of tartar or other acid when the eggs are frothy. Add sugar gradually. Beat until the peaks stand up straight. If the sugar does not dissolve, add a little water.

*

The meringue is baked by first placing it on brown paper in the desired shape to make meringue shells, rings, rounds, or kisses. If the meringue is to be used as a pie shell, spread it in a greased pie plate for baking. Baking should be in a 250°F oven to dry out the meringue to a crisp, tender, very pale product. Too high a temperature gives a tough, gummy meringue that may weep. Gumminess can also result from too much sugar, underbeating, or overbaking.

Hard meringues are baked at **high/low** *temperatures.*   *Low.*

*

Hard meringues can be used to make *gateaus* or tortes of many kinds. These are desserts in which layers of hard meringues are alternated with fruits, fillings, whipped cream, or other products. Other meringue-like products also can be prepared. An Italian meringue is made by boiling a sugar syrup to 240°F and beating this into egg whites beaten to the third stage. This is used to make frostings and divinity. Sometimes an Italian meringue is blended carefully into a freezing mixture of fruit and syrup to make a frozen dessert. A very delicate frozen dessert, made by carefully folding equal quantities of vanilla-flavored Italian meringue and whipped cream together and freezing the mixture, is called angel parfait. It is a mousse. The familiar seven-minute icing is actually a meringue-like product made by combining sugar, egg white, cream of tartar, salt, and flavoring in a double boiler and beating vigorously while heating for seven minutes over boiling water. Occasionally this type of meringue is called a Swiss meringue, but usually

Swiss meringue refers to the hard meringue described above.

Which definition fits an Italian meringue?
A. *Egg whites and sugar combined and beaten over boil-*
    water.
B. *Egg whites and sugar combined by heating.*
C. *Egg whites combined with a cooked syrup.*

C. Egg whites combined with a cooked syrup.

*

Egg whites also are used frequently to make whips. These whips are made by folding pureed fruit into a soft meringue. Since they are not baked, whips should be served soon after they are made. Volume is lost on standing. We also can bake or steam some whips to make desserts similar to dessert souffles. Overbeating, excessive heat, or overbaking can reduce the quality of a baked or steamed whip. One popular baked whip is a prune whip which can be served either hot or cold. Another type of whip is one in which eggs (usually whites) are beaten to a stable foam and then combined with gelatin to make a light-textured product. A very popular dessert made this way is apple whip in which applesauce, egg whites, and gelatin are combined. This dessert is sometimes called "apple snow" or "apple sponge." A pineapple snow is also an excellent dessert.

Whips may be of three types: _____, _____, and _____.

Unbaked, baked, steamed.

*

White foams are unstable unless they are baked or stabilized. Heat coagulates the protein to give permanency to the foam. However, some recipes do not require heat to keep the foam from collapsing. Gelatin can be dissolved and combined with an egg white foam while the gelatin mixture is still slightly fluid. When the gelatin sets, the foam is permanently stabilized as long as the gelatin is kept too cool for it to melt. Chiffon pie fillings are flavored egg white foams stabilized with gelatin.

Where would you keep a chiffon pie until you are ready to serve it? Why?

Keep it in the refrigerator to keep the gelatin firm.

*

When beating yolks, whites, or whole eggs for the various products described above, use your knowledge to create stable foams with maximum volume. Careful attention to

305

technique is important in making a good foam. Watch the foam continuously as you beat so that you will beat it to just the right end point.

Would you say that the preparation of egg products is simple or takes some skill and knowledge.

When you have the knowledge and skill, the preparation of egg products will seem simple.

*

REVIEW

1. *True or false.* A hard-cooked egg is more difficult to digest than a soft-cooked one.

   False.

2. As an egg ages, the air sac gets *smaller/larger* and the egg gets more *acid/alkaline.*

   Larger, alkaline.

3. The size of the egg *is/is not* correlated with grade.

   Is not.

4. If you purchased small eggs for 40 cents a dozen, how much should you pay for medium eggs?

   One-seventh of 40 cents is about 6 cents; since medium eggs are one-seventh larger than small ones, you should pay about 46 cents for them.

5. If an egg has a shiny shell, is it a sure sign that it is an old egg?

   Not since they started dipping eggs into oil to preserve them.

6. If an egg broken out of the shell has a flat yolk and a runny white, is it apt to be a fresh egg?

   No, these are indications of an old egg.

7. Sulfides form more easily in an *alkaline/acid* medium.

   Alkaline.

8. The dark ring around a hard-cooked egg and the strong flavor it has are caused by:
   A. *Ferrous sulfide.*
   B. *Hydrogen sulfide.*
   C. *Rancid fatty substances.*
   D. *None of the above.*

   A. Ferrous sulfide.

9. *True or false.* An egg fried "sunny-side up" has a coated yolk.

   False.

10. Scrambled eggs should be in *fine/large* segments.

    Large.

11. To softcook an egg at 3000 feet elevation, you need to cook it *less/more* than three minutes.

    More.

12. If you simmer an egg, you make it *tougher/more tender* than when you boil it and *tougher/more tender* than when you coddle it.

More tender, tougher.

13. To coddle an egg, use a _____ of boiling water per egg.

Pint.

14. When you poach an egg, add _____ and _____ to the cooking water.

Salt, vinegar.

15. A *cold/warm* egg beats up into the most stable foam and a foam with the most volume.

Warm.

16. *True or false.* Eggs beaten to stage 4 are frequently used in food preparation.

False.

17. Acids make eggs *more/less* extensible in making a foam.

More.

18. *Old/fresh* eggs beat up into the best foam.

If an egg is either too fresh or too old, it will not beat up into a good foam.

19. There are two types of omelets: plain (French) and _____.

Foamy (or fluffy).

20. *True or false.* When you burn an omelet, you brown it to quite a dark color.

False.

21. To make a souffle, you fold egg whites beaten to stage 3 into a *starch-thickened base/stiffly beaten egg yolks/hot sugar syrup.*

Starch-thickened base.

22. A dessert souffle *is/is not* made in much the same manner as a regular souffle.

Is.

23. The four basic ingredients of a baked custard are: _____ _____, _____, _____, and _____.

Whole eggs, milk, sugar, flavoring.

24. If a custard breaks down in baking and curdles, we call this _____.

Syneresis.

25. When eggs coagulate, they *absorb/give off* heat.

Absorb.

26. A stirred custard is apt to break down above _____°F.

185.

27. A Spanish cream is made from a stirred custard and is set with _____.

Gelatin.

28. *True or false.* Soft meringues are used for toppings for pies.

True.

29. The base of a casserole fondue is a *cream sauce/custard.*

Custard.

30. A timbale is made by adding chopped or pureed vegetables or other foods to a _____ base and then baking it.

Custard.

31. To prepare a shirred egg, we *steam/bake/fry/poach* it.

Bake.

11

Baking Basics

You may wonder why it is necessary for you to understand the basic information that supports the art and science of baking. After all, there are many mixes for baked products on the market today. In fact, there is probably a mix available for just about any baked product you can name. Since modern food technology has made mixes so readily available, what is the rationale for studying the individual substances common to most baked products and determining their behavior in batters and doughs? It is simply this. Without an understanding of the science underlying the production of batter and dough products, you are handicapped in your ability to evaluate the quality of baked products. Unless you can assess the strengths and shortcomings of baked products, whether prepared with or without a mix, you cannot expect to improve your baked goods.

Do you think there is some skill required to make a good baked product from a mix?

Despite the fact that mixes have been tailored to make them as foolproof as possible, people who understand baked products usually can make a better product from a mix than can someone with no knowledge of the science of baking.

*

To understand the total baking process, it is necessary to consider each basic ingredient used in batters and doughs. There are important choices to be made in selecting the best possible ingredient form for a specific product. Then it is necessary to trace the changes that take place when these ingredients are mixed together and when they are baked as a complex mixture. Batters and doughs are complicated systems wherein the quantity and form of one ingredient will influence the behavior and contribution of another ingredient. Throughout this chapter, keep in mind that the interrelationships of ingredients are significant factors in determining the quality of the finished product.

Batters and doughs are **simple/complex** *mixtures; the quantity and form of ingredients* **will/will not** *significantly influence the quality of the baked product.*

Complex, will.

*

A good recipe for baked products will carefully state the form of a product that will give the best result possible in that particular mixture. It also will clearly outline the basic procedure for mixing and will give concise directions for baking, including specification of type of baking container to use. If you add the unspecified ingredient, knowledge of baking, to a good recipe, you will produce a good product. Without that knowledge, even a good recipe often will not be sufficient to help you make a high-quality product.

The ingredient that cannot be added simply by using a good recipe is _____.

Knowledge.

*

Let us begin to increase your knowledge of baking by examining in some detail the basic ingredients in batters and doughs. Then you can begin to make good decisions regarding the purchase and selection of ingredients. You will better understand the techniques involved in mixing batters and doughs as you gain knowledge of the contributions each ingredient makes to the finished product.

Will the specific function an ingredient is to perform in a baked product influence the type of ingredient you may wish to use?

Yes. For example, consider whether you should choose brown sugar, granulated sugar, superfine granulated sugar, or powdered sugar in preparing a white cake batter.

*

FLOUR

For baked products to rise and hold together satisfactorily, it is necessary to have a structural network that is flexible and rather stretchy during baking, yet is reasonably strong after the product is removed from the oven. Some proteins have the ability to provide this network in batters and doughs. The ingredient that is most important in forming this framework is flour. Flour contains several proteins which are significant in baked products. Two of these proteins form a complex called gluten when flour is manipulated after water or some other liquid is added. Gluten is primarily responsible for the elastic character of batters and doughs during baking and for their semirigid structure after baking.

_____, the protein complex in flour, is developed when flour is mixed with _____.

*

Actually, flour is a complex mixture of protein, carbohydrates, and some fat. When the term "flour" is used alone, it is understood to mean the finely ground product made from wheat. However, flour can be used in a more general sense to refer to any finely ground product from a cereal. For cereal flours other than wheat, it is customary to use the name of the grain to designate the specific flour product. For example, you can buy rye flour or rice flour.

If a recipe simply called for flour, would you use any type of flour you happened to have in the kitchen or would you be careful to use only flour made from wheat?

Recipes that simply say flour assume that wheat flour will be used.

*

Flour is prepared from the cereal grain by a process known as milling. The first step in the manufacturing of flour is cleaning the grain to eliminate dirt, insects, and other contaminants. Steam is then used to temper the grain to prepare it for the first break. This process toughens the grain so that the outer bran layers and the germ of each grain may be separated readily from the desired endosperm portion. Breaking is the name of the process used to crush the grain. This is accomplished by feeding the grain into steel rollers. The broken wheat is then bolted or sifted to remove the fine flour particles from the coarser material known as middlings. This process is repeated to crush the coarse particles further. The best flours come from the first siftings and are called patent flours. These streams of flour are higher in protein and somewhat lower in mineral content, whereas the product resulting from the last bolting or sifting of the flour, referred to as clears, is higher in mineral and lower in protein content.

The patents are **high/low** quality flours that are produced in the first break; the clears are **high/low** quality flours that come off in the last breaks.

High, low.

310

If all the streams of flour produced during milling are mixed together and blended into one flour, the product is called a straight flour. If all the milled streams except the last 5 per cent are used, the flour is said to be a 95 per cent patent. When only the first 65 per cent of the flour streams are used, the flour is a 65 per cent patent and is sometimes called a fancy patent flour.

Flour produced by blending all streams of flour is called **fancy patent/straight** *flour.*

Straight.

*

Some of the bran layer may be included to give a coarse-textured flour. This flour, known as graham or whole wheat, is a medium-brown color because of the presence of the bran particles. The keeping quality of whole wheat flour is greatly improved if the germ or embryo of the grain is removed because this portion of the grain contains some fat. This fat will become rancid during long-term storage and result in an off flavor in the flour. Of course, this problem limits the shelf life of any flour containing a portion of the germ. Ideally, the germ will be completely removed not only from whole wheat flour but also from white flour. The storage life of all types of flour and of whole wheat in particular is also limited by the possibility of insect contamination. Such contamination is more of a problem in whole wheat flour than in white flour because of the inclusion of part of the bran in whole wheat flour.

Storage life of flour is limited by what two possible problems?

Rancidity developing in the fat (if germ is not completely removed) and insect infestation.

*

Sometimes you see a cereal product that is labeled "old-fashioned" or "stone ground." This means that the product is ground quite coarsely, as would be the case if the cereal had passed between stone rollers rather than the more closely meshed metal rollers used today. These cereal products often contain the germ. In hot climates the shelf life of stone ground flour can be extended by keeping it tightly sealed in refrigerated storage. This minimizes the likelihood of the development of rancidity in the fat and of insect infestation.

Coarsely ground wheat flour is often referred to as _____.

Stone ground (or old-fashioned).

The white flour that is so popular today contains none of the bran and very little of the germ. As a result of the removal of these portions of the grain (a process known as refining), the vitamin content of the flour is significantly reduced. To offset this nutritive loss, most refined flours are fortified or enriched with thiamine, riboflavin, niacin, and iron. They may also have calcium added. The addition of the B vitamins that are normal components of wheat is important, because this enrichment once again makes bread and other products made with enriched flour useful sources of these important vitamins.

*When the bran layer is removed during the refining of flour, the vitamin content is **raised/lowered**; refined flour is commonly enriched with _____, _____, _____, and _____ to offset this loss.*

Lowered, thiamine, riboflavin, niacin, iron.

*

At the end of the milling process, refined flours will be a creamy color and will possess poor baking quality. If the flour is allowed to age, it will lighten in color and will begin to develop better baking properties. Storage of flour to gain these improvements is expensive and time consuming. These changes can be accomplished by the use of bleaching and maturing agents. Commonly used bleaching agents to lighten the color of the flour are nitrogen peroxide and benzoyl peroxide. Chlorine, nitrosyl chloride, and chlorine dioxide are maturing agents that modify the baking quality of the flour.

True or false. Desirable changes in flour that used to be accomplished by aging are now accomplished by the use of chemical bleaching and maturing agents.

True.

*

Flours from cereal grains other than wheat are produced by a similar process. These flours are valued because they add considerable variety to baked products, especially to the realm of breads. They add variety in color, flavor, and texture. However, if you study recipes for breads containing these other types of flour, you will quickly note that they almost invariably require a combination of wheat flour and the other type of flour, such as rye flour. The inclusion of wheat flour is necessary to give the desired structure to baked products, because these other types of flour lack the strong gluten complex of wheat flour which is necessary to bind and hold baked products together. Some true southern cornbread recipes contain only cornmeal (no wheat flour is used); these products typically have a very heavy, tight structure and they are very crumbly and difficult to spread with butter. You will note similar characteristics if you attempt to make cookies and other baked products with rice flour, oat flour, or other flours unless you also include some wheat flour.

It is wise to add some wheat flour to cornbread, rye bread, and other types of bread made from flours other than wheat, because these flours lack the _____ found in wheat flour.

Gluten.

*

The characteristics of gluten and the amount of gluten in wheat flours vary with the type of wheat used to make the flour and with the locale in which the wheat was grown. These differences in gluten content can be used to advantage to produce different types of wheat flour with specific baking characteristics. Flour made from hard wheat (also known as spring wheat) has a relatively high gluten content and the gluten is a strong gluten. Hard wheat flour with the strongest gluten is marketed as bread flour. The strongest gluten for home use is in all-purpose flour, but this is slightly weaker than bread flour. In products such as yeast breads and quick breads, the strong gluten in all-purpose flour withstands the pressure generated by the carbon dioxide released from the fermenting yeast and the bread has a sound structure.

*Flour from hard wheat is **high/low** in gluten content.*

High.

*

For more tender and delicate baked products such as cakes, it is desirable to use a flour with a weaker gluten. Flour from soft wheat contains less gluten and the gluten is weaker than that found in hard wheat. This weaker gluten promotes the fine yet tender structure characteristic of high-quality cakes. The protein content of cake flour (flour made from soft wheat) is approximately 7.5 per cent, whereas the protein content of all-purpose flour is approximately 10.5 per cent. You can see that there is a significant difference in the amount of protein in these two types of flour that are available for home use. The differences in

313

the behavior of the two gluten complexes are more subtle to observe but are nevertheless very significant.

Flour from hard wheat for home use is marketed as **cake/ all-purpose** *flour, and that from soft wheat is called* **cake/all-purpose** *flour.*

All-purpose, cake.

*

Commercial bakers have considerable choice in wheat flours to obtain the flour with just the desired characteristics. They can buy bread flour containing as high as 16 per cent protein to meet the specific demands of large-scale yeast bread production. For baking pastries commercially, pastry flour is the usual choice. This flour is lower in protein than all-purpose flour and provides a more tender gluten for the pastry.

True or false. The flour available to commercial bakers that is highest in protein is bread flour.

True.

*

When baking, select the type of flour needed to produce the item. For instance, yeast breads require a strong gluten, but you may wish to add some pastry flour (if available) to a yeast dough for sweet breakfast rolls or raised dough-nuts to promote the desired tender, delicate structure. Cakes are customarily made with cake flour to create the characteristic fine texture and tender structure, but pound cake is sometimes made with some all-purpose flour to support the structure more adequately. Cream puffs and pop-overs are examples of products that undergo extreme structural stress during baking. A strong gluten is essential to withstand the pressure and maintain an intact finished product. When a pastry chef makes a puff paste, he uses some bread flour to provide the structural properties needed to permit the dough to puff up without rupturing.

For baked products that undergo extreme increases in volume during baking, it is necessary to select a flour **high/low** *in gluten.*

High.

*

In commercial bakeries, it is possible to use the flour specifically designed for the products being prepared. Thus, a bakery will have available bread flour for yeast breads and quick breads, pastry flour for pastries and cookies, and cake flour for cakes. In the home, it is practical to keep two types of flour: all-purpose for all types of bread, pastry, and cookies, and cake flour for cakes.

Do you think that a package of yeast roll mix or a package of cake mix would contain all-purpose flour as one ingredient?

No, it would not. The manufacturers of mixes have available to them the specific type of flour best suited to their product.

*

In baked products, it is necessary to understand the factors that influence gluten development during mixing and use these factors to your advantage. Of course, it is important to start by selecting the appropriate type of flour for the product you are making. It is also essential that you measure all ingredients carefully and correctly so that you are working with the best possible ratio of ingredients. Gluten development is significantly influenced by the ratio of flour to liquid, to sugar, and to shortening. In commercial production, flour is weighed. In the home, flour should be sifted once before being lightly spooned into a Mary Ann measuring cup and gently leveled. There is more chance for variation in the quantity of flour by measuring than there is by weighing, but you can achieve excellent results by measuring, if you are careful.

The ratio of flour to ____, to ____, and to ____ is important in the development of gluten during mixing.

Liquid, sugar, shortening.

*

Gluten begins to develop when flour and liquid are manipulated or mixed together. If this is an extremely dry mixture, the gluten will not develop as quickly as it does when slightly more liquid is added to make a sticky mixture. A very fluid mixture does not develop gluten quickly because the gluten strands tend to slide right past each other as they are mixed. But in a sticky mixture, the gluten strands tend to cling to each other and the gluten develops quickly into a strong, cohesive structure. In bread, it is desirable to develop a rather strong gluten network, but in pastry and cakes the gluten should be developed just enough to give cohesiveness and some flexibility to the mixture during baking. Extensive development of gluten results in tough pies and cakes.

*Gluten develops quickly in a   **very fluid/sticky**   batter.*

Sticky.

315

The total amount of mixing is directly related to the extent of gluten development. In any baked product, some gluten development is necessary so that there will be a structural network that holds the food together. If gluten development is not adequate, the food will be very crumbly and will fall apart too readily. As you would expect, the stronger and more cohesive the gluten network is, the less tender will be the final product. The amount of mixing that is appropriate varies with the type of product being prepared. Quick breads become tough very readily and are usually mixed only enough to blend the ingredients completely. Yeast breads need a relatively strong network because of the pressure exerted by the gas produced by the yeast during fermentation. Yeast breads, therefore, are kneaded vigorously to develop the necessary tenacity and strength in the gluten network. Cakes are mixed more than quick breads because the large quantities of shortening and sugar delay gluten development and necessitate the additional mixing. However, even in rich cake mixtures, the amount of manipulation is much less than that required for high-quality yeast breads.

True or false. Gluten development is undesirable because it causes baked products to be tough.

False. Too much gluten development is undesirable for this reason, but adequate gluten development is essential if the product is to have a structure strong enough to hold together.

*

Gluten develops more readily in a warm mixture than in a cold one. This is the reason that pastry often is made with ice water or chilled fat. Muffin batters also may be kept rather cool to avoid overdevelopment of the gluten. The desired development of structure in yeast bread doughs is encouraged by the addition of the warm milk.

Chilling a dough _____ gluten development.

Slows.

*

SHORTENINGS

The primary reason for using shortenings in baked products is to promote tenderness of the structure. Either fats or oils tend to coat the strands of gluten as they develop and cause the strands to slide across each other rather than cling together. This slippery quality of shortenings slows gluten development, thus increasing the tenderness of the product.

Shortenings **speed/retard** *gluten development in batters and doughs.*

Retard.

*

Some fats have a greater tenderizing effect than others. Free fatty acids in shortenings contribute to the tenderizing ability of a shortening. The ability to tenderize a product also is influenced by the softness or spreadability of the fat. A very hard fat will not coat the gluten easily due to its resistance to spreading. A softer, more plastic shortening spreads readily during mixing to coat the gluten effectively. An oil is particularly effective because of its flow qualities. Sometimes, batters or doughs become too warm during mixing, and the shortening may become so soft that it begins to separate from the mixtures. Less shortening or tenderizing results when this happens.

True or false. A hard shortening is less effective than a plastic shortening for tenderizing a cake.

True.

*

Shortenings in bakery goods also give what a baker calls "slip." Fats or oils give a slip or pliability during mixing because of their lubricating effect. This quality is also important as the product stretches during baking.

Will an oil give more slip to a batter than a hard fat?

Yes.

*

The other contributions that shortenings make to batters and doughs depend upon the particular product used. Butter, margarine, salad oils, lard, and various hydrogenated vegetable shortenings are all available for use in baking. Butter, margarine, and vegetable shortenings that are colored yellow all contribute to the color of the finished product. The color from these fats will appear to be just a delicate yellow. Remember that egg yolks also frequently are present to add to the yellow color of the finished product. The choice of fat also influences the flavor. Each of the above-mentioned fats has somewhat different flavor characteristics. Butter is preferred by many for use in cakes and breads because of its pleasing distinctive flavor. This

317

flavor preference has led to the recent introduction of an oil to which a synthesized butter flavor has been added. Some persons can detect the very subtle flavor of lard in a pic crust and select lard for that reason. Although the flavors of the fats vary, any of them will give a richness of flavor to the baked product.

Fats contribute sometimes to the _____ of a product; they always add to the _____ and promote tenderness.

Color, flavor.

*

LIQUIDS

Liquids add fluidity to a batter or dough. Some ingredients will be dissolved in the liquid and thus be readily dispersed uniformly throughout the mixture. This solvent action is of particular importance in relation to leavening. Baking powders or baking soda will not react to release carbon dioxide unless they are dissolved. You can readily observe the importance of dissolving baking powder by making a muffin batter that is mixed so little that dry flour is still conspicuous. Place a muffin, mixed to just that extent, in the muffin pan. Continue mixing the remaining batter until all ingredients are moistened. Place the remaining muffin batter in the pan, making each muffin the same size as the undermixed muffin. After these are baked, you will readily notice the smaller size of the undermixed muffin. This smaller volume is due, in part, to failure of some baking powder to react because it was not moistened by the liquid.

Baking powders **must/need not** *be dissolved if they are to react during baking.*

Must.

*

Liquids are necessary in any batter or dough to develop the gluten. As has been pointed out already, some gluten development is essential if the product is to hold together at all. The actual ratio of liquid to flour varies widely in different batters and doughs, but in all cases there is some liquid present. Pie crusts and many cooky doughs have a minimum amount of liquid in proportion to the flour. These doughs are quite stiff; popovers have a very high ratio of liquid to flour and are extremely fluid. In pastry,

318

the proportion of liquid to flour by measure may be as little as 1 to 10; in some quick breads it may be as high as 2.4 to 1. This increase in liquid dilutes the protein of the flour and helps to promote the tenderness of the product if the amount of mixing is properly controlled.

True or false. Liquid must be added to a batter or dough before gluten development will begin.

True.

*

During baking, liquid interacts with the flour. Starch in the flour is gelatinized in the latter part of the baking period as the mixture becomes hot enough for gelatinization to occur. Liquid is drawn into the starch granules, causing some swelling of the starch that is embedded in the protein matrix of the product. This binding of some of the liquid by the starch granules accounts for part of the change in the flow characteristics of a batter during the baking process.

True or false. A batter will be heated sufficiently during baking for starch to gelatinize (absorb water).

True.

*

Many different liquids may be included in baked products. Water and milk are commonly used, but some recipes include other types of liquid to provide variations in color or flavor. A burnt sugar syrup in a chiffon cake contributes both color and flavor, as does molasses in cookies and gingerbread. Orange juice may be used in a pie crust to give a very subtle color change as well as a flavor variation. Buttermilk and sour cream are two dairy products that add pleasing flavor notes in a recipe.

Few/many *different liquids are used in a variety of recipes for baked products. These liquids provide variations in _____ or _____.*

Many, color, flavor.

*

Eggs may be considered as liquid during the mixing of batters and doughs. A whole egg is about 75 per cent water, the white is 87 per cent water, and the yolk is almost 50 per cent water. The moisture in eggs functions during mixing in the same way as water, that is, as a solvent and an aid in gluten development. During baking, however, the protein in the egg binds the liquid.

319

What other functions do you think eggs would play in a batter or dough besides serving as a liquid?

1. The yolks contribute color.
2. Eggs contribute flavor.
3. Eggs foams, particularly egg white foams, add to the volume of a baked product.
4. The protein in eggs is a part of the protein framework of a baked product.
5. Eggs aid in forming emulsions in cake batters and other baked products. This promotes good volume and a fine texture.
6. Eggs are important sources of nutrients.

Had you realized that eggs could do so much?

*

SUGAR

The obvious role of sugar in baked products is to add sweetness to the flavor. Sucrose (regular granulated sugar) is the type of sugar most commonly used, but other sugars are available and may be used in baking. The sweetness of sugars varies from one sugar to another. The sweetest sugar commonly used is fructose (also called levulose). The next sweetest is sucrose, followed in order of decreasing sweetness by glucose (dextrose), maltose, and lactose (milk sugar).

Rank these sugars in order from least sweet to most sweet: maltose, sucrose, fructose, lactose, and glucose.

Lactose, maltose, glucose, sucrose, and fructose.

320

It may seem surprising that sugar contributes to the color of baked products. Of course, you would expect brown sugar to add to the color, but do you think granulated sugar does? You may have melted some sugar in a skillet and observed that it gradually began to brown or to caramelize. This same transition takes place on the crust of a baked product containing sugar. The sugar on the surface becomes very hot and gradually begins to change color, with the result that the baked crust takes on a golden-brown hue.

Why does sugar only caramelize on the crust of a baked product?

The interior of the product does not become hot enough to caramelize the sugar.

                    *

Sugar has the possibility of increasing volume by working in two different ways. First, during creaming of sugar and fat as the preliminary step in preparing a cake batter, the sugar crystals aid in trapping air in the fat to form a heavy foam. This trapped air contributes in a limited way to the total volume of the finished product. Sugar also contributes to the volume by raising the final coagulation temperature of the gluten network so that the product has more time to expand before the structure is set.

What do you think would happen if you put too much sugar in a cake batter?

The cake probably would rise well for awhile and then fall as the gluten strands stretch so far that they break.

                    *

Tenderness is increased in batters or doughs when sugar is added. The presence of sugar appears to delay the development of gluten. This factor promotes tenderness. In addition, the larger volume due to the addition of sugar means that the cell walls are stretched thinner and thus are more delicate and tender.

Is it possible to mix cakes containing sugar to the point where the gluten will be overdeveloped?

Yes, but it will require more stirring to over-develop the gluten in a product containing sugar than in a similar mixture that does not contain sugar.

                    *

LEAVENING AGENTS

Leavening agents are any ingredients or components in a batter or dough that causes expansion in the mixture.

With this broad definition, you will recognize that there are two leavening agents available naturally in any batter or dough. These two natural leavening agents are steam and air. Yeast is a leavening agent that has been used for many, many years to lighten baked products. A relatively recent newcomer is baking powder, a chemical leavening agent that has been available for just a little over a century.

_____ and _____ are natural leavening agents.   *Steam, air.*

*

Leavening agents are effective because gases expand dramatically when they are heated. This expansion causes a stretching of the gluten framework of dough and batter mixtures. Baking must be continued until the protein in the framework is denatured to form a relatively strong structure that can hold up the entire baked product without the pressure from the gas trapped in each air cell. Rigidity of the structure is essential when the item is removed from the oven because the cooler outside air immediately begins to cool the product. As the gases within the item cool, they contract and no longer exert pressure against the cell walls. The cell walls then contract and the product shrinks in volume if the protein in the cell walls is not strong enough to hold up the entire mass.

*When gases are heated, they **contract/expand;** when they are cooled, they **contract/expand.***   *Expand, contract.*

*

Air is trapped in any batter or dough during the mixing process. In products such as pastry, there is very little air trapped and little leavening occurs during baking. In an angel cake which, of course, includes an egg white foam, there is a great deal of air held within the foam structure. This air expands significantly and the finished product has a very noticeable increase in volume. All batters and doughs also contain some water; water is converted to steam during baking, and the steam exerts considerable force to expand the cell walls. The combination of steam and air is operating in all items when they are baking. Their effectiveness in increasing the volume is controlled by the amount of air and steam potentially available in the product.

If air and steam are effective leavening agents, why are other leaveners commonly added to batters and doughs?

By using chemicals and yeast as leavening agents, it is possible to achieve great increases in volume and to make a wide variety of baked products. It is no longer necessary to rely on using an egg white foam to encompass enough air to give a light product.

*

Yeast is a one-celled fungus. The particular yeast used in baking is *Saccharomyces cerevisiae*. This simple chlorophyll-free plant feeds on sugars in the batter or dough to produce carbon dioxide. As yeast multiply in a dough fermenting at room temperature, more and more carbon dioxide is produced and the dough starts to expand because of the pressure of the gas. Yeast fermentation is temperature-dependent. If yeast plants are too cold, they will not multiply at all. On the other hand, yeast are living things and they will die if they are heated to too high a temperature. When the dough is held between 85° and 95°F, the multiplication process is very rapid and much carbon dioxide is produced. The production of carbon dioxide by the yeast continues throughout the fermentation period and the baking period until the yeast are killed by the oven heat. This accelerated production of carbon dioxide in the early phase of baking explains the phenomenon known as oven spring. Oven spring is the abrupt increase in volume of the baked product just prior to the death of the yeast plants.

The increase in volume of yeast breads is due to **an increase in the volume of yeast plants because of reproduction/production of carbon dioxide by the yeast.**

Production of carbon dioxide by the yeast.

*

You will need to be careful of the temperature of any ingredient that comes in contact with yeast during the prepa-

323

ration of yeast-leavened doughs. If the temperature is too high at any time during production, the yeast will be killed and no more carbon dioxide will be produced. Yeast is easily distributed in a dough if it is first softened in lukewarm water. Be sure that this water is not too hot. It should be just about body temperature. (You may elect to mix dry granulated yeast directly with the other dry ingredients now as a result of improved yeast.) The second possible problem point in preparation is when the yeast are added to the mixture containing the scalded milk. It is imperative that you do not add the yeast until the milk has cooled to lukewarm. If these precautions are observed and if adequate opportunity is given for the dough to stand while the yeast ferment, the finished product should have a good volume.

Any ingredient or mixture that comes in contact with yeast should not be hotter than _____.   *Lukewarm.*

                                                           *

Chemical leavening agents are popular because it is not necessary to wait a long time for the gas to be generated before the product can be baked. Chemical leavening agents may produce almost all of their gas at room temperature, or they may be formulated to give only a small fraction of the gas at room temperature and to give off most of the carbon dioxide at baking temperatures. Long ago it was noticed that a batter or dough containing an acidic product such as sour milk and an alkaline material such as baking soda would increase in volume if it were baked soon after the mixing was completed. Many recipes in use today still require an acid ingredient. This may be cream of tartar or such liquids as sour cream or a fruit juice. To neutralize this (react with the acid), soda is added. Actually, this is comparable to adding a baking powder in two parts. As soon as the soda is dissolved, it reacts with the acid ingredients, and carbon dioxide is released into the mixture. If the product is baked immediately in a preheated oven, most of the gas will still be trapped in the mixture and considerable expansion will take place during baking. However, if there is a delay, the volume in the finished product will be poor because much of the carbon dioxide will escape before the mixture is placed in the oven.

324

True or false. Soda plus an acid ingredient makes an effective leavening combination that will regularly give you a predictable volume in the finished product.

False. Gas is quickly lost when this combination is used for leavening.

*

Use of baking powders is very similar to the use of an acid and soda in baked products. Soda is always used as the alkaline ingredient in baking powders. All baking powders contain an acid salt to react with the soda to release gas. Use of baking powders sounds as though it would be as unpredictable in its gas production as is the simple addition of soda and acid separately. This, however, is not true. By choosing the type of baking powder best tailored to your baking abilities, you can get reasonably reproducible results each time you bake. Baking powders have starch added to them to absorb moisture that may creep into the can and this prevents loss of gas in the can before the powder ever gets into a product.

Baking powders always contain _____, _____, and _____ _____.

Soda, starch, acid salt.

*

Tartrate salts may be used as the acid ingredient in baking powder. Cream of tartar is a familiar example of a tartrate salt that can be used in a baking powder formula. Baking powders containing tartrates give off much of their carbon dioxide at room temperature and are, therefore, recommended only for fast workers. If you work rapidly and also are careful to preheat the oven, you can obtain excellent results with a tartrate baking powder.

Cream of tartar can be used as the **acid salt/soda** *in a baking powder.*

Acid salt.

*

Phosphate salts also were tried in baking powder formulas as the acid. Although these react somewhat more slowly at room temperature than tartrate baking powders, they still release almost all of their carbon dioxide at room temperature. This may result in a product with poor volume. It seemed desirable to develop an acid salt that would retain most of its potential carbon dioxide until the product was in the oven. Sodium aluminum sulfate (also called

SAS) proved to be just such an acid salt, but it did have two distinct problems. Reaction of the powder was so delayed that the structure of the baking product might set before sufficient gas had been generated to produce a good volume. Secondly, the residue from the reaction had a strong bitter flavor.

True or false. Phosphate salts react too quickly in a baking powder and SAS reacts too slowly.

True.

*

To overcome the shortcomings of the phosphate baking powder and the sodium aluminum sulfate baking powder, it seemed wise to use a combination of these two acid salts in one powder. This combination was termed sulfate-phosphate or double-acting baking powder. The phosphate portion produced some gas relatively quickly and the sulfate salt produced the desired gas in the oven to give reliable leavening from this baking powder even when the worker was slow. There was a slight residue flavor, but it did not seem to be too objectionable for most people's tastes. The result was the widespread marketing and acceptance of today's numerous double-acting baking powders.

What type of baking powder do you think would be most suitable to use in a class where food preparation is taught?

Double-acting (sulfate-phosphate) baking powders usually give the most reliable results in a class of beginning students.

*

Baking powders and other leavening agents are important in batters and doughs because they contribute significantly to the volume, texture, and tenderness of the finished product. With the correct amount of leavening, you will be able to prepare a product of optimum volume, yet one with a good texture rather than one that is too porous or too compact. The stretching of the cell walls as a result of leavening action contributes significantly to the tenderness.

What do you think would happen to biscuits that were accidentally made with soda in place of baking powder?

The volume would be poor because biscuits do not contain enough acidic ingredients to react completely with the soda. You also would find that the biscuits would be distinctly yellow because of the alkaline dough. The flavor would be very objectionable.

The amount of gas given off by baking powders is standardized so that any type of baking powder can be used in a recipe. Baking powders for home use must give off a minimum of 12 per cent carbon dioxide. Manufacturers usually formulate the mixture to give off 14 per cent. This allows some small loss to take place before use and still results in a reaction exceeding the required minimum. Baking powders for commercial or institutional use are formulated to give off 17 per cent carbon dioxide. Starch is used as the filler to standardize the different types of baking powders.

Why are baking powders for institutional use designed to give off more gas than those marketed for home use?

The large batches made in institutions or commercial bakeries have to be handled vigorously and they may have to stand for a long period of time before they are baked. It also takes longer to heat the large quantities sufficiently to set the protein structure during baking. For these reasons, there is considerable opportunity for loss of gas during the preparation of baked products in quantity. Consequently, it is necessary to have more gas available originally.

*

In some very old recipes or in some institutional recipes, you may find that baking ammonia is the leavening agent. This is an alkaline material and is used in a manner similar to soda. Baking ammonia may be used to make thin items (or thin-shelled ones) such as cookies or cream puffs. It can only be used successfully in thin items. Thicker items trap some of the ammonia, and this flavor permeates the product. When baking ammonia is heated, its vapors expand and leaven the product.

Baking ammonia is an **acid/alkaline** *ingredient.*

Alkaline.

*

In some instances, you might wish to substitute baking ammonia for soda. This can be done by using three-fourths teaspoon of baking ammonia for every teaspoon of soda in the recipe. Baking ammonia leaves no aftertaste in cookies and thin-shelled products. It is of value in a cooky dough because it helps the dough to spread. It is possible to drop cookies containing baking ammonia on a sheet and have them spread out as if they were rolled and cut. The labor saved is of obvious significance. You will also discover that baking ammonia adds to the volume of cream puffs and eclairs.

If you were making rolled chocolate cookies and the recipe called for half a teaspoon of soda, could you use three-eighths teaspoon of baking ammonia?

> *Yes. The cookies would probably be better and would have more spread than if the soda were used.*

*

The use of soda is also recommended sometimes to give an alkaline reaction to the mixture. For instance, soda is used in devil's food cakes to give a reddish-colored crumb. In gingerbread, you may wish to use dark molasses to gain the desired dark color, but you may object to the strong flavor. By adding soda to the gingerbread and using a lighter molasses, you will get a milder flavor and the color of the molasses will be darkened by the soda.

Soda is an **acidic/alkaline** *ingredient.*

> *Alkaline.*

*

CHANGES DURING BAKING

Anyone who has ever watched a batter or dough during its transformation in the oven is aware that some impressive changes take place during the baking process. Batters change from relatively fluid mixtures into a fixed structure. Doughs lose their flexibility and become firm. These changes are caused in part by loss of free water in the mixture. Some moisture evaporates in the oven during the baking period. Part of the moisture will be bound by the starch granules when the starch gelatinizes. This bound water will not flow.

Do you think that loss of free water explains entirely the change that takes place from a fluid batter to a rigid baked product?

> *As you have probably concluded, these changes do not entirely explain the change to the finished product.*

328

Protein changes as a result of heating. The coagulation of protein causes a change from the native state to the denatured rigid state. The gluten in flour is coagulated during baking and the structure is no longer so flexible. The egg and milk proteins also change to the denatured state, accounting for part of the strength of the structure.

Three common sources of protein in baked products are _____, _____, *and* _____.

Flour (gluten), eggs, milk.

*

The increase in volume that occurs during baking is always exciting to watch. If chemical leavening agents have been used, a considerable amount of carbon dioxide is generated during baking as the heat in the oven supplies the necessary energy for the chemical reaction to take place. If yeast has been used, the oven heat will accelerate yeast growth briefly to cause oven spring. The carbon dioxide that is present in any batter or dough will begin to expand as the oven heats it, and you can readily observe the expansion of the product. Air in the item also accounts for some expansion. As was mentioned above, some water is evaporated during baking. As the interior of the batter becomes heated, some water will be converted to steam and this also increases the volume. Leavening by steam is particularly significant in popovers, which combine a high oven temperature with a very thin fluid batter.

_____, _____, *or* _____ _____ *will cause an increase in volume during baking because of their expansion when heated.*

Air, steam, carbon dioxide.

*

Flavor changes develop in the baking process. The flavors blend together into a more homogeneous flavor. Some very volatile substances evaporate during baking and are lost from the total flavor impression. For example, alcohol formed during yeast fermentation vaporizes during baking of bread. Sugar caramelizes a bit on the crust to give a slightly different flavor to that region of the finished product.

True or false. The flavor of a baked product is unchanged from that of the batter or dough.

False. Some volatile substances are lost during baking and other substances may be formed.

329

The crust of baked products changes in color during the baking process. The pleasing golden-brown color is due to changes taking place in the carbohydrates and proteins during baking. Sugar caramelizes a bit to produce part of the color change. Starch contributes some browning as it is dextrinized to a limited extent on the crust. The heat also encourages a browning reaction between the protein and the sugar. All these factors combine to give a pleasing color to the crust when baking is completed.

Browning of the crust is due to **a single/several** *reaction(s).*

Several.

*

REVIEW

1. If the cost of a homemaker's or baker's time is considered, mixes may frequently be *more/less* expensive than making the product from the beginning.

    Less.

2. The main purpose of flour in a baked product is to furnish _____ to the product.

    Structure (or gluten).

3. The main protein in flour is: (a) gluten, (b) albumen, (c) vitelline, or (d) gliadin.

    Gluten.

4. Hard wheat flour has *more/less* protein in it than does pastry flour, and pastry flour has *more/less* protein in it than does cake flour; hard wheat flour has *more/less* starch in it than does pastry flour, and pastry flour has *more/less* starch in it than does cake flour.

    More, more, less, less.

5. The germ (embryo) in cereals *promotes/retards* onset of off flavors during storage.

    Promotes.

6. *True or false.* Bleaching of flour is done to improve its baking quality.

    True.

7. Gluten develops more quickly in a *warm/cold* dough.

    Warm.

8. Gluten develops more quickly in a *stiff dough/sticky batter.*

    Sticky batter.

9. Increasing sugar in a recipe *speeds/retards* gluten formation.

    Retards.

10. Increasing fat in a recipe *speeds/retards* gluten formation.

    Retards.

11. Check the reasons why liquids are helpful in batters and doughs.
    A. *Act as solvent for ingredients.*
    B. *Provide moisture for gelatinization of starch.*
    C. *Give moistness to product.*
    D. *Aid in gluten development.*

    Check them all.

12. *True or false.* A liquid can be used in baking to give a product a desirable pH. — *True.*

13. *True or false.* Sucrose is the sweetest sugar. — *False. Fructose is.*

14. *True or false.* Sucrose contributes to the color of a baked product. — *True. It adds a golden-brown color to the crust.*

15. Yeast are one-celled *plants/animals.* — *Plants.*

16. Sulfate-phosphate baking powder is commonly called a _____ _____ baking powder. — *Double-acting.*

17. Baking powder leavens by producing _____ _____. — *Carbon dioxide.*

18. Yeast leaven by producing _____ _____. — *Carbon dioxide.*

12

The Making of Bread

Most breads are leavened primarily in two ways: either with yeast or with a chemical leavening agent. Yeast breads require approximately three hours from the beginning of mixing until baking is completed. A rather small portion of this time is consumed in mixing and shaping the dough. The majority of the time is needed for the yeast to multiply and produce carbon dioxide to leaven the dough. To avoid this long preparation time and still have leavened products, chemical leavening agents were developed over 100 years ago. When baking powder is used instead of yeast in a bread, the product is called a quick bread. Much less time is needed to prepare quick breads because baking powders produce their carbon dioxide without a fermentation period.

Although yeast breads and quick breads are both leavened by air and steam, their greatest leavening is caused by ____ ____, which is produced by ____ in yeast breads and by ____ ____ in quick breads.

Carbon dioxide, yeast, baking powder (or chemical leavening agent).

*

YEAST BREADS

Yeast doughs may be classified as rich doughs or lean ones. Rich doughs usually are used for sweet breakfast rolls, yeast coffee cakes, Danish pastries, and sometimes dinner rolls. These doughs contain more fat, more eggs, and sometimes more sugar than do lean doughs. A rich dough may have the shortening and sugar creamed, the eggs added, and creaming again occur, ending up with a product that is almost as rich as a rich cake. Some rich doughs are made simply by combining the sugar, salt, melted shortening, and scalded milk and adding the beaten eggs. When this mixture has cooled to body temperature, the softened yeast is added and flour stirred in to make a dough stiff enough to be kneaded. A third method for making rich doughs is the rolled-in method. This procedure involves rolling pieces of hard butter, margarine, or fat into the layered dough to produce a flaky, layered, very rich product such as is desired in French *croissantes.*

Rich doughs are **yeast/quick** *breads containing extra _____, _____, and sometimes _____.*

Yeast, fat, eggs, sugar.

*

Bread doughs are usually lean doughs; this means that they have less fat, fewer eggs, and probably less sugar than rich doughs. In fact, some lean doughs contain only yeast, salt, water, and flour. French bread and other hard-crusted breads and rolls are made from such very lean mixtures. Ordinary breads and rolls are usually made with a moderate amount of shortening, eggs, and sugar. The liquid is milk, rather than the water used in the very lean French bread doughs.

Classify the following as rich or lean yeast dough products: coffee cake, cinnamon roll, Danish pastry, Italian bread, hard rolls, and regular bread.

The first three would be considered rich doughs, the last three lean doughs.

*

The liquids used to make bread include water, milk, potato water, and whey. The carbohydrate in potato water may be used by yeast as food. Milk and whey add richness and softness as well as nutrients to the bread. Dried milk solids can be used in a bread formula; water is the liquid used when dried milk solids are used.

True or false.   The liquid in bread does not furnish any nutrients to the bread.

False.

*

The temperature of the liquid when it is added can be adjusted to give a dough temperature suitable for good fermentation. Normally, if all ingredients are at room temperature, the addition of lukewarm (100°F) water gives a dough of 80° to 84°F. This is a safe temperature for the yeast and they will produce carbon dioxide at a satisfactory rate. Actually, the dough will be slightly warmer at the end of kneading because of the friction developed in such manipulation. If the bread is to ferment in an area where the temperature is extremely high, it is wise to cool the liquid with some ice before it is added to the bread dough. However, this precaution is unncessary in most kitchens.

True or false. Liquid is usually warm when added to a yeast dough, so the dough will be within the temperature range recommended for fermentation (80 to 95°F).

True, but if the kitchen is unusually warm, you may need to use cold liquid.

*

Bread and roll formulas are carefully developed to give a final product that is tender and flavorful and has a uniform, bread-like texture. The best final results are obtained when you are aware of the role of each ingredient and use each of them to advantage. Yeast, the one-celled plant used for leavening, can produce the necessary amount of gas in a practical length of time if you use the amount recommended in the recipe and if you are careful to add it only to ingredients warmed to a temperature between 80° and 95°F. Higher temperatures (110°F and above) kill the yeast; lower temperatures make an unnecessarily long fermentation period.

If you made a yeast dough and it failed to rise, what might be the cause?

You may have had the ingredients so warm that the yeast was killed when added, or you may have omitted the yeast.

*

Compressed yeast cakes and active dry yeast are the two forms in which yeast (*Saccharomyces cerevisiae*) is marketed for home baking. Either type works more satisfactorily in breads if it is softened to a fluid paste in some lukewarm water before being added to the dough. In this way the yeast can be distributed uniformly throughout the dough to give consistent leavening in all portions of the bread. The water used to soften the yeast is considered part of the total liquid in the formula. Recently, it has been shown that active dry yeast can be mixed with the dry ingredients rather than softening it in the liquid.

Why should compressed yeast and active dry yeast be placed in lukewarm water until they form a rather fluid paste?

This is done to enable you to distribute the yeast uniformly.

The proportion of liquid to flour in a dough influences tenderness, but the exact amounts needed vary with the type of flour available. An approximate ratio is three cups of flour to one cup of liquid. However, this proportion has to be adjusted to meet individual situations. For instance, if eggs are added to the dough, these will contribute approximately three tablespoons of liquid per egg; therefore the amount of milk or other liquid will need to be reduced accordingly. Add just enough flour to make the dough easy to manage during kneading. Too much liquid makes a sticky dough. Excess flour makes a tough bread or roll.

_____ *helps to make rolls and bread tender, but excess* _____ *makes them tough.*

Liquid, flour.

*

Salt is added to yeast doughs to improve the flavor and to regulate yeast growth. Salt and sugar act in opposition to each other in yeast doughs. Salt slows down yeast growth; sugar serves as food for the yeast and speeds up its production of carbon dioxide gas. Of course, sugar in excess of the nutrient needs of the yeast also contributes to the flavor and the browning of the crust that occurs during baking.

Salt regulates yeast growth by **speeding growth/retarding growth.** *Sugar has* **a similar effect/the opposite effect.**

Retarding growth, the opposite effect.

*

Fat is added to yeast doughs to increase tenderness and help soften the crust. If butter or margarine is the fat used, flavor and color are added. A lean dough usually contains approximately one tablespoon of fat per cup of flour. Added richness of flavor and increased tenderness in rich doughs are due to the addition of three to four times as much fat as is included in a lean dough.

Approximately how much butter or other fat would you use per cup of flour in a rich dough?

Three to four tablespoons.

*

Normally, about a fourth to a third of a cake of compressed yeast is used per cup of flour. Rich doughs require slightly more yeast than lean ones. Also, if you plan to either freeze

335

or refrigerate a yeast dough during some stage prior to baking, the quantity of yeast should be increased slightly. There may be times when you need to reduce the total preparation time. At such times, you will need to add more yeast and shorten the fermentation and proofing times. Slightly more sugar is needed when more yeast is added. If you double the yeast, fermentation is faster but the quality of the bread will not be as good because the texture will be less uniform and more porous.

If you add more yeast, **decrease/increase** *the quantity of sugar slightly.*

Increase the sugar to provide more food for the extra yeast plants.

*

Doughs can be retarded or delayed during processing by refrigerating or freezing them. Doughs may be retarded at any time during their preparation. They may be refrigerated after they have doubled in volume and have been punched down, they may be refrigerated before rising the first time, or they may be made up and panned before they are frozen. It is advisable to cover any dough that is to be retarded so it will not dry out at the surface. A retarded dough, after it is warmed again to room temperature, is handled the same as an ordinary dough and the remaining production steps are completed. Doughs retarded by refrigeration should be processed and baked within two or three days. If they are retarded by freezing, they can be held for a week or more.

Which of the following would you **not** *do to a retarded dough?*
A. *Increase the yeast.*
B. *Increase the salt.*
C. *Increase the sugar.*
D. *Cover well during the retarding period.*

B. You would not want to reduce yeast activity further by increasing salt.

*

If fluid milk is used in making bread, it is usually warmed before being added to the dough mixture. Scalding is not necessary to destroy enzymes because they were inactivated when the milk was pasteurized. The solid fat used for rolls can be added to the warm milk. The warm milk melts the butter so it can easily be mixed uniformly throughout the dough. The melting fat helps to cool the milk to lukewarm so that it can be added to the remaining ingredients.

What would happen to a yeast bread if the scalded milk were not cooled to lukewarm before the yeast were added?

The product would be a heavy, compact failure because hot milk will kill the yeast and no leavening will occur. This mistake is commonly made when the cook does not understand the importance of controlling the temperature of ingredients that come in contact with yeast.

*

The easiest way to make a yeast dough is to combine sugar, salt, melted fat, and lukewarm scalded milk. Then add the softened yeast, beaten eggs, and about one-third of the flour. This gives a batter that is thin enough to mix all ingredients thoroughly and develop the necessary gluten network. Sometimes this batter is fermented until it doubles in volume before the remaining flour is mixed in. This is called the sponge method. The more common method used today is called the straight dough method. This method starts the same as the sponge method, but after mixing to develop the gluten, the remaining flour is added before fermentation begins. Either of these methods can be modified to shorten production time. To do this, double the yeast and ferment and proof the dough at approximately 95°F. The dough made by this fast method is sometimes called a "no-time" dough.

Match the method in column B with the name of the method in column A.

A	B	
1. No-time	A. All the flour is mixed in before the bread is fermented	
2. Sponge	B. More yeast and higher temperatures are used	1. B.
3. Straight dough	C. Only part of the flour is mixed in before the first fermentation	2. C. 3. A.

Bread dough should be mixed until the flour is thoroughly incorporated with the other ingredients. Then take it from the bowl and knead it well on a lightly floured board. Avoid working an excessive amount of flour into the dough from the board. Kneading is done vigorously for about five minutes to finish developing the necessary gluten. If you knead well, the rough spots and cracks in the dough will gradually disappear and the dough will be a smooth round ball when the kneading is completed. To knead, lift the far edge of the dough with your fingers and fold it over to meet the edge of the dough nearest to you. Press the dough away from you firmly with the palms of both hands. Turn the dough a quarter of a turn. Lift and fold the far edge of the dough to the front edge, press with the palms, and again rotate the dough a quarter of a turn. Keep repeating this process until the dough is smooth on the surface and reveals small blisters under the surface when stretched. Work for a good rhythm when kneading.

*During kneading, the dough is **pounded and turned/folded and pressed** with the hands.*

Folded and pressed.

*

When kneading is finished, the dough should feel more cohesive than it did when you began, but it should still be rather pliable. The ball of dough should soften a bit on the board rather than being perfectly rigid. This desired softness of the dough can only be maintained if you avoid working extra flour from the board into the dough. Use just the minimum of flour needed to keep the dough from sticking. Extra flour in the dough causes tough breads.

True or false. Dough should be firm enough to hold a rigid ball rather than softening slightly when it is resting on the board.

False. The finished bread will be more tender if the dough softens slightly.

*

When the dough has been kneaded, it is placed in a bowl and the surface is lightly but thoroughly greased with salad oil to prevent the crust from drying during fermentation. Solid fats can be used to grease the surface of the dough, but avoid excessive greasing since this may show later as streaks in the bread. The dough is covered with aluminum foil, a damp cloth, or other cover and set in a warm place to ferment. If no suitable place is warm enough, the covered bowl can be put into a pan of lukewarm water to give a good fermenting temperature.

Putting a fat or oil on the surface of a dough to be fermented prevents _____ of the surface; warm air surrounding the dough during fermentation will **promote/retard** *fermentation.*

Drying, promote.

*

Several changes take place in the dough during fermentation. Yeast multiply and produce alcohol, water, and the carbon dioxide needed for leavening. The gluten becomes a bit more extensible as a result of limited chemical changes causd by acid and enzymes present in the dough. When a dough is well fermented, it will feel soft, pliable, and smooth. A dough that has not yet doubled in volume is not completely fermented and it will feel tighter and less pliable.

During the fermentation period, the dough **tightens/ relaxes.**

Relaxes.

*

Fermentation should be allowed to continue just until the dough doubles in volume. At that time, the dough is ready to be made-up or shaped. If the bread dough is not fermented enough, that is, it is not allowed to double its volume, the cells are too compact for good palatability. The texture is apt to be tough and soggy.

Underfermenting causes:
A.   A sour flavor.
B.   Toughness.
C.   A white crumb.
D.   A good volume.

B. Toughness.

*

Too much fermentation causes the gluten strands to stretch so far and become so weak that the final product has poor volume and a coarse thick cell. The crumb will seem dark and soggy, the bread will have a sour odor and flavor, and the crust will brown readily.

339

Overfermenting gives:

A. *A coarse, heavy cell.*

B. *A sour flavor.*

C. *A dark crust.*

D. *An acid flavor.*

All of these happen when too much fermentation occurs.

*

If you are making a rich dough product or a bread dough using graham, rye, or other flours that do not have the strong wheat gluten, do not allow the dough to ferment until it doubles in volume. Instead, allow it to go to only about three-fourths of its full fermentation. This is said to be "taking the dough young." Then punch the dough and proceed to shape it for proofing and baking.

"Taking a dough young" means to punch it and begin to shape it **after full fermentation is complete/when fermentation is about three-fourths complete.**

When fermentation is about three-fourths complete.

*

The actual fermentation process continues until the yeast is killed by the oven heat during baking. If you expect to take a considerable amount of time making up the dough into the desired shapes, start shaping the dough before the dough has quite doubled in volume. This precaution keeps the dough from fermenting too much before it is proofed and baked.

True or false. If you are a slow worker or are planning to shape your dough into complex shapes, it is wise to begin shaping it before fermentation has caused the dough to quite double in volume.

True.

*

You can tell when a dough has fermented sufficiently by inserting your fingers into the dough. If it puckers up and withdraws, the dough is ready to be punched and shaped. However, if the fingers leave an impression but the dough still has some resistance and does not withdraw, allow the fermentation to continue. Usually, a bread dough that has risen to twice its original bulk is said to be adequately fermented. One that has not quite reached that is said to still be "young."

A dough that is fermented adequately will **withdraw and pucker up**/still have some resistance in it *when pressed with your fingers.*

Withdraw and pucker up.

*

To punch down a dough, gently press with your fist and then fold the sides of the dough into the center of the dough. Punching down the dough helps to redistribute the yeast and carbon dioxide to give more uniform texture in the finished product. The dough will tighten up a bit when punched because of loss of gas from it. It will be easier to make up the dough if you allow the dough to rest upside down on a lightly floured board for about 10 minutes until the dough relaxes.

After the dough is punched down, it is allowed to rest 10 minutes so that the dough has time to **tighten up/relax.**

Relax.

*

Make-up methods should be simple. Pan rolls are simply rounds of dough put into a pan. Parkerhouse rolls are made by folding these rounds of dough in half. You can roll dough into pencil-sized strips and leave them straight for bread sticks or tie them into many different shapes. Cinnamon rolls are made by rolling the dough a fourth to a third inch thick, brushing heavily with melted butter or margarine, spreading with a mixture of sugar and cinnamon, and then rolling into a roll. Make individual rolls by cutting this large roll at one-inch intervals and then turning the small rolls on their sides in the baking pan to reveal the cinnamon pinwheel. Numerous other shapes can be done as easily as these.

After fermentation, punching, and then a short rest, the next step is the _____ or shaping of the rolls.

Make-up. Good make-up takes practice. You might like to try the shapes described above. What other shapes can you make? We have suggested only a few.

*

The rolls are now ready for proofing—the final step before baking. The rolls are proofed on the pans used for baking so that they can be inserted directly into a preheated oven when they have doubled in volume. Since the rolls are not covered during proofing, it is helpful if they can be proofed in a warm, moist place. Proofing must continue until the dough doubles in volume if the baked roll is to be light and tender.

341

How would you describe a roll that was not proofed long enough before baking?

*

Rolls are baked at 425°F for about 15 minutes, but bread is usually baked at 400°F for 30 or more minutes to allow time for the heat to penetrate and bake the center of the loaf. Rich doughs, sweet rolls, or dough containing raisins should be baked at 350°F to prevent them from burning before they are baked through.

Why is it necessary to bake a loaf of bread at a slightly lower temperature than is recommended for plain rolls?

*

During the first part of baking, there will be a final increase in volume; the heat accelerates the production of carbon dioxide by the yeast and also causes a large increase in the volume occupied by the carbon dioxide as this gas is heated. These factors combine to cause "oven spring." Oven spring is the increase in volume that occurs as the result of heating the dough in the oven during baking. Fairly quickly the yeast will be killed by the oven heat (when the dough reaches 140°F) and no additional gas will be produced in the dough. However, the carbon dioxide already generated in the dough will expand and hold up the structure of the rolls for some time. It is necessary to heat the dough quickly enough to coagulate the egg and gluten protein network in the cell walls while the carbon dioxide is still holding the structure in an extended position. Then, the rolls will be a good volume and light and tender. When the heat coagulates the protein structure and gelatinizes the starch, the product becomes rigid enough to hold its shape and will not collapse when removed from the oven.

The increase in the volume of a loaf of bread during baking is known as _____.

*

Bread is done when it is a pleasing brown color and when it sounds hollow when tapped on top. The hollow sound indicates that there are no doughy unbaked places remaining in the loaf. If you have a keen sense of smell, you may also notice that there is little or no trace of alcohol in the bread vapors because most of the alcohol produced during fermentation evaporates during baking.

Check the true statements:
A. *Yeast action speeds up to 140°F and then stops as the yeast is killed.*
B. *Plain rolls are baked at a higher temperature than is a loaf of bread.*
C. *Rolls take less time to bake than loaf breads.*
D. *Baked rolls weigh less than they weighed before baking.*

They are all true statements.

*

After baking, yeast breads and rolls should be lifted from the pans onto wire racks to cool. This permits steam to escape and gives a drier, lighter crumb. Otherwise, the bottom portion seems soggy and heavy because moisture condenses in the pan rather than escaping as it does when air circulates around all parts of the bread. When the bread has cooled to room temperature, it can be placed in polyethylene bags and stored for later use.

Couldn't we take breads from the oven and serve them hot?

Certainly, you can serve bread hot from the oven. The above procedure is outlined because people rarely eat a whole recipe of bread or rolls straight from the oven.

*

Now, let us review the essential steps in the making of yeast breads. First, the ingredients must be measured and mixed. If you have a warm place for the dough to rise, a small batch of dough should double in volume in one to one and one-half hours. Large batches take longer. After this, the bread is punched down and allowed to rest for 10 minutes. After resting, it is made-up into the desired shape. Now proofing, the last fermentation period, occurs. When the shaped dough doubles in volume, it is baked and then cooled on racks and stored.

Place in their proper order the following steps in bread making: measuring, proofing, fermentation, resting, mixing, make-up, punching, cooling, baking, and storing.

Measuring, mixing, fermentation, punching, resting, make-up, proofing, baking, cooling, and storing.

*

Homemade breads stale and mold rapidly. Commercial breads contain chemical additives that retard staling and molding of bread. Bread containing such compounds lacks the firm texture of homemade bread and readily becomes soggy when used for sandwiches. When these commercial breads are toasted, the center remains moist and a bit gummy while the outside is crisp.

Can you purchase bread that does not have these softeners added to them?

Yes, but most bread from large commercial bakeries does contain mold inhibitors. Smaller bakeries are likely to bake bread without such additives.

*

Freezing halts staling in bread, but refrigerator storage speeds staling because of great evaporation loss. Bread held at room temperature will stale less rapidly than when refrigerated. Breads should be wrapped in moisture proof polyethylene bags for frozen storage. Rolls can be warmed in the oven for serving directly from the frozen state by wrapping them in foil.

Bread stales fastest in the **refrigerator/freezer** *and keeps longest in the* **refrigerator/freezer.**

Refrigerator, freezer.

*

QUICK BREADS

The name "quick breads" includes a variety of items such as biscuits, muffins, hot cakes, waffles, popovers, cake doughnuts, and fruit breads. They are made from batters or doughs leavened by substances other than yeast. Some quick breads are batters before baking and others are doughs. Doughs are flour mixtures thick enough to be kneaded; batters are those mixtures thin enough to pour or drop.

Would the main difference between a dough and a batter be the quantity of liquid in proportion to the flour in them?

Yes, this is the major difference.

*

Most quick breads are made with flour, baking powder, salt, sugar, shortening, eggs, and liquid. By varying the quantity of these ingredients and the techniques used in making them, we can get many different types of quick breads. A hotcake batter and a waffle batter are much the same, except the waffle may be richer and have just slightly less

344

liquid in it. Date-and-nut loaf bread, cornbread, bran muffin, and steamed brown bread are variations of a muffin batter.

What are the seven ingredients used in most quick breads?

Flour, baking powder, salt, sugar, shortening, eggs, and liquid.

*

The ratio of shortening and sugar to flour in a muffin batter is low, and there is a good chance for too much gluten development unless mixing is carefully controlled. With lean muffin batters, mix just enough to incorporate and moisten completely the sifted dry ingredients with the liquid ingredients. If the batter is mixed too much, the muffins will be tough. Other indications of overmixing a muffin batter include tunnels in the baked product, a peaked top, and a smooth crust. A good muffin is tender and has a rounded top, a uniformly coarse interior, no tunnels, and a cauliflower-like surface. Occasionally, someone is so concerned about overmixing a muffin batter that the product is actually undermixed. Then the volume is poor and the muffins are very crumbly.

Match the terms in column A with the amount of mixing indicated in column B.

A	B	
1. Cauliflower-like top	A. Undermixed	1. B.
2. Poor volume	B. Mixed the	2. A.
3. Tunnels	correct amount	3. C.
4. Very crumbly	C. Overmixed	4. A.
5. Peaked top		5. C.
6. Rounded top		6. B.

*

The muffin method of mixing is a simple process. First, all the dry ingredients (flour, baking powder, salt, and sugar) are sifted together. Then the eggs are beaten until well blended and the milk and melted shortening or oil are blended thoroughly with the eggs. These liquid ingredients are poured all at once into the dry ingredients and the mixture is stirred just enough to blend and moisten the ingredients completely. The batter should still look pebble-like and rough; it should be transferred gently into the muffin pan. Use one large spoonful of batter for each muffin. Do not stir the batter as each muffin portion is removed from the bowl or the last muffins will be mixed too much.

345

If a muffin batter is smooth, it has been mixed **the correct amount/too much.**

Too much. Muffin batter mixed the correct amount is still rather lumpy.

*

Some muffins are very high in sugar and shortening. They may be so rich that they are called cake muffins. These can receive much more stirring before they become tough. The sugar and shortening slow down the development of the gluten in rich muffin batters. Consequently, more mixing needs to be done to develop the gluten network to support the muffins.

Extra _____ and _____ in a rich muffin batter delay the development of gluten, thus making it necessary to mix this batter **more/less** *than a regular muffin batter.*

Sugar, shortening, more.

*

If Teflon-lined muffin pans are used, they need not be greased. Otherwise, grease the bottoms and sides of muffin pans lightly. Use enough fat to keep the muffins from sticking, but do not use so much that a crisp hard crust develops. If batter happens to spill on the muffin pan while being spooned into the cups, wipe the pan clean before baking to avoid burning these spills.

Why is a Teflon-lined muffin pan useful?

Muffins can be removed quickly and easily without breaking. Cleanup time also is reduced.

*

Muffin batters should be baked at 400° to 425°F. These small quick breads will bake through to the center quickly at this temperature before the outside burns or dries out. However, some rich muffin batters are better if baked at 350° to 375°F. Usually, muffins are served immediately. If it is necessary to serve them later, they should be lifted from their cups when baked so that steam can escape from the bottom.

Would you also remove muffin-like fruit bread loaves from their pans so steam can escape?

Yes, this should be done for all quick breads. However, be careful with rich loaves of fruit and nut breads for they are fragile when very hot.

*

Muffins can be varied easily by the addition of different ingredients. Blueberries, chopped nuts, dates, crisp bacon bits, or different flours are all effective means of varying muffins. Besides being used for muffins and fruit-or-nut loaf breads, muffin-like batters can be used to make quick coffee cakes, cake doughnuts, dumplings, fritters and even some desserts.

If you were making a quick coffee cake, cake doughnuts, dumplings, or fritters, should care be taken to avoid overmixing and should the ingredients be reasonably cool?

Yes. These items become tough and less desirable when they are overmixed. Warm ingredients speed gluten development and promote toughness.

*

Biscuits are made by cutting solid shortening into a sifted mixture of salt, flour, and baking powder until the mixture is in pieces about the size of uncooked rice grains. Milk or liquid is then added all at once and mixing is gently done just until a soft pliable dough is formed. The biscuit dough is kneaded on a lightly floured board or pastry cloth for about 30 to 60 seconds to develop a flaky texture and to mix the ingredients thoroughly. Biscuit dough is kneaded lightly with the finger tips. This kneading is much less vigorous than the kneading of yeast doughs. You can split the baked biscuit and peel off flaky sheets of it. This is considered to be evidence of good technique in preparing biscuits. Overkneaded biscuits are compact and tough.

About a half minute of kneading develops **desirable/ undesirable** *flakiness in biscuits.*

Desirable.

*

After kneading, the dough is lightly rolled to about half the thickness desired for the final biscuit. Press the cutter straight down through the dough without twisting it. Twisting is apt to give a misshapen biscuit. Cut as conservatively as possible, because the biscuits from the first rolling will look better and be more tender than those cut from subsequent rollings. Some people prefer to dip a knife into melted butter or flour and cut the dough into squares and bake square rather than round biscuits. In this way the quality of all biscuits is the same because no scraps are left for a second rolling. The biscuit quality is lowered each time the dough is rerolled for cutting.

Biscuit dough will **double/triple** *its volume during baking.*

Double.

Place biscuits on an ungreased baking sheet. For biscuits with crisp sides, set them apart. Even more crisp biscuits can be obtained if the biscuit is dipped into melted margarine or butter and then placed apart on the pan. If you prefer biscuits with soft sides, place them on the sheet so that they touch each other. Bake in a 425°F oven for about 15 minutes, until the top is a golden brown and the interior is done. Well-prepared biscuits are tender and flaky. The golden-brown top should be flat and the sides straight. The flavor should be delicate and pleasing.

Are canned refrigerator biscuits exactly like these baking powder biscuits?

No. Canned refrigerated biscuits are not kneaded, hence they are not flaky.

*

Drop biscuits are not kneaded but are spooned from the mixing bowl and dropped onto a baking sheet. This biscuit has a rough, uneven shape and a coarse texture. The crumb is more tender than that from a rolled biscuit because the gluten is not developed as much. The baking is the same as for regular biscuits.

True or false. Drop biscuits are usually flaky.

False. Since drop biscuits are not kneaded, the dough is not folded into layers as it is during the kneading of rolled biscuits.

*

There are many variations that can be made from biscuit dough. You can use the dough to make quick cinnamon rolls or other types of rolls. Chopped bacon, chopped raw apples, shredded cheese, a mixture of honey, chopped pecans, and raisins, or maple sugar and butter, jam, or jelly can be added to the dry mixture or put on the bottom or top of the biscuit before baking. Mashed sweet potato biscuits are made by adding the potatoes with the liquid; reduce the liquid somewhat to make up for the moisture in the potatoes. Some biscuits have acid liquids in them. If soda is added to the dry ingredients, it reacts with the acid, and carbon dioxide is released to leaven the biscuits. In such recipes, the baking powder is greatly reduced or omitted. Buttermilk or sourmilk biscuits are made with soda instead of baking powder. Tomato juice or chopped tomato biscuits are excellent with a beef stew. Soda is added to react with the acidic tomatoes to leaven these unusual biscuits.

If your regular recipe called for two cups of flour, four tablespoons of shortening, one-half teaspoon salt, three teaspoons double-acting baking powder, and two-thirds cup of milk, how would you vary it if you wanted to make chopped tomato biscuits?

Sift three-fourths teaspoon soda with the flour instead of the baking powder and use three-fourths cup of chopped tomatoes with their juice in place of the milk. The slight addition (over two-thirds of a cup) of liquid compensates for the solids that are in the tomatoes.

*

A rich biscuit dough is used for shortcakes. Sugar and a bit more shortening are added to a regular dough. Some recipes call for eggs and reduce the liquid. The shortcake dough is baked in two layers either in portion sizes or in a pie or cake pan. The layers are split while hot, filled with fruit and whipped cream, topped with fruit and whipped cream, and served.

Could you mix up a very large batch of dry ingredients for biscuits, not add the liquid, and then put this aside and use it as a packaged mix? If you wanted to make a short-cake from this, what would you do?

This is done with good success, but use the mix within several weeks. Add some sugar and melted butter for your shortcake dough.

*

Cobblers are fruit mixtures topped with a rich biscuit dough and baked. They are usually served hot. Biscuit dough also can be rolled out, filled with a thickened fruit mixture, cut into rounds like cinnamon rolls, and baked. These also are good served hot with a sauce, cream, or rich milk. This sliced product is called a *roly-poly*. If you just make and bake a roll of the biscuit dough filled with thickened fruit, the product is called a baked fruit roll.

349

Indicate the name in column B that goes with the description in column A.

A	B	
1. *Fruit mixture topped with biscuit and baked*	A. *Roly-poly*	
	B. *Fruit roll*	1. C.
2. *Roll filled with thickened fruit*	C. *Cobbler*	2. B.
3. *Baked slices from a roll*		3. A.

*

Scones are another type of bread made from rich biscuit dough. Eggs are usually added, and frequently dried fruits such as raisins or currants are included in the recipe. The dough is patted out onto a floured board and then rolled out to a thickness of about a half inch. A mixture of milk and whole egg or undiluted egg white is brushed over the top as a glaze; sugar may be sprinkled over this. The dough is cut into rectangular shapes about two inches long and an inch wide or into triangles of the same approximate size. A knife, dipped into melted butter or margarine, is used to make a deep cut in the center of the scone running along the long end. These are then placed onto greased baking sheets and baked. They are an excellent bread for breakfast or for teas.

Scones are rich biscuit dough cut into _____ or _____ shapes; they usually contain dried fruit such as _____ or _____; they are glazed over the top by a wash made of milk and _____ _____ or _____ _____.

Rectangular, triangular, raisins, currants, whole eggs, egg whites.

*

Hotcakes are made from a thin batter baked on a griddle. The ratio of ingredients is usually one cup liquid to one cup flour. This is a very fluid batter which permits a reasonable amount of mixing with less likelihood of developing the gluten enough to make the pancakes tough. However, if the ratio of liquid to flour drops below this, then the gluten in the hotcakes develops faster and they can be toughened by overmixing. The method of mixing is the same as that for muffins; that is, sifted dry ingredients are blended into the combined liquid ingredients.

The ratio of liquid to flour in a hotcake batter is usually _____ _____ _____ or _____ _____ _____.

One to one, cup for cup.

If two tablespoons of shortening are used per cup of liquid, the griddle or pan on which the hotcakes are baked needs only to be greased lightly at the start of baking. The griddle should be preheated to the point where a drop of water dances on it when dropped onto the hot surface (about 425°F). If the water spits and disappears, the griddle is too hot. Pour or spoon the batter onto the griddle to get the desired size and then bake until bubbles appear on top and the edge *just* begins to show some sign of drying. The bottom of the hotcake should then be a pleasing brown. Turn and bake the second side until the pancake is baked through. If you have followed these pointers and regulated the heat correctly, it should not be necessary to turn the hotcake more than one time.

Do/do not   *preheat the griddle for pancakes.*   Do.

*

Sweetmilk wheat cakes, buttermilk or sourmilk griddle cakes, cornmeal or buckwheat cakes, or regular sweetmilk batter to which blueberries, huckleberries, other fruits, or nuts are added all can be used for variation. You also can get wide variation with hotcakes by using different kinds of syrups or serving them with assorted whipped toppings. The names flapjacks, griddle cakes, pancakes, or hotcakes are all used to identify these products. Hotcakes, when leavened by yeast, are sometimes called old-fashioned or sourdough cakes. Sometimes, the term flapjacks means these cakes. They should be served very hot and fresh from the griddle.

If sweetmilk is used for hotcakes, what leavening agent is used? What is added to get leavening if sourmilk or buttermilk is used?   *Baking powder, soda.*

*

The hotcake is a popular item all over the world, but in many places it appears as an unleavened thin item that is filled and then rolled or folded. The Swedes love their *plattar* made into a large cake filled with lingonberries or sauce (the lingonberry somewhat resembles our cranberry). The Russians love their *blinis* or *blintzes* filled with cottage cheese or sour cream. The Frenchman likes *crepes* filled with a rich buttercream and topped with flaming brandy. The Mexican has a large flat cake called a *tortilla* that is

351

made largely from cornmeal or wheat flour. The Norwegian eats a cake called *lefse* and the Scotsman his heavy *oatcake*. The Germans have potato cakes with their sauerbraten roast and eat apple cakes sprinkled with sugar and cinnamon for dessert.

You are in a restaurant and see on the menu the following foods. Can you match the item with the nation?

A	B	
1. France	A. Pancake served with lingonberries	1. F.
2. Sweden		2. A.
3. Mexico	B. Yeast-raised flapjacks served with beefsteak	3. E.
4. Germany		4. D.
5. Russia	C. Blintzes served with sour cream	5. C.
6. Western cowboy dish		6. B.
	D. Potato cakes served with pot roast	
	E. Tortillas served with chili beans and ground meat	
	F. Crepe served with orange buttercream and flamed	

*

The unleavened type of hotcake is usually high in eggs. These need to be baked in a well-greased or a Teflon-lined pan. Pour in the thin batter and quickly tilt the pan to all sides so the bottom of the pan is completely covered with batter. Allow the cake to bake, browning well underneath, and then turn and bake on the top side. Fill as desired. Such cakes can be made ahead, rewarmed, filled, and served.

An unleavened hotcake is **thick/thin.**   *Thin, very thin.*

*

Waffles are usually made with a moderately rich batter containing more shortening, sugar, and eggs than a hotcake batter. If the ratio of flour to liquid is one to one, the batter is a little thin and too much leavening gas is lost in baking. Usually, waffle batters contain a little more flour than milk or cream. The ratio may run as high as one and one-half cups flour to one cup milk. Many waffle recipes use cream or sour cream and a large amount of sugar. Such waffle batters are almost as rich as cakes and may be served as dessert topped with ice cream and a rich sauce.

The batter of a waffle is much like that for hotcakes but is _____ and richer.

Thicker.

*

Tenderness is important in a waffle. To get tenderness, be careful not to overmix the batter, especially if you are using an all-purpose flour rather than pastry or cake flour. More mixing is necessary if the batter is a rich one or if it contains cake flour. If the ingredients are cool, more mixing can be done because gluten develops more slowly in a cool batter than in a warm one.

Waffles **can/cannot** *be easily toughened.*

Can (unless the batter is quite rich).

*

Crispness is usually a desirable characteristic of a waffle. This quality is related to the ratio of flour to liquid, the amount of sugar and shortening, and the time and speed of baking. Waffles baked from thicker batters or from richer batters and those that are baked at elevated temperatures (about 425°F) and for a longer time are more apt to be crisp. Limp waffles may be the result of underbaking, baking at too low a temperature, or using too much liquid in proportion to flour. If you allow a waffle to bake until the excess moisture is driven off, it will be fairly crisp. One way to tell when a waffle is done is to bake it until you cannot see anymore steam coming from the waffle baker.

True or false. You can increase the crispness of a waffle by baking it until it is golden brown and steam stops coming from the waffle baker.

True.

*

Popovers are a unique quick bread because they contain no baking powder but are leavened by steam developed inside them. No baking powder or other leavening agent is added. For this reason, popovers must be heated through very quickly to get rapid steam production inside so the leavening will be available before the structure is set. If the outside is set too firmly before steam develops inside, the popover cannot rise because the steam will not have enough pressure to force the coagulating mixture upward. It is best to use a heavy pan for popovers. Preheat it while you are mixing the batter, quickly pour in the batter, and immediately put the pan back into a 450°F oven. Well-seasoned, cast iron popover pans do not need to be greased, but other popover containers need to be greased. After 15 minutes at 450°F, the oven heat is reduced to 350°F; the popovers are then baked at this heat for another 30 minutes.

The leavening agent in a popover is _____; to get a rapid development of steam in a popover, we put the popover into a **warm/hot** *oven.*

Steam, hot.

*

Popovers are made from milk, eggs, salt, and flour. The ratio of liquid (milk and eggs) to flour is even higher than in a hotcake batter. Many recipes do not call for shortening. If the milk is preheated to lukewarm before being added to the eggs, the steam development occurs more quickly and a better volume is obtained. Care must be taken to be sure that the milk is not allowed to become hot enough to begin to cook the egg or gelatinize the starch during mixing.

In making popovers, would you add the hot milk to the flour or would you mix it well with the eggs and then blend in the flour?

The latter method would be preferable because the eggs would cool and reduce the likelihood of gelatinizing the starch in the flour.

*

When a popover bakes, the protein in the flour and eggs coagulates and the starch gelatinizes. This gives a solid structural framework surrounding the large hole in the center created by the steam. To avoid having popovers collapse as they cool, it is essential to have this framework dry and firm before removing the popovers from the oven. Turning off the heat, opening the oven door, and allowing the popovers to stay in the oven a short time before removing them help to dry them out. Additional rigidity is obtained by using a flour with strong gluten.

Would you prefer to use all-purpose or pastry flour for a popover?

Use the all-purpose flour because it will give a stronger framework.

*

A good popover should have an irregular peaked shape and a very large volume. The dark brown top should be rather shiny and smooth. Ideally, there will be a very large hollow center surrounded by walls that are crusty and crisp but not tough. The bottom should be intact and the sides should be shiny and crisp. You can tell a good popover when you take it from the pan because it will be very large and light and feel crisp.

What is the probable explanation for the collapse of a popover when you remove it from the oven?

You probably did not bake it long enough to set the protein in the walls to give the necessary rigidity and strength.

*

Cream puffs and eclairs are used as desserts rather than hot breads, but they are so closely related to popovers that they can logically be discussed here. To make a cream puff or eclair batter, water and shortening, butter, or margarine are brought to a boil and the flour and salt are added to make a smooth heavy paste. This is cooked only until a firm paste forms and then it is cooled slightly. Then, one by one, whole eggs are added, with each egg being beaten in well after the addition. One secret of making a good cream puff batter is to see that the beating is sufficient after adding each egg so the batter is smooth and glossy. When the eggs have been added, the batter is still quite thick and will stand by itself. If it does not, then it must be poured into muffin tins and baked. This will not give a top-quality product. About a fourth to a third of a cup of paste makes a good size cream puff or eclair shell. Normally, baking sheets need no greasing for these products.

Which of the following is necessary to have a good cream puff or eclair shell?
A. *Cooking the water, shortening, flour, and salt for a long time.*
B. *Blending all the eggs in well at one time.*
C. *Having a thin batter.*
D. *Beating each egg in separately until the paste is smooth and glossy.*

D. Beating each egg in separately until the paste is both smooth and glossy.

*

Cream puff or eclair shells are baked much like popovers. The oven is preheated to 450°F and the panned items are put in. Rapid expansion of steam is needed. Full expansion takes place in about 15 minutes, after which you reduce the heat to 375°F and bake for about 30 minutes. The shells should bake until a firm hard frame is obtained.

355

Would you think when the shells are baked that leaving them in the oven a few minutes with the door open would be a good idea, just as it is for popovers?

Yes. This allows them to dry out a bit in the declining oven heat and adjust slowly to cool air.

*

The batter used for cream puffs and eclairs can be used for other products. Tiny little puffs can be baked and then filled with cold crab, cheese, or some other tangy filling and then reheated and served as a hot *hors d'oeuvre*. About a teaspoon of filling is enough for one of these small puffs. The paste can also be used to make French doughnuts. These are put in rings onto a piece of greased paper and then this paper is slid carefully into hot fat. Soon the steam developed in the paste causes it to leave the paper and the doughnut rises to the top, cooking first on one side and then on the other as it is turned. As soon as the doughnuts are free from the paper, the paper is removed. Danish dumplings cooked in rich chicken stock are nothing but a cream puff paste.

Cream puff paste is called by the French "choux" paste— "choux" in French means cabbage. Why do you suppose they named it this?

Compare the shape of a head of cabbage with a cream puff shape and you can see why.

*

REVIEW

1. *True or false.* Yeast is the only leavening agent used to make breads and rolls.

   False. Baking powder, soda and acids, steam and air are used to leaven quick breads.

2. What technique is used to punch down bread dough and redistribute nutrients for the yeast?

   Lift the dough up from the sides of the bowl, folding over to the center, and then turn the dough upside down.

3. Match the type of dough in column A with its description in column B.

A	B	
1. Italian hard bread	A. A lean-dough product	*1. A.*
2. Danish pastry	B. Butter in thin layers between thin layers of dough	*2. D.*
3. Rolled-in dough		*3. B.*
4. Retarded dough	C. Dough refrigerated for late processing	*4. C.*
	D. A rich-dough product	

4. If you wish to make French bread or hard rolls, what kind of liquid would you probably use?   *Water.*

5. Compressed yeast *is/is not* softened in lukewarm liquid before being added to a bread dough; active dry yeast may be softened in this manner but *may/may not* be added with the dry ingredients.   *Is, may.*

6. Scalding of fresh milk for bread may be done to:
   A.   Inactivate enzymes.
   B.   Condition the milk.
   C.   Change the lactose for fermenting.
   D.   Concentrate the milk.
   E.   Melt shortening and aid in warming the dough mixture.   *E.*

7. Kneading is done to develop the *starch/gluten.*   *Gluten.*

8. Yeast ferments and produces _____ _____, water, and alcohol.   *Carbon dioxide.*

9. Fermentation *tightens/relaxes* a bread dough.   *Relaxes.*

10. *Too much/too little* fermentation gives a sour flavor to the bread.   *Too much.*

11. If you have a rich bread dough or one made of a flour that lacks a strong gluten, would you let it ferment completely or take it "young?"   *Take it "young."*

12. *True or false.* Loaf breads are baked at higher temperatures than plain rolls.   *False.*

13. Homemade breads stale *more/less* easily than do commercial breads.   *Usually more because they do not contain bread softeners or mold inhibitors.*

14. *Freezing/refrigerating* bread stops staling.   *Freezing.*

15. *True or false.* Bread flour is usually used for quick breads.   *False. All-purpose flour is most commonly used.*

16. If sour milk or buttermilk is used in the making of a quick bread, *soda/double-acting baking powder* is usually added to react with the acid in the milk and produce carbon dioxide.   *Soda.*

17. *True or false.* A muffin batter that contains a fairly large quantity of sugar and shortening can be mixed more than one that does not have these ingredients.

True.

18. If a recipe said to keep a doughnut dough cold and also to work in as little flour as possible in rolling and cutting, this would be done to:
    A. Prevent excessive gluten development.
    B. Give a more tender doughnut.
    C. Give a doughnut that fries quicker.
    D. None of these is the reason.
    E. Make the doughnuts more chewy.

A and B.

19. If you were asked to judge a muffin with a pointed top and smooth surface, what could you say about the preparation technique?

The muffin was mixed too much.

20. Are muffins or biscuits being made: shortening is cut into flour, baking powder, and salt until the mixture looks like uncooked rice grains; liquid is added and a dough made?

Biscuits.

21. *True or false.* Biscuit dough will never have eggs added to it.

Most biscuits do not, but scones are an exception for they are a variation of a biscuit dough made with eggs.

22. What is the topping on a cobbler?

Biscuit dough.

23. The ratio of flour to liquid in a hotcake batter is _____ cup to _____ cup.

One, one.

24. A waffle batter usually contains *more/less* liquid in it than does a hotcake batter.

Less.

25. To help prevent hotcakes and waffles from sticking to the griddle or waffle iron, use _____ tablespoon(s) of shortening per cup of liquid.

Two.

26. What is the American version of the following items: tortilla, lefse, crepe, blini, and plattar?

Hotcakes.

27. To get a crisp waffle, which of the following is done?
    A. Bake a long time.
    B. Add more sugar and shortening.
    C. Add more flour.
    D. Reduce the liquid.
    E. All of the above will help.

E.

28. Popovers are a mixture of _____, _____, _____, and _____ with melted shortening or oil sometimes added.

Flour, eggs, milk, salt.

29. If popovers are made with hot milk, they rise *faster/slower.*

Faster.

30. In making a cream puff paste, the water and shortening are brought to a boil and *flour/flour and eggs/eggs* are added; this is cooked until a smooth paste forms.

Flour only; salt may be added at this time too.

31. Bake cream puff shells first in a _____°F oven for 15 minutes and then in a _____°F oven for _____ minutes.

450, 375, 30.

13

Desserts

The kinds of desserts that can be served are almost endless. You can make cakes, cookies, pies, fruits, mixtures of fruits, baked fruit dishes, gelatins, egg-thickened mixtures (souffles, meringues, custards), puddings, crepes (pancakes), frozen desserts, and fried desserts. Some are simple to prepare and others are difficult. Most are sweet. A dessert is something that finishes or "tops off" a meal.

Are all desserts sweet, or is a better definition something that "tops off" a meal?

Most are sweet but some may be only slightly so; some fruit desserts can be quite tart. If you consider cheese and crackers a dessert when served at the meal's end, then perhaps "topping off" a meal is a better definition.

*

Many desserts have been studied elsewhere in this text. For instance, frozen desserts were discussed under milk products, fritters and doughnuts under quick breads, egg-thickened desserts under eggs, and some fruit desserts under the cooking of fruits. The topping for a cobbler or the base for a shortcake is made from a biscuit dough, discussed under quick breads. A filling for a pie may be a pudding mixture, a custard, or even a fruit sauce set with gelatin. Therefore, a discussion of desserts will range across a wide variety of preparation techniques.

True or false. This chapter will be limited to one basic type of cookery, as was done in the previous chapters.

False. We shall discuss batters and doughs, different pudding mixtures, gelatins, and other preparations used to make desserts.

*

The number of dessert mixes available is so numerous and these are so well accepted and widely used that it might seem somewhat academic to discuss the making of desserts from scratch. However, some mixes are fairly expensive, and not all types of desserts can be made from today's mixes. Moreover, if you are to be able to judge the mixes you purchase, you need to know something about the preparation of the original product and to compare accurately the quality of the original with that of the mix. Just as important as the practical aspects of comparing homemade products with mixes are the pleasure and satisfaction you get when you prepare your own superior creations without having to credit a mix with your achievement.

If a custard prepared from a mix was slightly rubbery and tough and had a pasty texture, how would you compare this with the texture of a custard made from scratch?

A true custard is delicate and soft and, when cut, breaks into tender segments; there is no suggestion of pastiness or toughness when eaten.

*

CAKES

Cakes are best classified into two categories according to the ingredients used in their preparation. On this basis, the categories are foam and shortened cakes. The foams created from eggs can be used to make three foam cakes: angel, sponge, and chiffon. Pound cakes, which were the forerunners of shortened cakes, may be classified as true or modified. A true pound cake is made from a ratio in which a pound each of flour, shortening, sugar, and eggs is used. Air incorporated in creaming the shortening, sugar, and eggs, and steam provide the leavening for a pound cake. In modified pound cakes, the ratios are adjusted, liquid may be added, and some leavening may be provided by baking powder. Other shortened cakes are usually classified according to their method of mixing. Thus, we have conventional, conventional-sponge, one-bowl, and muffin-type butter cakes.

There are two main classifications of cakes; these are _____ and _____. The three types of foam cakes are _____, _____, and _____. There are two types of pound cakes: _____ and _____. In addition to pound cakes, the other types of shortened cakes are _____, _____, _____, and _____.

Foam, shortened (butter). Angel, sponge, chiffon. True, modified. Conventional, conventional-sponge, one-bowl, muffin.

*

The differences between foam and shortened cakes or between pound and other shortened cakes are not clear-cut.

A chiffon cake resembles a shortened cake in that oil is added as a shortening and baking powder contributes to the leavening. In modified pound cakes, the liquid may be increased while the eggs and shortening are reduced and baking powder added until some modified pound cake formulas are very similar to those for rich, conventionally made butter cakes.

Are the only differences between types of cakes the kinds and proportions of ingredients?

No. As we shall see, the technique of mixing is also a factor in making cakes different from each other.

*

Shortened Cakes

The ingredients for a shortened cake usually are flour, baking powder, salt, sugar, eggs, liquid, shortening, and flavoring. Butter cakes today are more tender, sweeter, and richer than those our grandmothers made. More sugar and shortening can be added to our cakes, because today's cake flour is particularly well suited to the development of a tender cake with fine texture and good volume. Commercial shortenings available today have been modified by the addition of emulsifiers so the cake batters form more stable emulsions and ultimately result in fine-textured cakes. These shortenings also help to increase volume. Because of these changes and also the production of sugar with much finer crystals, today's cakes may have as much as 125 per cent more sugar than flour. Previously, cakes contained, as a maximum, equal amounts of sugar and flour. Today's granulated sugar is fine and readily goes into solution in a cake batter.

*The flour, sugar, and shortening available today are **the same as/modified from** that our grandmothers used. Today the ratio of sugar and shortening to flour in our cakes can be **higher/lower** than the ratio used in cakes 50 years ago.*

Modified from, higher.

*

Because butter was the shortening most commonly available for cakes, the term "butter cake" was used originally to indicate the group of cakes made by first creaming butter and sugar, then adding eggs, and ending with alternate additions of sifted dry ingredients and liquid. Baking powder, or soda in conjunction with sour cream or buttermilk, was used to create carbon dioxide gas to leaven the cake. However, many cake recipes today are formulated for hydrogenated shortenings rather than butter. Some people insist that butter or margarine gives a better-flavored cake and prefer to use either of these in place of hydrogenated shortening. When this substitution is made, more butter or margarine must be used to make an equally tender cake because these contain only 80 per cent fat while hydrogenated shortning is 100 per cent fat. The extra water in butter and margarine and the lesser quantity of fat account for the reduction in tenderness of a cake that has been made with butter being substituted for the expected hydrogenated shortening.

If you were going to make a cake that used butter or margarine and replace this with hydrogenated shortening, would you add more or less hydrogenated shortening?

Slightly less.

*

While the ratio of ingredients can vary quite widely in a shortened cake, usually a well-balanced recipe will contain the following proportions:

Ingredient	Weight of Ingredient Compared with Weight of Flour	Measure of Ingredient Compared to 1 Cup Flour
Shortening	Slightly less than ½	¼ cup
Sugar	Exceeds (usually 1¼)	⅝ to ¾ cup
Liquid	Equal	½ cup
Eggs and liquid	1¾	¾ to ⅚ cup

If you set up a rule for recipe balance for butter cakes, would the following be correct? The shortening will be slightly less than half the flour by weight, but the liquid weight will equal that of the flour; both the weight of the sugar and the combined weight of the eggs and liquid will be greater than the weight of the flour.

Yes, this would be correct.

*

Now, let us see how the preceding rules help in evaluating a recipe. Check the following recipe, assuming that one egg equals one-fourth cup.

Cake flour, sifted	2 cup
Butter or margarine	½ cup
Sugar	1¼ cup
Milk	1 cup
Eggs, whole	½ cup (two large eggs)

As you check through, you will notice that all ingredients are within the recommended range. To this basic formula, you would need to add two teaspoons of baking powder, one-half teaspoon salt, and some flavoring.

In this formula, could hydrogenated shortening be substituted for the butter satisfactorily? How would the cake made with the hydrogenated shortening differ from the cake made with butter?

The butter cake would be more yellow and have a more buttery flavor, but the cake made with the hydrogenated shortening would be more tender.

*

Sugar has other important functions in cakes besides giving a sweet flavor. The tenderizing effect of sugar is due to two factors: (1) sugar slows gluten development and (2) sugar raises the coagulation temperature of protein in the cake so the cake can have a longer time to stretch cell walls. This means that the volume of the cake is increased, the cell walls are thinner, and the cake is more tender. In addition to the effect of sugar on flavor, tenderness, and volume, sugar contributes to a uniformly fine cell structure because its fine granules help to create many tiny air pockets when sugar is creamed with shortening. Finally, much of the eye appeal of cakes is attributed to the attractive browning that occurs when the sugar in the crust of the cake browns during baking.

What function(s) does sugar perform in a cake?
A.   Colors the crust.
B.   Sweetens the cake.
C.   Increases volume.
D.   Increases tenderness.
E.   Reduces the air brought into the batter during creaming.

All but E are contributions made by sugar.

*

If a cake is low in sugar, it lacks sweetness. The texture is apt to be tough, the volume low, and the color of the crust will be pale. If too much sugar is added to a cake, the gluten will develop very slowly during mixing and, during baking, the gluten strands forming the cell walls will be stretched to the point where they cannot support the cake, causing the cake to fall. This problem is also accompanied by excessive browning of the surface of the cake and a slightly sticky, crystalline-looking crust.

363

Match the result in column B with the factor in column A.

A	B
1. *An excess of sugar in a shortened cake*	A. *Pale crust*
2. *Too little sugar in a shortened cake*	B. *Gummy, heavy crumb*
	C. *Hard, flinty crumb*
	D. *Small volume*
	E. *Dark crust*
	F. *Sticky crust with crystalline look*
	G. *Cake collapses in baking*

1. *B, E, F, and G.*
2. *A, C, and D.*

*

Eggs help to give moistness, flavor, richness, color, and volume to a shortened cake. If they are beaten before being added to the creamed sugar and fat, they help incorporate air into the batter and increase the volume. During the mixing of a cake, eggs are considered to be a liquid. In an angel or pound cake, the eggs are the only moisture the cake receives. In the baked cake, eggs contribute strength to the structure because of the coagulated proteins added to the cell walls.

If you use yolks in place of whole eggs, should you increase the moisture slightly in the cake recipe? What if you replaced whole eggs with egg whites?

Yes, because yolks are only 50 per cent moisture while whole eggs are about 75 per cent. If you use egg whites in place of whole eggs, reduce the moisture slightly because egg whites are 87 per cent moisture.

*

A butter cake should be fine-grained and have a soft, velvety, open crumb that is delicate and tender. These characteristics are influenced greatly by the type of flour used. The flour used for cakes should be cake flour, because it has less gluten; the gluten in cake flour forms a delicate and tender structure. Although pastry flour can be used, a less tender cake will result. All-purpose flour causes a cake to be slightly tough and bread-like in texture. More shortening can be added to help offset the decreased tenderness of a cake made with all-purpose flour. Strong bread flours give a cake with a tough crumb and a tight, heavy texture.

*A quick review: Cake flour has **less/more** starch but **less/more** gluten than does pastry flour.*

More, less.

*

If not enough flour is used, the structure of the cake is weak, resulting in a compact, gummy product with poor volume. If too much is used, you get a tough and compact cake with a tight, heavy grain. A cake containing too much flour is apt to split in the center when it bakes. A cake with too little flour is apt to fall during baking.

Match the condition in column A with the result in column B.

A	B	
1. *Too much flour is used*	A. *The grain is heavy and compact*	
2. *Too little flour is used*	B. *The top splits in baking*	
	C. *Volume is slightly more than expected*	
	D. *Volume is slightly less than expected*	
	E. *Cake falls in baking*	1. *A, B, and C.*
	F. *Poor, gummy texture*	2. *D, E, and F.*

*

Baking powder is normally used in butter cakes for leavening, but soda may replace it if an acid liquid such as buttermilk, fruit juice, or even perhaps mashed fruit is used to react with the soda. You need to work rapidly to benefit from the carbon dioxide that is produced when the soda comes in contact with the acid ingredient. This reaction will go to completion at room temperature. This means that all the gas may be formed in the batter before you finish mixing a cake batter. Much of this gas will then be lost during the remainder of the preparation and panning of the batter, with the result that the cake may have poor volume.

Is soda a leavening agent?

Soda alone is not a leavening agent, but when an acid is present, the soda and acid react readily to form a gas (carbon dioxide) which leavens the cake.

*

Our grandmothers' recipes often recommended that soda be added to the liquid being added to the cake batter. This was done because it was difficult to distribute the coarsely ground soda uniformly in the cake batter unless it was dissolved in the liquid first. As you would expect, considerable leavening gas is lost when soda is dissolved in the liquid. Some recipes even state that soda should be added to hot water, buttermilk, sourmilk, or other acid liquid with which it reacts immediately, causing great loss of carbon dioxide. Today's soda is in fine particles that can readily be distributed in the batter; many recipes today could be revised by simply having the dry soda sifted with the dry ingredients. If this procedure is followed, one-fourth teaspoon soda per cup of flour is usually adequate.

A spice cake recipe called for a teaspoon of soda added to three-fourths cup of boiling water; two cups of flour are used in the recipe. How much soda would you use if you sifted the dry soda with the dry ingredients rather than adding it in the boiling water?

Reduce the soda to one-half teaspoon, using one-fourth teaspoon for every cup of flour.

∗

Baking powder is a more consistent leavening agent than is a mixture of soda and acid. Baking powder is actually a dry mixture of soda and one or more acid ingredients. The potential gas production from these powders is standardized by the addition of a starch filler. This starch is also important because it absorbs moisture that may penetrate into the can and greatly slows the reaction that might take place in the can. Baking powders containing tartrate salts as the acid ingredient are preferred by some cooks because they do not create an unpleasant after-taste. Unfortunately, tartrate baking powders generate their gas fairly quickly at room temperature once they are moistened. A double-acting baking powder is the choice of the majority of cooks; this type gives most of its gas only at oven temperatures and is therefore reliable for slow workers as well as fast ones. The residue flavor from double-acting baking powders may be slightly objectionable to some people. One teaspoon of baking powder per cup of flour is the customary ratio.

What do you think might be the result of too much baking powder in a cake?

The answer depends upon how much extra baking powder was added. The volume might be increased and the texture might be rather coarse. However, if enough extra baking powder were added to stretch the gluten strands to the breaking point, the cake would fall and have a very small volume.

∗

The first step in making a butter cake by the conventional method is to cream the shortening, flavoring, and sugar together until they form a fluffy (but heavy) fat foam. If the fat is very hard, it is sometimes creamed briefly before the sugar is added. Thorough creaming is important because the many fine air bubbles in this fat foam help to give the finished cake a very fine texture. On warm days, you may need to reduce the creaming time so the shortening does not become too fluid. Overcreaming leads to a broken emulsion later in mixing and causes a smaller volume.

What is the purpose of creaming sugar and fat together when making a butter cake?

This creates a fat foam with many small air bubbles and helps to produce a fine-grained, high-quality cake.

*

For best creaming action, a plastic shortening that is workable, yet not too fluid at room temperature, is needed. If cake shortenings, butter, or margarine are in the temperature range from 60° to 75°F, they will usually cream well. A shortened cake batter should not rise above 80°F in its mixing.

What happens if the shortening is too cold or too warm?

It will not cream well. Cold shortenings require an unnecessarily long time to cream with the sugar. Warm fats become too fluid during mixing, resulting in a curdled cake batter and a cake with small volume.

*

The manner of adding eggs to butter cakes varies from one recipe to another. Sometimes the well-beaten whole eggs are blended thoroughly into the fat-sugar foam, thus increasing the air incorporated into the cake. In other cakes, the unbeaten yolks may be added one at a time into the creamed mixture. If this is done, the whites are usually beaten to a foam, preferably a foam stabilized with some sugar, and the egg whites are folded in carefully as the final step in preparing the batter. This is a good way of increasing the quantity of air in a cake and obtaining a slightly greater volume. Whether the whole eggs or just the yolks are added to the creamed mixture, it is important to blend these completely with the fat-sugar mixture, because this forms the basis of the emulsion in the cake batter. A stable emulsion produces a better finished product.

The beaten eggs are added to a cake batter **at the same time the sugar is added/after the sugar and fat have been creamed.**

After the sugar and fat have been creamed.

*

The flour, salt, leavening agent, and any other dry ingredients are sifted together preparatory to their addition. Approximately one-third of the dry ingredients are stirred in

367

after the eggs have been added. Then, about half the liquid is added. Next, the second third of the sifted dry ingredients is added, followed by the last half of the liquid. The last third of the sifted, dry ingredients is then stirred in completely. The batter should now be fairly thick and uniformly blended.

Arrange the following steps in their proper sequence for mixing by the conventional method.
A. *Fat and sugar are creamed together.*
B. *Half the liquid is added.*
C. *One-third of the dry ingredients is added.*
D. *Beaten eggs are added.* $A, D, C, B, C, B, C.$

*

If a cake is low in sugar and shortening or is made with a soft shortening, the conventional-sponge method is a particularly good method to use. This method is very similar to the conventional method of mixing. The prinicpal difference is in the addition of the sugar. In the conventional-sponge method, only half the sugar is creamed with the shortening. The remainder is beaten into whole eggs or the egg whites to make a stable foam which then is carefully folded in during the last 50 strokes of mixing. The sugar-stabilized foam helps to give cakes made by the conventional-sponge method a good volume. The quality of the cake is as good as can be obtained with the limitations imposed by having a cake low in shortening and sugar or having a shortening that is not the best for cake making.

If the shortening is excessively soft or the cake recipe is low in sugar or shortening, the _____ method is a particularly suitable method to use in preparing the batter.

Conventional-sponge.

*

Cakes made by the one-bowl method are sometimes called high-ratio cakes because this method allows higher ratios of sugar and liquid to flour than do other methods of mixing. To make a cake by this method, the flour, salt, sugar, leavening agent, and other dry ingredients are sifted into the bowl and soft shortening and part of the liquid are added. These ingredients are mixed with an electric mixer until a smooth mixture is formed. Subsequently, the beaten eggs and remaining liquid are added and mixed in thoroughly. There are some minor variations of this one-bowl method, but such methods produce cakes of moderate rather than fine grain, and these cakes tend to stale fairly quickly.

Outline the steps in making a one-bowl cake.

Sift the dry ingredients into the mixing bowl and add the soft shortening and part of the liquid. Mix well before adding the beaten eggs and remaining liquid. Again, mix thoroughly for about three minutes after adding the eggs and liquid.

*

Machine mixing produces a more satisfactory cake by the one-bowl method than does hand mixing. The more vigorous machine mixing seems to be suited particularly well to manipulation of this rather large volume of materials in the initial mixing of the batter.

For best results in making a cake by the one-bowl method, mix by **hand/machine.**

Machine.

*

The muffin method of making a cake is suited only for cakes that will be eaten soon after they are made, because the keeping qualities of this cake are limited. In the muffin method of mixing, liquid ingredients are added to the sifted dry ingredients, as in making muffins, and these are mixed together. This allows liquid to reach the flour before the gluten in the flour is coated with shortening. The better coverage with the conventional method appreciably slows the onset of staling and, consequently, is well worth the extra time involved in preparing cakes by the conventional method.

When you are able to coat flour particles with shortening, you **delay/speed** *staling.*

Delay.

*

When a cake is made by the muffin method, the shortening, of necessity, is oil or melted fat. This oil is thoroughly blended into the milk, eggs, and any other liquids prior

369

to addition of the liquids to the dry ingredients. The muffin method has one virtue in that it is a very quick way to make a cake, but the texture is coarse and the keeping qualities are poor.

Would you mix a cake made by the muffin method more or less than you would a muffin batter?

More. The amount of shortening and sugar used is usually as high as that used in other cakes; this permits mixing without the formation of an excessively strong structure.

*

A pound cake gets its name from the fact that the ratio of ingredients is a pound of flour, a pound of shortening, a pound of eggs, and a pound of sugar. The shortening is creamed, the sugar added, and the mixture creamed again. Eggs are added gradually and creamed well to obtain a light fluffy mixture in which sugar granules are not readily apparent. The eggs act as liquid for the cake. The flour is carefully folded in as the last step in preparing a pound cake. The key to success in a pound cake is thorough creaming. Without adequate creaming, a pound cake will appear to be a ton cake. The air incorporated in creaming, along with some steam developed during baking, will leaven the cake.

The basic recipe for a pound cake required a _____ each of flour, eggs, sugar, and shortening.

Pound.

*

A pound cake has a rather fine and compact grain. It is not as open and velvety as other shortened cakes. A pound cake can use a stronger flour than most cakes because of the high ratio of shortening and because of the need for a firm structure. Commercially, bread flour sometimes is mixed with cake or pastry flour to make this cake. But in the home, cake flour is customarily the choice for pound cake.

True or false. A typical pound cake has a fine and compact grain.

True.

*

Many pound cake recipes used today are not true pound cakes. They are modified. Frequently, liquid and baking powder may be added to the basic formula. In a modified pound cake, the ratio of eggs and shortening to flour is reduced, and the cake becomes more like other shortened cakes. Pound cakes were made in medieval times. They could be made, even though chemical baking agents were not known, because air could be incorporated to give a light texture. The development of the other shortened cakes undoubtedly came about as chemical leavening agents became available and modifications in the true pound cake formula became possible.

If you add liquid and baking powder to modify a pound cake, you can **increase/decrease** *the ratio of shortening and eggs to flour.*

Decrease.

*

All cakes should be baked in baking pans designed for the job. Bright metal pans reflect heat away from the surface and, consequently, the cakes bake less well in these pans than in darker pans which tend to absorb the heat. Uneven, warped pans give misshapen cakes. For easy removal, grease the pan bottoms or line them with waxed paper. Leave the sides of the pan ungreased so that the cake can cling to the sides and help pull itself up to the maximum volume.

True or false. It is unnecessary to grease waxed paper used to line the bottom of a cake pan.

True.

*

Most cake recipes requiring two cups of flour make about a quart of batter. This can be divided to fill two eight-inch layer pans just over half full. During baking, the cake will just fill the pan because most cakes do not quite double in volume during baking. Overfilling the pan will give lower volume and may result in the cake running over. Approximately one-third more batter is required to fill an eight-inch square pan than an eight-inch round pan.

True or false. A cake batter more than doubles in volume when it bakes.

Usually false. Most cakes do not quite double in volume.

*

Shortened cakes should be positioned in the center of an oven preheated to the desired temperature. The usual baking temperature is 350°F, but this may vary with the recipe. Place the cake pans at least one inch from the walls and door of the oven to allow good air circulation and

371

even heating during baking. Space should also be left between pans to avoid interrupted air flow. If it is necessary to use two oven racks, try to stagger the pans on the two shelves rather than placing one directly above the other. Without a staggered arrangement, the surface of the lower cake will not brown well. It is important to avoid the lowest and the highest rack positions in the oven; the former creates too dark a crust on the bottom of the cake and the latter leaves the upper crust too light.

For the best baking results **do/do not** *preheat the oven and position the cake* **in the center/toward the bottom** *of the oven.*

Do, in the center.

*

Pound cakes need to be baked slowly. The cake is usually put into a loaf or angel cake pan. This gives a thick mass for the heat to penetrate. Pound cakes often show a crack on the top, indicating that the outside portion of the cake has been set by the heat before the inside portions have completely risen. This inside pressure causes the crack to appear. If the outside walls are too firmly set before all the leavening inside is completed, a misshapen cake with a poor texture will be obtained. Fruit cakes are pound cakes which have a large quantity of fruit and nuts added to them.

The leavening agent(s) for a true pound cake are: (1) air, (2) soda, (3) baking powder, and (4) steam.

1 and 4.
Air and steam.

*

Shortened cakes are done when a toothpick inserted in the center comes out clean. If some of the cake clings to the toothpick, the cake needs to be baked longer. A cake is overbaked if it draws away from the sides of the pan.

Select the best test for doneness for a shortened cake:
A. The cake draws away from the sides of the pan.
B. A toothpick inserted in the center comes out clean.
C. A toothpick inserted halfway between the center and edge comes out clean.
D. The cakes come to the top of the pan.

B.

*

When shortened cakes are done, remove them carefully from the oven and gently set them upright on the counter to cool. The structure of these cakes is extremely delicate when they are hot. However, the structure becomes less fragile as the cake cools; it is possible to remove shortened cakes from the pan onto a cooling rack while the cake is still fairly warm but not steaming hot. Removal is easy because the fat or the waxed paper will not be firm when warm and the cake is readily persuaded to leave the pan. Then it is a simple matter to peel off the waxed paper quickly. Allow the cake to finish cooling undisturbed on the rack. The rack permits steam to escape and helps prevent a soggy cake.

True or false. Remove a shortened cake from the pan as soon as it is removed from the oven.

False. Let the cake cool sufficiently to be handled easily before removing from the pan. You are less likely to break it.

*

When cooled, a cake is ready to be frosted. Before you frost the cake, gently brush the crust free of crumbs. This precaution reduces the problem of a rough-looking icing generously garnished with crumbs. Frequently, the bottom layer of a layer cake is turned upside down so that there will be a smooth level surface on the bottom layer. However, the top layer is usually placed in an upright position on the inverted bottom layer. This gives the best-shaped cake possible and one in which the top layer is unlikely to slide from the bottom layer.

True or false. When preparing to ice a layer cake, put both layers right side up.

False. The bottom layer is inverted and the top layer is right side up.

*

If a layer cake is to have a filling, the cake can be assembled with the filling while the cake is still warm if the filling is not affected by this warmth. However, fillings or icings containing whipped cream or butter (or other fats) should only be placed on cool cakes. Warm cake and whipped cream or a butter icing spell culinary trouble, because these toppings quickly soften and run off the cake. When the cake is cool, any type of icing can be applied.

If you are using a thick frosting, frost the sides first and the top last to make an attractive product. With a thin frosting or glaze, the frosting is poured over the top of the cake and allowed to run down the sides while it is spread over the top. Cakes glazed with thin icings have a more casual, less studied appearance.

373

*If you use a thin frosting, frost the **sides/top** first; if you use a thick frosting, frost the **sides/top** first.*

Top, sides.

*

Usually, a spatula is used to spread an icing. When spreading the sides and applying the finishing touches, keep the spatula straight and move around so that you get a smooth, even finish. When applying thick frostings, you can give a feeling of depth to the icing on top by lifting up the spatula to form some peaks or by forcing the spatula down to leave ridges in the surface. If a cake is to be frozen, it is best to freeze the cake without frosting and then frost the cake after it is thawed.

True or false.   Usually a cake frosted with a thick frosting should have straight, even sides and a rough top which gives the appearance of depth on the top surface of the cake.

True.

*

Foam Cakes

The leavening agent in foam cakes is largely air captured in the egg foam, but steam also assists in the leavening. Angel cakes and true sponge cakes have no baking powder added to them; chiffon cakes and some sponge cake variations use baking powder in addition to air and steam for their leavening.

Match the cake with the sources of leavening in each.
A. Angel	1. Air
B. Chiffon	2. Steam
C. Sponge	3. Baking powder
D. Pound cake	

A. 1 and 2.
B. 1, 2, and 3.
C. 1 and 2.
D. 1 and 2.

*

The functions of ingredients in foam cakes are the same as in shortened cakes, although the proportions, and even some of the ingredients, are different. The flour used for foam cakes should be high-quality cake flour. Good-quality eggs are also needed, but extremely fresh eggs whip to a foam of somewhat lower volume. The sugar sweetens, tenderizes, and aids in the browning of the cake. Increasing sugar increases the tenderness of a cake by raising the coagulation temperature of the protein structure. A small excess of sugar produces a cake with a slightly larger volume and a rather moist, soggy crumb. A large excess causes the cake to fall.

If you made a foam cake with all-purpose flour, the texture would be **finer/coarser** *and the cake would be* **more/less** *tender than if you had used cake flour.*

Coarser, less.

*

Angel cakes. Angel cakes are made from a very basic formula containing simply an egg white foam stabilized with cream of tartar and sugar, a mixture of sugar and cake flour, and flavoring. The egg white foam encases a large quantity of air which expands significantly during baking to give the light, airy, and tender product known as angel cake. Sometimes other ingredients such as cocoa, chopped maraschino cherries, or nuts may be added to make flavorful variations.

The basic ingredients in an angel cake are _____ _____, *_____, and _____ _____. Cream of tartar is commonly added to _____ the egg white foam, and vanilla or other extract may be added for flavoring.*

Egg whites, sugar, cake flour, stabilize.

*

Angel cake mixes are very popular today because they make a very acceptable cake and they eliminate the problem of creatively attempting to use the dozen or more egg yolks remaining when fresh egg whites are used. Dried egg whites in the angel cake mixes behave a bit differently than do fresh egg whites. The dried egg whites need to be beaten until the peaks stand up perfectly straight when you pull the beaters up from the foam. By beating the dried whites to this point, you will achieve maximum volume in your cake. Fresh whites, of course, should not be beaten quite this much because they break into chunks when the flour is folded in at this stage of beating.

When beating the whites in an angel cake mix, beat the whites until the peaks **just bend over/stand up perfectly straight.**

Stand up perfectly straight.

*

When making an angel cake with fresh egg whites, beat the egg whites with half of the sugar until the egg whites can be pulled into peaks that just bend over (stage 3). Begin adding the sugar at the foamy stage and continue

375

adding it gradually until half the sugar in the recipe is beaten into the whites. The remaining half of the sugar should be sifted with the cake flour. By combining some sugar with the cake flour, you decrease the tendency of cake flour to "ball up" when it is folded into the cake batter.

Why is sugar added to the egg white foam? To the flour?

Sugar is used to stabilize the egg white foam and to keep the flour from balling up.

*

Tenderness is extremely important in an angel cake. This characteristic is achieved by very careful mixing of the flour-sugar mixture with a properly beaten egg white foam. Sift approximately one-fourth of the flour mixture over the beaten whites, and gently fold in the flour with a flexible metal or rubber spatula. Efficiently draw the spatula across the bottom of the bowl and drag the egg white meringue up and across the upper surface where the flour has been dusted. Repeat this process about 10 times. Then, add the next one-fourth of the flour mixture, even though some flour may still be visible from the previous addition. Continue to add the flour, folding 10 strokes after each addition. After the last addition of flour, continue folding until you can find no trace of dry flour in the cake batter.

Why is it necessary to limit the amount of folding that is done to mix the flour into the egg white meringue?

A minimum amount of folding is necessary to combine the flour and egg whites; beyond this point, continued folding only causes the cake to lose volume and tenderness.

*

Angel cakes are baked in tube cake pans, preferably ones with legs on them and of two-part construction for easy removal. Do not grease the pan at all if you want to have maximum volume. Very gently pan the angel cake batter. Large air bubbles in the finished product can be avoided by simply running a spatula once around the center of the batter in the pan. Baking should begin immediately in a preheated oven. Delay at this stage allows air to escape from the foam and causes a smaller volume.

Why are all these precautions necessary when making an angel cake?

Leavening of the cake is done by steam and air incorporated in the egg white foam. Anything that causes or permits loss of air from the foam will produce a smaller, less desirable cake.

*

You can tell when an angel cake is adequately baked by touching its surface gently with your finger. When it springs back, it is done. If your finger leaves a small impression, quickly close the oven door and continue baking. Avoid slamming the oven door or any other disturbance of the cake when checking to see if it is done. The structure of a baking angel cake is quite delicate, and the cake will fall if cold air touches it or if it is shaken when it is almost but not quite done.

Cold air causes a baking angel cake to fall because it causes the expanding hot air in the foam **to shrink/to explode.**   *To shrink.*

*

Angel cakes should be cooled in an inverted position to keep the very delicate cell walls stretched out as far as possible. Legs on a tube pan keep the cake from resting on the counter and thus help to keep the cells stretched until the structure cools sufficiently to become more rigid. A second advantage of legs is that they allow air to circulate under the cake and thus prevent a soggy crust.

Angel cakes, **like/unlike** *shortened cakes, should be cooled in an* **inverted/upright** *position.*   *Unlike, inverted.*

*

You can judge your angel cakes by comparing them with these standards: (1) large volume, (2) great tenderness (almost seems to melt in your mouth), and (3) uniform but rather coarse air cells. These criteria can be used to evaluate angel cakes made with fresh eggs or with dried egg whites in a mix. Of course, these cakes should also be a pleasing brown on the surface and have a pleasant taste. Overfolding of the flour into the egg white mixture will cause a less tender cake with a poor volume. Underbaking will cause poor volume. In fact, an underbaked cake frequently will fall from the pan when it is inverted to cool.

Select the desirable characteristics for an angel cake:
A.   Light and tender.
B.   Very fine, uniform texture.
C.   Relatively large air cells.
D.   A few traces of cake flour.   *A and C.*

*

Sponge cakes. Sponge cakes, another type of foam cake, are very closely related to angel cakes. The chief difference between these two types of foam cakes is that sponge cakes are made using both the yolks and the whites, whereas only

377

the whites are used in an angel cake. In sponge cakes, you can trap air in a fluffy egg yolk foam as well as in the egg white foam. This means that you will be able to get a considerable amount of leavening from air and steam during baking.

Angel cake contains egg **whites/yolks and whites;** *sponge cakes contain egg* **whites/yolks and whites.**

Whites, yolks and whites.

*

The success of a sponge cake hinges on adequate beating of the egg yolk foam. You will find that egg yolks will gradually become lighter in color and begin to pile softly when they are beaten hard with an electric mixer. At this point, the water should be added and the beating resumed. It is essential to beat the resulting foam until it begins to hold a soft peak. This egg yolk foam, into which flour has been folded, is sufficiently stable that it can be held safely a few minutes while the whites are beaten with sugar to form a meringue. The whites should be beaten until the peaks just barely bend over. The final step preparatory to baking is to combine the two foam mixtures all at once and gently, but efficiently, fold them together. It is very important that you continue to fold until there are absolutely no streaks remaining in the batter.

True or false. The egg yolk foam in a sponge cake will be a little softer than the egg white foam.

True, but be sure to beat the yolks sufficiently.

*

The yolk foam is heavier than the egg white foam and thus has a tendency to drain to the bottom of the cake pan when a sponge cake is baking. A heavy layer in the bottom of a sponge cake can indicate that the yolks were underbeaten, because underbeaten yolks are more fluid and can drain rather easily from the egg white foam. Another cause of this rather rubbery layer can be inadequate folding of the yolk and white mixtures. This situation also makes it rather easy for the yolks to settle gradually to the bottom of the cake. This layer is also more likely to form if the cake has to stand on the counter for awhile before it is baked or if it is started in a cold oven.

Describe a sponge cake that was made with an underbeaten egg yolk foam.

The cake will be quite light in color near the top and will be increasingly more yellow toward the bottom. There may be a rubbery layer on the bottom. The volume will not be as great as expected.

Sponge cakes, like angel cakes, are baked in tube pans until the structure is set and the cake springs back when touched gently with the finger. The hot cake has a very delicate structure, even when it has baked to the correct end point. To maintain maximum volume of this delicate cake, invert the pan and allow the cake to hang suspended until it is cool.

For maximum volume, cool a sponge cake in an **inverted/ upright** *position.*

Inverted.

*

While angel cakes have relatively few variations, there are many ways in which a sponge cake recipe may be varied. A hot milk sponge cake is made by adding boiling hot milk to the mixture just after the flour has been folded in. Careful folding and immediate baking are essential to the success of this sponge cake. In other variations of a sponge cake, you may discover baking powder being used for part of the leavening, even though no chemical leavening agent is used in a true sponge cake. Chemically leavened sponge cakes usually have an increased ratio of liquid. In still other sponge cake recipes, you may find that melted butter is an ingredient, despite the fact that no true sponge cake ever contains any fat or shortening. This hot butter sponge cake is a very delicate product in which the melted butter is folded into a warm sponge batter after the flour has been folded in. Fat, of course, readily destroys an egg white foam so the folding must be done with unusual care. You might also like to try the variation of a sponge cake in which a syrup of sugar and water, boiled to 245°F, is slowly added to the egg white foam as it is beaten. The yolk foam is combined with the flour and folded into the egg white meringue in the usual manner.

More/less *variation is found in sponge cake than in angel cake recipes.*

More.

*

Several popular desserts use a sponge cake batter as their base. Jelly rolls are varied by filling them with fresh fruits in season, flavored whipped cream mixtures, or various jams and jellies. To make the cake portion of a jelly roll, simply prepare a sponge cake batter and bake it in a jelly roll pan lined on the bottom with waxed paper. The baked cake is cooled briefly; while it is still slightly warm, it is

379

removed from the pan onto a slightly dampened cloth that more than covers the cake. The warm cake is then very gently rolled in the cloth and allowed to cool. The cooled cake can be unrolled, filled with the desired filling, and re-rolled without the cloth. This dessert is then ready to be served.

To prepare a jelly roll, remove it from the pan **while it is warm/after it cools.**

While it is warm.

*

Chiffon cakes. Chiffon cakes are classified as foam cakes, but their ingredients are similar to those found in a butter cake. For instance, chiffon cakes are leavened with baking powder plus air and steam. They are tenderized somewhat by the salad oil used in the recipes. But chiffon cakes are baked in tube pans and are cooled in an inverted position like other foam cakes. Chiffon cakes are made by combining all the ingredients, except the egg whites and part of the sugar, in a mixing bowl and beating them together until they are satin smooth. This mixture contains egg yolks, salad oil, some sugar, baking powder, salt, flavorings, and flour. The mixing develops the gluten in the flour a bit to contribute to the final structure of the cake. However, the end product of this mixing is a rather fluid mixture that has a strong tendency to drain from the egg white foam during the first part of baking.

Chiffon cakes, unlike true sponge and angel cakes, contain _____ _____ and _____ _____.

Baking powder, salad oil.

*

Good technique is essential if you are to make a good chiffon cake. Your product will have good volume and be a uniform consistency throughout if you are careful in folding the egg white meringue with the other ingredients and if you form a proper egg white foam stabilized with some sugar. Since it does take time for the cake batter to heat up to the point where coagulation of the protein structure begins, stability of the egg white foam is very important. The addition of two tablespoons of sugar for each egg white to the foam during its formation is an important means of stabilizing the foam. Any sugar in the recipe in excess of this amount should be added with the other ingredients. The egg white foam should be beaten until the peaks just stand up straight. Such a foam can adequately bind in the fluid mixture.

What do you think would be the result if you made a chiffon cake that contained an underbeaten egg white foam? An inadequately folded batter?

Either inadequate folding or underbeating of the egg whites will cause a chiffon cake to have a rather heavy, almost rubber-like layer in the bottom of the pan. This layer is caused by drainage of the fluid mixture from the egg white foam before the structure is set.

*

Chiffon cakes are popular because many different liquids and flavorings can be used to give variety to them. The salad oil in them helps to retard their staling. Properly prepared chiffon cakes make showy desserts which you can take pride in making. Your individual technique will greatly influence the quality of your chiffon cake. A properly made chiffon cake will be very tender, will have pleasing color and flavor, and will be moderately fine textured.

True or false. Your technique is important in making a chiffon cake of high quality.

True.

*

Altitude Adjustments in Cake Formulas

A cake recipe that is suited for making a cake at sea level will not work satisfactorily at 5000 feet. The difficulties are caused by a change in atmospheric pressure. As the elevation increases, atmospheric pressure decreases. As a result, cake formulas must be modified if they are to be successful. Most cake formulas will work satisfactorily to an elevation of approximately 2000 feet, but above this elevation modifications must be made.

If someone who lived in an area with an elevation of 2000 feet sent you a cake recipe, could you expect to use it successfully at an elevation of 4000 feet?

No.

With an increase in altitude, it is necessary to reduce the amount of leavening agent used. It is also important to increase the strength of the structure by increasing the quantity of eggs and flour slightly. Some mixes for shortened cakes recommend the addition of two tablespoons of flour if they are to be baked at elevations above 3500 feet.

True or false.   Cake mixes need no adjustment for altitude.

False. They need to be modified just like any other cake.

*

If a shortened cake recipe is balanced for use at sea level, it is necessary to reduce the baking powder by approximately 15 per cent at 2000 feet and by 50 per cent at 6000 feet. To make the structure a bit less delicate, the quantity of sugar is reduced slightly in cakes prepared above an elevation of 3000 feet. Sometimes it is helpful to raise oven temperatures 25°F for cakes baked at an altitude of 4500 feet or more. The baking time is kept constant. In the home there is a limited opportunity to control the tendency for baking cakes to dry out, but a slight addition in the liquid in the cake formula helps. In commercial bakeries, oven dampers can be closed to minimize the drying during baking.

Match the formula changes in column B with the elevation changes described in column A.

A	B
1. *Elevation increases*	A. *Liquid is increased*
2. *Elevation decreases*	B. *Flour is increased*
	C. *Shortening is reduced*
	D. *Leavening is increased*
	E. *Eggs are reduced*
	F. *Sugar is increased*

1. A, B, and C.
2. D, E, and F.

*

Foam cakes also need to be adjusted to compensate for changes in altitude. At 3000 feet, the sugar in foam cakes should be reduced approximately 10 per cent and the cream of tartar increased about 20 per cent. Although these changes reduce tenderness, the resulting cakes have good volume and an acceptable texture. The liquid in sponge cake recipes needs to be reduced somewhat at elevations above 3000 feet. A slightly higher oven temperature is appropriate for all foam cakes baked above 4500 feet.

You should increase/decrease the amount of baking powder used in the formula when preparing a chiffon cake at 5000 feet. Should this change also be made for angel and true sponge cakes? Why?

Decrease. No. There is no baking powder in the formula for either angel or true sponge cakes.

*

COOKIES

Cookies are dessert items that can be made successfully even by inexperienced cooks if a few basic items of information are understood. Since cookies are so popular and are relatively easy to make, they are a good selection to introduce young people to the art of baking. Success with cookies of different types will encourage new interest in other types of baking.

*Cookies are **easier/less easy** to make successfully than are other baked products.*

Easier.

*

Cookies may be classified into three categories on the basis of the consistency of the dough and their form. Rolled cookies are baked from the dough mixtures sufficiently viscous to be rolled and cut into desired shapes for baking. The other two categories are bar cookies and drop cookies. Wide variations can be achieved in any of these categories with the use of different ingredients in varying proportions. However, the stiffest mixtures usually are rolled cookies, intermediate mixtures are spread and baked as bar cookies, and softer dough mixtures can be prepared as drop cookies.

The three categories of cookies are _____, _____, and _____.

Rolled, bar, drop.

*

Rolled Cookies

Rolled cookies are usually a rich mixture of sugar, butter (or other shortening), and flour coupled with vanilla or other flavorings. This basic cooky is often referred to as a sugar cooky. Its high ratio of sugar and butter to flour permits you to mix and roll this cooky without causing it to become excessively tough. However, even with the high quantity of fat and sugar, sugar cookies and other rolled cookies will become tough if they are worked excessively. They may also become tough if extra flour is worked into the dough when it is being rolled out.

A sugar cooky is classified as a _____ cooky.

Rolled.

*

For attractive, tender rolled cookies, the dough should be as soft as possible and still be rolled and handled easily. By chilling the dough before rolling, it is possible to work slightly less flour into the dough, thus keeping the cookies more tender. Efficient cutting is important; cookies from the first rolling will be more tender than those from reworked dough which is rolled out again. The cut dough needs to be gently and carefully transferred to flat cooky sheets for baking. Flat sheets permit good circulation of air over the entire surface of the cookies with the result that the cookies are uniformly browned.

Since cookies from the first rolling are the most tender, should you throw away the scraps of dough remaining after the cookies have been cut?

No. The cooky dough can be reworked and rerolled for subsequent cuttings. Although these are not as tender, they are still quite palatable if the dough is handled as little as possible.

*

Rolled cooky dough can be rolled as thin as one-eighth inch if you wish to have crisp cookies. If you prefer a somewhat softer cooky, roll the dough one-fourth to three-sixteenth inch thick. Thickness is generally a matter of personal preference, although sugar cookies are traditionally thin and crisp and gingerbread men are rolled less to make a thicker, more bread-like cooky. Select the thickness appropriate for the type of cooky you wish, and then uniformly roll the dough to that thickness. It is important to have the dough a uniform thickness so that the baking will proceed evenly.

Crisp rolled cookies should be rolled _____; to make softer rolled cookies, roll the dough _____ to _____ inch thick.

Thin (about one-eighth inch thick), three-sixteenth, one-fourth.

*

Refrigerator cookies are a variation of the rolled cookies that are made using a dough. The dough for refrigerator cookies is mixed and then shaped into long rolls and stored tightly covered in the refrigerator until you wish to bake them. The convenience of having rolls of dough ready to bake makes this variation of the rolled cooky particularly useful. Numerous refrigerated rolled cooky doughs are presently being marketed. These simply need to be refrigerated until you wish to bake them. With this type of cooky, you can easily serve freshly baked cookies whenever you wish. It takes only a few minutes to slice the desired number of cookies from the long roll and bake them.

Refrigerator cookies are prepared by mixing the dough and then shaping it into **long rolls/individual cookies.**

Long rolls.

*

Bar Cookies

Some bar cookies are made from doughs and others are baked from batters. The general similar feature in all bar cookies is that they are baked in a large baking dish and are cut into bars afterwards. Bar cookies are often the choice when there is limited time available for preparation, because it is much faster to spread the batter or dough into a pan than it is to either drop or roll the product into individual cookies.

Bar cookies have one feature in common; they are all baked in a _____ _____ _____ and then cut into _____.

Large baking dish, bars.

*

There are many variations of bar cookies. You may frequently find recipes for layered bar cookies. These usually consist of a stiff dough that is pressed out on the bottom of the baking dish to form a bottom crust. Buttered graham cracker or gingersnap crumbs may be used to form this bottom layer. Over this bottom crust there will be spread a batter which may often contain grated coconut, chopped nuts, maraschino cherries, or candied fruits for added interest and flavor. The cookies are then baked. The finishing touches include adding a layer of frosting and cutting the cookies into bars of the desired size.

Are all bar cookies made from a batter?

No. Bar cookies, such as those just described, sometimes are made from a thicker dough.

*

Layered cookies are more time consuming to prepare than are bar cookies made from a single batter. Brownies are a very popular example of a bar cooky that can be easily prepared and baked in a large low pan. Accurate measurements are important in preparing these and all other types of cookies. Although cookies do not contain quite such

385

critical ratios of ingredients as do cakes, the best product can only be obtained by careful workmanship throughout. If you measure and mix carefully, you should consistently produce good cookies. Good bar cookies should be uniformly mixed. They should be tender and should cut easily. Be sure to bake them until they are done throughout, but avoid burning them.

True or false. Recipes for bar cookies vary from a thick dough to pourable batters.

True.

*

Drop Cookies

Drop cookies are made from doughs that are moderately thick. These cookies need to be viscous enough to hold their shape or just soften slightly when dropped onto the cooky sheet. If the mixture is softer than this, the cookies will spread out a great deal on the sheet before the structure sets. A wide variety of cookies may be classified as drop cookies. The familiar chocolate chip, oatmeal and raisin, and chocolate drop cookies are popular in many homes.

For drop cookies you should have a _____ thick mixture so that they will not _____ too much during baking.

Moderately, spread.

*

Not all drop cookies are made from the traditional flour, shortening, egg, and sugar mixtures. Some drop cookies are made with egg white meringues that contain various other items such as walnuts, chocolate chips, or pecans. Coconut macaroons are also a variation of a drop cooky. These egg white mixtures contain twice as much sugar as is added to egg whites to make soft meringues for pies. Meringue cookies contain four tablespoons of sugar per egg white. To make these, add the sugar very slowly and continue beating until the whites will pull up into stiff peaks that do not bend over.

*To make meringue cookies, add **two tablespoons/four tablespoons** of sugar per egg white.*

Four tablespoons.

*

Although many drop cookies are made by dropping a spoonful of the dough mixture onto a baking sheet, other drop cookies may be made by rolling the dough in the palms of the hands to form balls or cylindrical shapes. This is done to make dream cookies, snickerdoodles, and several other cookies. Still other drop cookies are made by dropping the dough or rolling it lightly into a ball and then using a fork to flatten the dough and leave an impressed design. This is done in the preparation of peanut butter cookies.

True or false. Drop cookies are sometimes formed into balls or other simple shapes for baking.

True.

*

Preparation Precautions

Certain precautions need to be mentioned in the preparation of any type of cooky. Cookies need to be thoroughly mixed so that all ingredients are completely blended. This mixing distributes nuts, raisins, or other items throughout the dough as well as the main ingredients. Although it is essential to mix the dough, it is equally important to stop mixing after the ingredients have been uniformly dispersed. The high amount of sugar and shortening in cookies is a safeguard against overmixing, but it is possible to make tough cookies if mixing is prolonged. If the cookies are rolled, avoid unnecessary rerollings and reworking the dough. Extra manipulation decreases the tenderness.

*Excess mixing **tenderizes/toughens** cookies.*

Toughens.

*

It is important to use the specific type of shortening required in a cooky recipe; butter or margarine may not be interchanged satisfactorily with other shortenings. If you substitute margarine into a recipe that specifies shortening, the cookies will be soft and will spread too much during baking. This is also true when you substitute butter for shortening. You can correct this problem in a dough by using slightly more flour than is specified if you wish to substitute butter or margarine in the recipe.

True or false. Butter or margarine may be substituted for shortening in a cooky recipe without making any adjustments in the recipe.

False. It is necessary to add a little more flour to make up for the additional water provided by the butter or margarine.

*

Baking conditions for cookies vary, depending upon the type of cooky and its ingredients. Drop cookies and rolled cookies are baked on cooky sheets, which are flat baking

387

sheets without sides. Bar cookies, of necessity, are baked in shallow pans that hold the cookies in a rectangular shape and yet have sufficiently low sides so that browning of the upper surface is satisfactory. The temperatures recommended for baking cookies also are highly variable. Foam cookies are usually baked at 325°F and sometimes even lower. Bar cookies are satisfactory when baked at 350°F in most cases. Plain sugar cookies and other rolled and drop cookies that do not contain fruit may be baked at 375°F. If raisins or other candied fruits are in the cookies, it is necessary to reduce the oven temperature to approximately 335°F to prevent burning the fruit before the cookies are baked through. Cookies should be baked until they are a pleasing brown on the outside and no longer doughy on the interior. Their structures are sufficiently rigid so that they may be removed from the baking sheet without damage about a minute after they come from the oven.

Cookies should be baked until they are a _____ _____ on the outside and no longer _____ on the inside.

Pleasing brown, doughy.

*

PIES

What could possibly be a more typically American dessert than apple pie? Surely this fruit pie is an all-time favorite, but numerous other fruit, cream, custard, and chiffon pies are also very popular. When you master the knowledge and skill needed to make pies worthy of compliments, you can make desserts to suit any occasion and most palates. Pies can be varied by selecting different fillings or crusts and occasionally modifying the toppings.

How many different kinds of pies can you name?

You doubtless can name a long list, but it may surprise you to know that entire cook books are devoted to recipes for pies.

*

The crust most frequently used is prepared from a pastry dough. Most pastry recipes are very simple and contain only shortening, flour, salt, and water. The success of your pastry is closely linked to your knowledge and skill in manipulating the dough. Of course, the proportions of the ingredients also strongly influence your final product.

To make a good pastry, select a good recipe and develop your _____ in handling the dough.

Skill.

One of the chief problems in making pastry is overdevelopment of the gluten. A strongly developed protein network in pastry causes the crust to be tough. Of course, a tough crust is not desirable. The ratio of shortening to flour and the quantity of water greatly influence the rate of gluten development. A high ratio of shortening to flour delays gluten development because the shortening appears to coat the gluten strands, causing them to slide rather than to cling to each other as the dough is mixed. This coating action of shortening also makes it more difficult for the water in the crust to moisten the gluten.

Increasing the shortening in a pastry dough **increases/ decreases** *tenderness because it: (1) makes the _____ strands more slippery and (2) retards the moistening of gluten by _____.*

Increases, gluten, water.

*

You will find many variations of a basic pastry recipe. Some state that one cup of flour should be used for each half cup of shortening. This is a ratio of 2:1, that is, two parts flour to one part shortening by measure. It is easy to make a tender crust with such a high ratio. Unfortunately, this proportion results in a crust that is greasy and unnecessarily high in Calories. With a little practice and guidance, most people can make a very tender crust with a ratio of 3:1 (one cup flour to one-third cup shortening). Such a crust does not seem greasy and provides fewer Calories than a crust with a 2:1 ratio. A still leaner crust can be made with a 4:1 ratio (one cup flour to one-fourth cup shortening). This ratio, of course, is lower in Calories and less tender than the other crusts mentioned. However, with care and practice, you can learn to make a tender crust even with this lean mixture.

What ratio of flour to shortening do you think is the best choice for making pastry?

Most people prefer the 3:1 ratio.

*

The amount of water used in pastry is critical and must be carefully controlled. Enough water must be used to develop the gluten sufficiently to hold the crust together. However, too much water will quickly cause the crust to be tough due to excess gluten development. Approximately half as much water as shortening is an appropriate amount of water in pastry. This amount usually gives the necessary cohesive quality to the dough without making it sticky and tough. Flour does differ from one area to another in this country, so you may need to use a slightly different amount of water to make the crust cohesive yet not sticky.

How can you tell whether you have used the correct amount of water in a pastry dough?

The crust will roll out easily without cracking and will not tend to be sticky and soft when rolled.

*

The final ingredient in pastry is salt. It is added to heighten the flavor of the crust. In summary, a good pastry for a two-crust pie can be made by using one and one-half cups of flour, one-half cup of shortening, one-fourth cup of water, and three-fourths teaspoon salt. Note that this is a 3:1 ratio of flour to shortening and a 2:1 ratio of shortening to water.

If you are inexperienced in making pastry, do you think you can roll two crusts from the above dough?

Perhaps not. You might find it easier to use two cups of flour, two-thirds cup of shortening, one-third cup of water, and one teaspoon of salt until you develop your technique to the point where you can easily roll your dough into a well-rounded circle without unusual protrusions and indentations.

*

Pastry dough is made by thoroughly mixing the measured flour and salt with a fork. Then the shortening is added all at once and cut in with a pastry blender until most pieces are the size of rice grains. Of course, some particles will be much smaller and others will be larger. When cutting in the shortening, use a light flipping motion with the pastry blender. This same operation can be accomplished by cutting in the shortening with two table knives being used to cut against each other. This technique is more time consuming than is that using the pastry blender because the pastry blender has several cutting surfaces. After the shortening is cut in, the water is added very carefully. It works well to use a four-tined fork to flip the dough lightly while carefully dribbling the water dropwise throughout the dough. This technique is important, because it permits you to distribute the water fairly uniformly with an absolute minimum of mixing. Only a little stirring with the fork is necessary to make the dough cling together in a large ball after all the water has been added in the manner described above. Now, you deposit the ball of dough onto a rectangle of waxed paper and quickly work the dough with your hands until no pieces fall from the dough when the ball is unwrapped from the waxed paper. The dough is divided in half for a two-crust pie.

Describe a pastry dough made by adding all of the water at once and then mixing it in.

This dough is probably going to be very moist and sticky in some areas and dry and crumbly in others. If it is mixed enough to distribute the water more uniformly than this, the dough will be very tough. Be sure to sprinkle the water dropwise through-out the mixture rather than adding it all at once.

*

You will find that pastry for a pie is rather easy to roll out if you use a pastry cloth and a sock on the rolling pin. A pastry cloth is a hemmed, heavy muslin cloth. This is lightly floured by rubbing flour into the cloth over the area the rolled crust will occupy. Rub in flour until the cloth feels suede-like. Push any extra flour to one side where it will not contact the dough. A pastry sock is a heavy stockinette that is pulled over the rolling pin. This can be floured adequately by simply rolling the covered pin over the floured pastry cloth a few times. Now, take a ball of dough of the correct quantity to make one crust when rolled out. Gently, but quickly flatten this into a circular disc, using your right hand to flatten the dough and your left hand to shape the curve of the dough as it is flattened out. This step makes it simple to roll a circular crust because the crust has undergone preliminary shaping. The crust is rolled lightly in all directions uniformly to achieve a similar thickness throughout It is necessary to lift up on the rolling pin as you approach the edges if you are to avoid getting the edges too thin. Roll the entire crust until it is so thin that it just barely shows your fingerprint when the crust is touched with moderate firmness. A thick crust may make a pie give the impression of doughiness, but a very thin crust will become soaked too easily. Properly rolled dough is approximately one-eighth inch thick.

When rolling a pie crust, you should concentrate on maintaining a _____ shape and a _____ thickness.

Circular, uniform.

*

Dough can be easily transferred from the pastry cloth to the pie pan if you fold the dough in half and then in half again. This procedure makes a small package of dough which can be moved to the pan without stretching or tearing. Carefully unfold it and ease it into the pan without stretching the dough. If you are making a two-crust pie, roll out the second crust before putting the filling in the bottom crust. This method helps to avoid delay before baking and reduces soaking of the bottom crust.

Should you rub more flour into a pastry cloth before you begin to roll out the second crust?

Yes. Without the addition of a small amount of flour, the second crust may stick to the cloth.

*

Two-crust pies may be topped with a lattice top or with a regular crust equipped with simple ventilation vents. These two ways of finishing a pie require different techniques. For a lattice crust, trim the bottom crust approximately one-third inch beyond the lip of the pie plate. Then, arrange carefully cut strips of dough, cut to a uniform width, straight across the pie at regular intervals. Identical strips may then be placed across these or may be woven like a basket over and under the first strips. This latter procedure is rather laborious, but it does look somewhat neater than the top made by simply laying the strips across. After the strips are arranged, take kitchen shears and trim all the strips so that they end at the edge of the bottom crust. Now fold the portion of the bottom crust that is extending beyond the plate up and over the ends of the lattice strips and stand this folded crust upward to form a ridge all the way around the pie. To complete the pie, make an edging design. You can make a trim edge by resting your thumb and index finger of your right hand on the edge of the pie plate and pressing the dough between them into a small V-shape by pushing with the index finger of the left hand. Shift your right hand just to the right of the V and repeat the process uniformly all the way around the pie.

*To put a lattice crust on the top of a pie, trim the bottom crust **even with the edge of the lip of the plate/about one-third inch beyond the edge of the pie plate.***

About one-third inch beyond the edge of the pie plate.

392

The bottom crust of a pie that is to be topped with a conventional crust is trimmed by running a knife right along the outer edge of the pie plate. The filling is then added and the top crust adjusted into position. The top crust is trimmed, with kitchen shears, approximately one-third inch beyond the edge of the pie plate. Now lift the portion of the bottom crust that is resting on the lip of the plate and tuck the protruding portion of the top crust under it. Do this all the way around the pie; then go around again to put on the finishing V-shaped decoration as described above. To avoid a soggy upper crust due to trapped steam from the hot filling during baking, it is necessary to cut some steam vents in the top crust. Simple decorative designs can enhance the appearance of the pie while fulfilling this utilitarian function. Be sure to cut clear through the top crust, and cut large enough slits to be certain that they will not seal over with juice when the pie is baking. There is less likelihood of the filling spilling out through these vents if you confine your design to the center area of the pie.

The top crust of a pie needs to have slits cut for _____ to escape during baking.   Steam.

*

Single crusts are prepared by rolling them out as previously described and then gently easing them into the pie plate while being particularly careful not to stretch the crust at all. The pastry is then trimmed approximately one-third inch beyond the edge of the pie plate. This portion that extends beyond the plate is now turned under so that it just rests on the lip of the pie plate; an edging trim is made as before. Be sure that the V-shaped design rests firmly on the lip of the pie plate. This helps to hold the crust correctly in the plate during baking.

To make a single-crust pie, trim the crust **even with the edge of the pie plate/one-third inch beyond the edge of the pie plate.**   *One-third inch beyond the edge of the pie plate.*

*

Custard, pumpkin, and pecan pie fillings are placed in the unbaked pie crust, and the filling and crust are baked at the same time. For these pies that contain a filling when the crust is baked, do not prick holes in the crust. These holes would let the fluid filling run through and stick the crust to the plate as well as encourage soaking of the crust.

393

Cream pies, lemon meringue pies, and chiffon pies require that the filling be placed in a baked pie crust. Pie crusts baked without a filling in them are prepared as described previously. However, it is essential that you prick many small holes all over the crust before you bake it. These pricks can be made with a table fork; the holes should go clear through the crust to the pie plate. These holes prevent large blisters from forming in the crust as it bakes.

*Crust for custard pie **should/should not** be pricked before baking; for cream pies, you **should/should not** prick the crust.*

Should not, should.

*

To evaluate a pie crust, you will need to look at it carefully as well as to taste it. Most people value highly a flaky, tender crust. Flakiness is the layering that you can see when you look at a cross section of some crusts. This layered texture is promoted by cutting hard shortening into pieces the size of peas. Then, when the crust is baked, the shortening melts and leaves gaps in the gluten network. The steam formed from water in the baking dough causes expansion of these areas in the crust, resulting in a layered, flaky appearance. Sometimes crusts are made with salad oil rather than a solid shortening. The result is a crust that is described as mealy but not flaky. An oil crust will not be in flaky layers. However, it will usually be very tender because the oil spreads readily and unformly throughout a pastry.

If you wish to have a mealy pastry, use _____ _____ in place of solid shortening.

Salad oil.

*

In addition to looking at the flakiness or mealiness of a pastry, evaluate also on the basis of tenderness, crispness, and general overall appearance. A good crust is tender and easily cut; there is no sogginess in either the top or bottom crust. On the exterior, the pie should be neatly made and a pleasing golden brown.

Make a pie and evaluate it on the basis of the above criteria. How successful was your pie?

If you had any shortcomings in your pie, review the above material to determine what you can do to improve your next pie crust.

*

For a pleasingly golden-brown pie with a crisp crust, preheat the oven so that baking can proceed as soon as the pie is assembled. An oven temperature of 425°F quickly starts the baking and reduces the likelihood of juice from the filling soaking into the crust. For previously cooked fillings, the oven temperature may remain at 425°F until the pie is baked to a golden brown. Pies containing uncooked fruit fillings are usually baked at 425°F for 10 minutes and then at a setting of 350°F. This lower temperature for the remainder of the baking period allows time for the filling to become cooked before the outer crust is too brown.

Fruit pies with fillings that have been previously cooked are baked at _____°F; pies with uncooked fruit in the filling are baked at _____°F for _____ minutes, at which time the oven temperature is changed to _____°F.

425, 425, 10, 350.

*

If you search, you will find many different recipes for pastry. Some are made using boiling water as the liquid. Others are made with vinegar, while still others contain milk. The methods of combination are varied as well as the ingredients. In general, it appears that these variations produce different crusts, but that they are generally no more palatable than the simple basic crust described above.

True or false. Far better crusts are made by using exotic and unusual recipes rather than a basic pastry recipe.

False.

*

Crusts may be made from graham cracker or gingersnap crumbs as well as other types of cooky crumbs. The crumbs are usually mixed with enough melted butter or margarine to bind them together and a small quantity of sugar for flavoring. It is necessary to press this mixture together firmly in a pie plate so that the completed crust will hold its shape when cut. These crumb crusts may be baked in a 350°F oven for 10 minutes if desired; they are then chilled in the refrigerator until time to serve the pie. The cold butter or margarine will thus be sufficiently hard to hold the crust together.

True or false. It is necessary to bake a crumb crust.

False. Baking is not required for a crumb crust, but some people like the slight change in flavor that occurs during the baking period.

*

Other substitutes for the traditional pastry of a pie are available also. A coconut crust can be made by heavily

395

buttering a pie plate and then firmly pressing shredded coconut in a heavy layer all over the pie plate. This coconut crust is baked in a 350°F oven until it is a light golden brown. As you might expect, this crust is easier to serve if it is kept chilled until serving time so that the butter is firm. Angel pies are made by using yet another crust substitute. An angel pie crust is made by swirling a hard meringue mixture into a pie plate and baking it until it is dried out and very slightly browned. Sometimes soda cracker crumbs or chopped nuts are added to this type of pie shell for variety.

For an angel pie, make a **hard/soft** *meringue shell containing four tablespoons of sugar per egg white.*   *Hard.*

*

A pastry that is similar to the conventional pastry can be made by combining ground nuts such as filberts or almonds with flour and then continuing with the same procedure used in making a conventional crust. Slightly less shortening is used in this crust. You might also wish to try using grated orange rind and orange juice (in place of water) in a crust for a raisin or other fruit pie, or perhaps some grated sharp cheddar cheese in the crust for an apple pie might intrigue you.

True or false.   Many different foods may be added to basic pastry mixtures to individualize your pies.   *True.*

*

Now, let us look at the fillings that are used for pies. It takes approximately three cups of filling for a nine-inch pie. Slightly less than this amount may be used if you are going to put a meringue on top of the filling. It is important to have the correct amount of filling in a pie. If you are making a two-crust pie and have only a small quantity of filling in it, the upper crust will not brown well because it is too low in the pan for the hot oven air to reach it readily. On the other hand, too much filling is disastrous because the juices will bubble out over the edge of the pie. This spill-over will burn onto the oven and will result in an unattractive pie that may stick to the pan where the juices overflowed. When filling a pie, remember that uncooked fruit will shrink when baked. Therefore, apple and other fresh fruit pies need to be gently rounded before baking. Pack the fresh fruit in fairly firmly to reduce the distance the fruit will shrink during baking. If this precaution is not observed, you may make a pie with an arched upper crust that shatters when you cut through the crust and down into the filling resting lower in the pie plate.

Fresh fruit in a pie filling should be **gently packed/lightly placed** in the bottom crust.

Gently packed to minimize shrinkage.

*

Fillings for fruit pies may be made using fresh, frozen, or canned fruits. If canned or frozen fruits are used, the juice is usually drained from the fruit and thickened by heating it with a starch and sugar. This thickened mixture is then stirred into the drained fruit. This procedure eliminates unnecessary heating of the fruit itself and helps to retain the flavor. Frozen fruit, when prepared for a pie filling by this method, will have a flavor very closely approaching that of the fresh fruit. Fruit fillings should be high in fruit and contain only a limited amount of thickened juice. There is little satisfaction in eating a fruit pie where you have to hunt for the fruit through a sea of thickened juice. One cup of juice for each two cups of fruit is a maximum amount of juice to use, and you may prefer to use still less juice.

If you were making a fresh apple pie, would you cook the apples before putting them in the pie?

Apples and other fresh fruits that hold their juice well when cut usually are simply coated with a sugar and flour mixture and then arranged uncooked in the bottom crust. These fruits will soften and release some juice during baking, and the starch-sugar coating will sufficiently thicken the juice to give the filling the desired viscosity when baked and cooled.

*

Fruit fillings may be thickened with flour, cornstarch, or other starch products. Some people prefer to use tapioca as the thickener, while others prefer cornstarch. Cornstarch is generally preferable to flour because it gives the fruit juice a more translucent appearance. A modified cornstarch, called waxy maize, is an excellent starch for thickening fruit pies because it gives a pleasingly clear starch paste which is thickened but still flows slightly. Regardless of the type of starch selected, be sure to use just enough starch to thicken the juice sufficiently to avoid a runny filling and a soggy bottom crust. Too much starch gives a rigid, pasty filling that has little appeal. The proportions to use can be carefully measured when making fruit fillings with canned or frozen fruits, but the exact amount of thickener to use with fresh fruits is difficult to estimate because of the large variation in the juiciness of fresh fruit.

If a fruit filling is very thin and flows readily, you know that the problem could have been caused by:
A. *Using cornstarch instead of flour.*
B. *Not heating the filling to a high enough temperature.*
C. *Using too much juice in proportion to starch.*
D. *Using unusually juicy fresh fruit.*
E. *Using too little sugar.*   *B, C, or D.*

*

Cream pies are essentially cornstarch puddings to which egg yolks have been added. These pies are prepared by carefully stirring together cornstarch and sugar and then thoroughly stirring in milk. This mixture is heated at a moderate rate over direct heat and is stirred constantly until it comes to a boil and is thick enough to leave a path when a spoon is pulled through it. Then it is removed from the heat and four tablespoons of this thickened starch mixture are quickly but carefully stirred into beaten egg yolks. This egg yolk mixture is then stirred into the hot starch mixture and placed over boiling water for five minutes. This final five-minute heating period is done over boiling water to coagulate the egg yolks without overheating them and causing them to curdle. It is necessary to stir the mixture slowly during this final heating period because the egg yolk will coagulate unevenly and form lumps in the mixture if the pudding is not stirred. If you have a range with a sensitive enough heat control to prevent overheating and boiling of the mixture, you may cook the egg yolk mixture over direct heat rather than over boiling water. At the end of the five-minute cooking period, the egg yolk should have coagulated. This coagulation will cause a slight dulling in the glossy appearance of the filling and will increase its viscosity slightly. It is essential to have the egg yolk coagulated before the filling is poured into the baked crust. If the egg yolk is not coagulated, the filling will become very thin and runny as the pie cools.

If you were asked to cut and serve a banana cream pie and you discovered that it was so runny that you could only serve it with a spoon, what would you think was probably the cause of this failure?

Most likely the filling was not heated enough to coagulate the egg yolks after they were added.

*

Coconut cream pie fillings, butterscotch fillings, chocolate fillings, and lemon meringue pie fillings can be placed in a baked pie shell while the filling is still hot. The meringue is quickly added and the pie is placed in the oven to bake the meringue. The hot filling helps to bake the meringue by adding heat from beneath. However, for a banana cream pie it is necessary to let the filling cool almost to room temperature before it can be poured over bananas in the baked shell. A hot filling will cook the bananas and change their flavor.

Why do you cool the filling for a banana cream pie before pouring it over the bananas in the crust?

To avoid cooking the bananas and changing their flavor.

*

When you place the meringue on the filling, avoid any unnecessary manipulation of the meringue so that as much air will be retained as possible. However, it is very important to seal the meringue to the crust at every V-shaped indentation before it is baked. This can be done easily with a rubber spatula. With this same rubber spatula, make a few swirls on top to add an interesting appearance to the pie. The higher places will turn to a tempting golden brown that will contrast with the pale white of the lower portions of the meringue when it is baked. Avoid making high peaks because they will burn before the rest of the meringue bakes sufficiently.

*Use a rubber spatula to _____ the meringue to the crust and to make **swirls/peaks** in the meringue.*

Seal, swirls.

*

Lemon meringue pies are made in much the same way as the cream pies just discussed. However, one important difference should be noted. The lemon juice used in a lemon pie is, of course, quite acidic. If the lemon juice is cooked with the starch or with the egg yolk mixture, the filling will be thinner than you expected. The acid causes the starch to break down to a more soluble carbohydrate substance with less thickening ability. The protein in the egg yolks is also adversely affected by the lemon juice. To get maximum thickening of the filling, combine the starch, salt, sugar, and water and proceed as described above. After the starch mixture is thickened and has boiled, add some of the hot starch mixture to the egg yolks and recombine this with the filling. The heating of this mixture is the same as for the cream pies. After the egg yolk-starch mixture is heated for five minutes and the egg yolk proteins have coagulated, the butter and lemon juice are stirred in. At this point, the lemon juice does not cause a chemical change in the starch or protein.

The lemon juice in a lemon meringue pie recipe is added:
A. *At the beginning.*
B. *After the starch and sugar are mixed.*
C. *After the starch is gelatinized.*
D. *When the egg yolks are added.*
E. *After the egg yolk proteins are coagulated and cooking is completed.*

E.

*

Chiffon pies are yet another type of single-crust pie. Usually, chiffon pies are made by first coagulating an egg yolk-sugar-liquid mixture over boiling water and then softening gelatin in this hot base. This mixture is set aside to cool until the gelatin just begins to congeal it into a syrup-like liquid. Immediately, egg whites are beaten to stage 3 where the peaks just bend over. The slightly viscous gelatin mixture is folded into the egg white meringue until there are no streaks of the whites or gelatin. This filling is mounded into a baked pie shell and chilled until set. Chiffon pies are best if they are kept chilled until served and if they are served the same day they are made. Gelatin-stabilized chiffon pies develop a slightly rubbery character when kept a day or longer.

Why should a chiffon pie be kept chilled until it is served?

A chiffon pie is stabilized with gelatin and gelatin will become too fluid to be effective if the pie is allowed to remain long in a warm room.

*

A word of caution regarding the refrigeration of cream and chiffon pies is appropriate here. Both these types of pies contain eggs and they frequently also contain milk. Milk and eggs together are an excellent medium for bacterial growth. An unrefrigerated cream pie may look tempting to eat and yet may cause serious food poisoning. Therefore, avoid these pies when you can see that they have not been refrigerated adequately. In your own kitchen, refrigerate these pies when they have cooled almost to room temperature. Keep them refrigerated until you serve them. This simple precaution is all that is necessary to keep these pies wholesome.

True or false. Cream pies and chiffon pies can cause food poisoning if they are held at room temperature several hours.   *True.*

*

FRUIT FOR A DESSERT

Fruit may be served in many ways to provide a variety of pleasing desserts. One of the most satisfying ways of serving fresh fruit is to simply chill it. Often a piece of mellow, flavorful cheese will blend beautifully with such a dessert. Crisp apples, succulent pears, or sweet flavorful grapes are particularly pleasing with dessert cheese. Tropical and subtropical fruits such as mellow papayas, mangoes, and tart juicy pineapples provide an exotic finale to a meal. The natural acids in fruits contrast with their sugar content to give a stimulating flavor contrast that is very refreshing. Fresh fruits are not only simple to prepare but are also very good sources of various minerals and vitamins.

For a simple, yet pleasing dessert, serve **chilled/unchilled** *fruit accompanied by some* _____.   *Chilled, cheese.*

*

You may wish to prepare fruits in slightly more elaborate ways than just serving them whole or in slices. Pieces of various fruits may be combined to make a chilled fruit cup. Fresh berries, carefully sorted and washed, are a popular dessert when served with cream, whipped cream, or various sauces. Cherries, peaches, and some other fruits may be varied by serving them in a brandy sauce or other syrups and sauces. You may also wish to try combining puddings with fresh fruits. A tapioca cream pudding or a vanilla cream pudding topped with sweetened fresh berries or frozen berries makes a colorful and very nourishing dessert. Broiled grapefruit halves topped with a sprinkling of brown

401

sugar can be served successfully as either an appetizer or a dessert.

True or false.   Fruits may be served alone or in combination with sauces, syrups, cream, ice cream, or pudding for dessert.

True.

*

Some fruits discolor when they are pared or sliced. This discoloration occurs when air comes in contact with cut surfaces. Fruits that are particularly troublesome because of the darkening that occurs when the fruits are cut include bananas, apples, and peaches. Some other fruits also will darken, but some fruits show little or no tendency to change color or darken when cut. Citrus fruits, berries, and grapes are all good examples of fruits that retain their original bright colors. Obviously, it is important to avoid this darkening in susceptible fruits; this easily can be done by simply dipping the cut pieces in a little acidic fruit juice such as lemon or pineapple juice. This oxidative change can also be slowed by dipping the cut fruit in a solution of ascorbic acid and water. Ascorbic acid is an antioxidant which keeps oxygen from reacting with the pigments in the fruit. It is also wise to keep these fruits from contacting any copper or iron; these metals also cause the browning reaction.

Acidic fruit juices coat the fruit and keep the air from discoloring the cut surfaces.

If apples are sliced, they should be dipped in **cold water/ fruit juice (tart)** *to prevent darkening of the cut surfaces.*

*

When fruits are cooked, they may be served as sauces. A fruit is usually said to be in a sauce when it is served in a light sugar syrup; in a medium sugar syrup it is usually referred to as stewed fruit. Fruits may be mixed and cooked in a heavy sugar syrup to make a fruit compote for dessert. A fruit compote is most successful when you include tart as well as sweet fruits to give the necessary sparkle to the overall flavor combination. For apple sauce, it is customary to prepare the fruit so that it is broken into small pieces or pureed before serving. Rupturing of the fruit is encouraged by cooking the apples in a little unsweetened water and adding sugar after the cooking is completed. With most other fruit sauces, it is desirable to retain the shape of the fruit. The fruit is cooked in a sugar syrup containing two cups sugar for each cup water. This proportion helps to plump the fruit and to protect its shape. Canned and frozen fruits also may be served simply in their own syrup. The canned fruits are chilled to prepare them for a meal. Frozen fruits are best when served with just a few ice crystals still remaining in them. If thawing proceeds to completion, most frozen fruits become a bit mushy and lose their textural appeal.

Match column B with column A.

A	B	
1. *Stewed fruit*	A. *Mixture of fruits cooked in heavy sugar syrup*	1. B.
2. *Fruit sauce*	B. *Fruits cooked in medium sugar syrup*	2. C.
3. *Fruit compote*	C. *Fruits cooked in light sugar syrup*	3. A.

*

Dried fruits are excellent desserts when soaked and gently simmered. The high sugar content of dried fruits usually makes these stewed fruits sufficiently sweet without the addition of more sugar. To stew dried fruits, add a small quantity of water and gently simmer the fruit in a covered pan. A simmering temperature is sufficient to rehydrate the fruits without burning or breaking up their shapes. The covered pan is important because the lid traps the volatile, mild flavoring substances in the pan with the fruit and thus helps to retain the rather delicate flavor of stewed dried fruits. Dried fruits should be simmered until they are tender to cut. If they contain seeds, cook the fruit until it can be removed easily from the seeds. As you would expect, large fruits with no cut surfaces need to be simmered considerably longer than raisins. The exact length of time varies. You need to check the simmering fruit occasionally to see whether it is sufficiently tender. The amount of moisture in dried fruit also influences simmering time. Fruits naturally dried or artificially dried at atmospheric pressure by warm air currents usually contain approximately 25 per cent moisture; vacuum-dried fruits may have less than 5 per cent moisture remaining in them. The rehydration of these drier fruits takes slightly longer than for those dried at atmospheric pressure. A soaking period prior to simmering reduces the actual cooking time required to rehydrate and tenderize the fruit.

The purpose of cooking or stewing dried fruit is to _____ and _____ the fruit.

Rehydrate, tenderize.

*

Fruits are used to make several desserts that incorporate various quick breads. Biscuit doughs, shortcake, and cobblers combine well with fruit to make pleasing desserts. These have been described previously under the preparation of quick breads. Other fruit desserts of this nature are a betty and a crisp. A betty is a fruit dessert prepared by alternating layers of crumbs, sugar (often brown), butter, and fruit. The crumbs and fruit form the main portion of the dessert. The sugar and butter are sprinkled and dropped in primarily as flavoring items. This mixture is then baked. A crisp consists of a layer of thickened fruit topped with a crust consisting of approximately equal quantities of flour, sugar, and margarine or butter. When this mixture is baking in a baking dish, the dough mixture on top fuses into a solid, crisp crust. Sometimes a peach crisp is made with uncooked oatmeal, butter, and brown sugar as the top crust over a thickened peach sauce. Very often, fruit desserts of this type are served with whipped cream, a custard sauce, or a hot fruit sauce prepared by thickening fruit juice. They are particularly pleasing when served warm.

Can you remember what was discussed under quick breads and fruit desserts to complete this question by matching column A with column B?

A	B	
1. Shortcake	A. Fruit mixture with crisp, sweet topping	
2. Cobbler		
3. Fruit roll	B. Fruit mixture with a pie crust topping	
4. Crisp		
5. Betty	C. Fruit mixture with biscuit topping	
6. Deep-dish pie		
	D. Fruit mixture baked in a biscuit roll and sliced	1. F.
		2. C.
	E. Fruit mixture baked between layers of crumbs, sugar, and butter or margarine	3. D.
		4. A.
		5. E.
	F. Fruit mixture placed between baked rich biscuits	6. B.

*

PUDDINGS

Many puddings, especially those containing milk and eggs, may be desserts that add significantly to the nutritive value of a meal. One versatile and simple pudding is *blanc mange* or vanilla cornstarch pudding. *Blanc mange* is a mixture of milk, sugar, cornstarch (or other similar starches), and vanilla. Variations of this basic pudding are easily made by using brown sugar for butterscotch pudding, chocolate or cocoa for chocolate pudding, or a burnt sugar syrup for caramel pudding.

Butterscotch, caramel, and chocolate are variations of a _____ pudding.

Cornstarch.

*

A cream pudding is perhaps poorly named because it is simply a cornstarch pudding to which egg yolks have been added. Cream is not customarily used in the preparation of a cream pudding. However, *blanc mange* or vanilla cornstarch pudding is transformed from a stark white to a creamy color in a cream pudding because of the yellow color contributed by the egg yolks. This color change may explain the title selected for a "cream" pudding. Egg yolks may be added to any of the variations of cornstarch pudding. The addition of the egg yolks is made after the starch mixture has been completely thickened. As described in the section on cream pie fillings, the egg yolks are combined with the hot pudding by first blending some of the hot pudding into the beaten egg yolks and then stirring this mixture back into the remainder of the pudding. Again, it is essential to stir this mixture during the five-minute cooking time over boiling water. If this cooking period is not adequate, the pudding will be very thin. For a lighter pudding, you may beat the whites remaining after the yolks have been added to the cream pudding. Make a meringue of these whites and fold the cream pudding mixture into the whites. This type of pudding is sometimes called a chiffon cream pudding because of its very light and airy character.

*

Match column A with column B.

A	B	
1. *Blanc mange*	A. *Milk, sugar, cornstarch,*	1. *A.*
2. *Cream pudding*	*vanilla*	2. *B.*
3. *Chiffon cream*	B. *Egg yolks added to a*	3. *C.*
pudding	*cornstarch pudding*	
	C. *Pudding mixture folded into*	
	an egg white meringue	

*

Smooth texture is essential in any of these puddings. Lumps can be avoided by observing the following suggestions. Carefully blend the sugar and starch together before adding any liquid. Be sure to do this very thoroughly. If you are making a large quantity of pudding, you may wish to scald approximately three-fourths of the milk to reduce the cooking time required to gelatinize the starch. The remaining one-fourth of the milk (or all of the milk if you are making only a couple of cups of pudding) should be stirred carefully into the dry mixture without heating the milk at all. Make a perfectly smooth slurry with the cold milk before stirring in the scalded milk. Now, heat the pudding at a moderate rate while stirring conscientiously to avoid forming lumps while the starch is thickening the pudding. If you are making a cream pudding, care also must be taken to stir the small quantity of hot pudding into the egg yolks thoroughly and then stir while adding this mixture back to the pudding and while thickening the egg yolks. With these precautions, your pudding should be perfectly smooth.

Do you think that it is possible to stir a pudding too much?

Surprisingly enough, you can overstir a pudding. A pudding that has been stirred too hard will feel pasty and gummy on the tongue.

*

Tapioca pudding is actually a variation of cornstarch pudding; tapioca is the thickening agent used here so there is no cornstarch in these recipes. There are two basic types of tapioca that might be used: one is long cooking and the other is a quick-cooking product. The quick-cooking tapioca is broken into small particles that swell readily. The long-cooking variety is often called pearl tapioca because it is in large pearl-shaped pieces. For maximum thickening of a tapioca pudding, do not heat the pudding to boiling. Tapioca starts to lose its thickening ability when it is heated too much. Considerable care must be taken in

406

the stirring if you are to avoid the stringiness that develops in an overstirred tapioca pudding. Tapioca cream pudding is made by folding this hot thickened pudding into an egg white meringue. Tapioca puddings are excellent served with fresh sweetened berries or other fruits, either fresh, frozen, or canned.

The term used to describe a tapioca pudding containing egg yolks and an egg white meringue is: (a) tapioca pudding, (b) tapioca cream pudding, or (c) tapioca cream chiffon pudding.

(b)
Unlike the cornstarch puddings made with meringues, the term chiffon is not usually applied; hence this is simply called tapioca cream pudding.

*

There are many other types of starch-thickened puddings. In the New England states, early settlers learned to make Indian pudding, a pudding which is still popular in that region. Cornmeal, brown sugar, spices, and molasses form the base of this pudding. Cake, bread, or cracker crumbs, farina, or sago may be added as thickeners to make other puddings. Additional flavor and interest are often achieved by adding raisins, other fruits, and nuts to the basic recipe.

Indian pudding is thickened with _____.

Cornmeal.

*

Steamed pudding is a traditional dessert at Christmas time. These puddings are not difficult to make. Another advantage is that they can be kept satisfactorily for a long time so that they can be made ahead of the holiday season. They usually are made from flour or from a starch product such as bread, cracker, or cake crumbs. Frequently, eggs and a small quantity of liquid are included in the recipe. Some steamed puddings are cake-like mixtures such as a steamed chocolate pudding; others such as plum pudding contain a large quantity of fruit and nuts and closely resemble a fruit cake. The ancient origins of steamed puddings are evidenced by the recommendation in many recipes that suet be used as the shortening. Some steamed fruit puddings have no leavener. Others may use soda as an alkaline ingredient to react with the acidic fruits to produce a little carbon dioxide gas for leavening. The soda causes the pudding to be a distinctly darker color. Steamed puddings are served hot so they should seem reasonably tender and only moderately compact on the tongue.

407

*Steamed puddings are **easy/difficult** to make and keep well/poorly.*

Easy, well.

*

To steam a pudding, grease a metal container well and then lightly dust it with flour. A coffee can works very well for steaming a pudding. Drop the batter gently from a large spoon into the can. This pudding will expand so the cans should not be filled clear to the top. Cover the container with aluminum foil and place it on a rack in a covered pan containing about an inch of water. The water is maintained at a boil to keep the pan well filled with steam during the entire steaming period. The actual time required for steaming varies with the size of the container being steamed. An average time is approximately two hours. After steaming, the pudding is allowed to cool and is then tightly covered with foil or a tight plastic cover. It may be stored in a refrigerator or freezer for long-term storage. Steamed puddings are usually served warm with a hot fruit sauce or a hard sauce, although whipped cream may be used.

Actually, it also is possible to bake them at a very low oven temperature for a long time or to boil them.

Do you think that steamed puddings can only be cooked by steaming them?

*

GELATIN DESSERTS

Many gelatin mixtures are used as desserts. These may simply be prepared by combining boiling hot water (two cups) with approximately one-half cup of a prepared mix of gelatin, flavoring, and sugar and allowing this mixture to chill and set. You can get variety in gelatin desserts by adding fruits of many kinds or by cutting the gelatin into cubes or ricing it. You can blend various gelatin flavors and colors in layers by allowing one layer to set and then pouring another layer that has cooled to a syrupy consistency. After this layer is allowed to set, still another layer may be added. For variety, try putting some riced gelatin in a glass dish, adding chilled *blanc mange* or cream pudding, and then topping this with more of the riced gelatin mixture. For some desserts, it is desirable to use the flavored gelatin mixtures that are commonly available. In other recipes, you may need to use the unflavored gelatin. Unflavored gelatin usually needs to be hydrated in cold liquid before being used because it is in large particles and is difficult to dissolve if added directly to hot liquid. Desserts made with plain unflavored gelatin have other ingredients that contribute both color and flavor to the finished dessert.

True or false. *Flavored gelatin mixtures can be combined directly with boiling hot liquids, but unflavored gelatin should be softened in cold liquid before being combined with hot liquids.*

True.

*

Gelatin desserts are often prepared in molds to give added design interest. To unmold these products, it is necessary to melt, just barely, the gelatin that is in contact with the mold. This can be done by quickly dipping the mold in a container of warm water or by applying a towel dipped in warm water to the mold briefly. Do not linger over this process. Invert a plate over the mold and then turn the mold and plate right side up. Give a quick shake and the gelatin should slide easily from the mold onto the plate. If the gelatin remains in the mold, apply a little more heat and repeat the unmolding process. Do be careful to avoid overheating the mold or the gelatin will soften too much and the design from the mold will be rather indistinct.

Gelatin can be released from a mold by quickly dipping the mold in **warm/cold** *water or by applying a _____, damp _____ very briefly.*

Warm, warm, towel.

*

REVIEW

1. When shortening and sugar are creamed and then eggs are beaten in, you are making a cake by the _____ method.

Conventional.

2. The two main classes of cakes are _____ and _____.

Shortened, foam.

3. If you substitute butter for shortening in a cake recipe, you need to add *more/less* butter than the shortening required in the recipe.

More. Butter is only 80 per cent fat.

4. In a modern shortened cake, frequently the sugar *exceeds/ is less than* the amount of flour.

Exceeds.

409

5. *True or false.* During mixing, eggs are considered as a liquid in a cake recipe.

True. Whites are about 87 per cent moisture, whole eggs are about 75 per cent, and yolks are 50 per cent.

6. *True or false.* Foam cakes are made using cake flour; some pound cakes and other shortened cakes occasionally may be made with all-purpose flour.

True. However, most shortened cakes are made with cake flour.

7. Outline the procedure for the addition of the dry and liquid ingredients when making a cake by the conventional method.

One-third of the dry ingredients will be added after the sugar, shortening, and eggs have been creamed thoroughly. Next, one-half of the liquid is stirred in, followed by one-third of the dry ingredients, one-half of the liquid, and the last one-third of the dry ingredients. After each addition, the batter is mixed to combine the ingredients.

8. *True or false.* To make a cake by the muffin method, you must use either a melted fat or salad oil.

True. Shortening is treated as a liquid ingredient in the muffin method.

9. If a cake has a ratio of a pound of flour, a pound of butter, a pound of sugar, and a pound of eggs, it is a _____ cake.

Pound.

10. *True or false.* Angel cakes have shortening in them.

False.

11. *True or false.* Sponge cakes are leavened by steam and air.

True.

12. *True or false.* Chiffon cakes are leavened by carbon dioxide released from baking powder, as well as by air and steam.

True.

13. To make an angel cake with a mix containing dried egg whites, it is necessary to beat the whites until the peaks *just bend over/stand up straight.*

Stand up straight.

14. To make an angel cake with fresh egg whites, it is necessary to beat the whites until the peaks *just bend over/stand up straight.*

Just bend over.

15. Sugar added to an egg white foam as it is being beaten *delays/speeds* foam formation.

Delays.

16. *True or false.* Sugar beaten into an egg white foam aids in stabilizing the foam.

True.

17. *Warm/cold* eggs whip into a foam more easily.

Warm (room temperature).

18. Cake recipes need to be adjusted at any altitude above *2000/5000* feet.

2000 feet.

19. *True or false.* Angel, sponge, and chiffon cakes usually are baked in tube pans and cooled in an inverted position.

True.

20. Cookies may be divided into what three categories?

Rolled, bar, and drop.

21. To make a flaky pie crust, use *a solid shortening/an oil.*

A solid shortening. This should be cut into particles the size of uncooked rice kernels to obtain a flaky crust.

22. A crust made with salad oil will be *tough/tender* and *mealy/flaky.*

Tender, mealy.

23. A custard pie is baked in a *baked/unbaked* crust.

Unbaked. The filling and crust are baked at the same time.

24. Fruit pies are usually baked 10 to 15 minutes at _____°F; the remainder of the baking is done at _____°F.

425, 350.

25. It takes nearly a *quart/pint* of filling to fill a nine-inch pie.

Quart.

26. To get maximum thickening in a lemon meringue pie, add the lemon juice *before cooking/after cooking is completed.*

After cooking is completed.

27. If a cornstarch pudding has egg yolks added to it, it is called a _____ pudding.

Cream.

28. *True or false.* Soda darkens the fruit and molasses in a steamed pudding.

True.

14

Sugar
Cookery

Sugar serves several functions in food preparation. It is a sweetener in many foods. In baked products, sugar is important as a tenderizing agent. It also increases volume and aids significantly in the browning of baked goods. However, sugar is not used solely as a supplemental ingredient. In candies and icings, sugar is the most important single ingredient. Of course, sugar is responsible for the very sweet flavor of these items. You will learn in this chapter about the way in which the concentration of sugar influences the firmness of candies and icings.

In candies and icings, sugar contributes _____ and influences the _____.

Sweetness, firmness.

*

Several types of sugar are marketed for use in the various food products containing sugar. The discriminating consumer knows and selects the sugar product best suited for the item being prepared. The two practical food sources of sugar are sugar cane and sugar beets. Either of these crops can be treated to extract the sugar. The resulting syrup is refined and crystallized to yield granulated sugar.

Granulated sugar can be obtained from _____ _____ and _____ _____.

Sugar cane, sugar beets.

*

The color of brown sugar, obtained from unrefined cane sugar, is important because flavor and color are directly related. The darker the color, the stronger is the flavor of the brown sugar. The darkest brown sugar contains the most impurities and has the strongest flavor. Light-brown sugar has a flavor that is more subtle and less sharp because it contains fewer impurities. When selecting brown sugar for a recipe, pick the dark-brown sugar if you desire to have a dominant brown sugar flavor. For more subtlety, choose the light-brown sugar. Of course, the dark-brown sugar results in a product that is darker in color. Usually, however, the choice should be based on the flavor desired.

True or false.   Light-brown sugar has a milder flavor than has dark-brown sugar.

True.

*

During the manufacture of granulated sugar, the raw sugar syrup is filtered to remove impurities. This filtration process results in clarification of the syrup, and this syrup can be precipitated to form the familiar white granules of sugar used in so many food products. The size of the crystals produced can be controlled. Most granulated sugar in this country is fairly fine and dissolves quite easily. It is coarse enough to cream well in baked products and trap air in fats creamed for cake batters, yet it is fine enough to dissolve without difficulty when making candies. Meringues are most easily prepared when sugar of still finer crystal size is used. This small-sized sugar crystal is marketed as superfine or dessert sugar.

*For most purposes, **granulated/superfine** refined sugar is suitable, but for meringues **granulated/superfine** sugar is a better choice.*

Granulated, superfine.

*

Powdered or confectioners sugar is made from granular sugar by simply grinding the sugar into extremely fine particles and then mixing this sugar with cornstarch to prevent caking of the sugar. The three per cent cornstarch level is adequate to absorb moisture from the atmosphere that would otherwise cause the sugar to cake. This cornstarch does give a slightly cloudy appearance to products made with powdered sugar; you may sometimes also be aware of a raw starch taste in icings made with powdered sugar.

Powdered sugar contains _____ per cent _____ to absorb moisture and prevent caking of the pulverized sugar.

Three, cornstarch.

Corn syrup is another sweetener used in cookery, often in conjunction with another type of sugar. Corn syrup, as its name implies, is a corn product. It can be made either by the action of acid or by the addition of an enzyme to cornstarch. Acid or enzyme hydrolysis causes a breakdown of the starch to smaller molecules; eventually the majority of the products present will be sugars. Corn syrup contains a mixture of glucose, maltose, and other short-chain carbohydrate molecules. Although this product is a concentrated mixture of sugars, crystallization does not take place because of the variety of substances present in the solution. There is little tendency in such mixtures for any sugar to precipitate or form crystals.

When cornstarch is treated with an acid or an enzyme, the starch will be hydrolyzed and the resulting product mixture will be marketed as _____.

Corn syrup.

*

Molasses is a rather acidic sweetener that is a by-product of sugar production from sugar cane. This sweetener contains numerous impurities. Iron is nutritionally the most important of these contaminants. As with brown sugar, the flavor and color of molasses are interrelated. Very dark molasses has a flavor that penetrates any product in which it is used. The lighter molasses has a more mellow, less strident flavor.

True or false. You could use either dark or light molasses in gingerbread.

True. However, you will notice a strong molasses flavor and dark color if you use the dark molasses. The choice is largely a matter of personal preference.

*

PROPERTIES OF SUGAR

One physical property of sugars is their ability to attract water from the atmosphere. Because of their affinity for water, sugars are said to be hygroscopic. This property is useful in baked products because it retards moisture loss from products containing sugar. In candy cookery, this hygroscopic nature of sugar needs to be considered when making sugar in a humid environment. Candy can actually attract a significant amount of moisture from very damp air just in the short time required to cool the candy to be beaten.

Sugar is said to be **hygroscopic/hydrophobic** *because it does/does not attract moisture from the air.*

Hygroscopic, does.

*

Sugar will dissolve to form a true solution. Because of this property, sugar does influence the boiling point of a sugar solution. This particular property is extremely significant in the preparation of candies. At any given temperature just so much sugar can be dissolved. If additional sugar is added at that temperature, it cannot be dissolved. There will be a dynamic situation in which the undissolved sugar and the dissolved sugar are in equilibrium. That is, as a granule of sugar goes into solution, some of the sugar that was dissolved will precipitate. No more sugar can actually be dissolved without having some that was in solution precipitate from solution. This condition is described as being a saturated solution.

At a given temperature, a saturated solution **can/cannot** *dissolve additional sugar.*

Cannot.

*

If the temperature is raised, more sugar can be dissolved. In other words, a larger quantity of sugar can be dissolved at a high temperature than can be dissolved at a lower temperature. You can observe this when you make candy. When the ingredients are all placed in the pan and stirred, the mixture feels gritty because much of the sugar will not go into solution in the cold mixture. As the candy becomes warmer, the gritty feel gradually is reduced until you cannot feel any sugar crystals with the spoon as you stir the solution. This is because the hotter liquid is able to dissolve more sugar than the cool liquid. Therefore, all of the sugar gradually goes into solution as the temperature rises.

More/less *sugar can be dissolved at a hot temperature than at a cold one.*

More.

*

Sugar dissolves to form a true solution, which causes the vapor pressure of the solution to be reduced. This reduction in vapor pressure raises the temperature of the boiling solution above the temperature of boiling water. The greater the concentration of the sugar, the higher the boiling temperature of the sugar solution will rise. In candy

415

making, this increase in concentration of sugar in the boiling candy solution is caused by evaporation of water from the boiling candy. Of course, evaporation takes place gradually and, as a consequence, the concentration of sugar slowly increases. The increase in the concentration of sugar is paralleled by the gradual rise in the boiling temperature of the candy.

When making candy, you increase the concentration of sugar in the boiling solution by **evaporating some water/ gradually adding more sugar.**

Evaporating some water.

*

As the temperature of the boiling sugar solution rises, the solution will constantly be saturated. The increasing concentration of sugar is still all soluble in the remaining liquid. This is true throughout the boiling period. You will notice that this saturated solution has undergone changes since it was first mixed together. The solution no longer has a gritty feel because all of the sugar is now in solution. The solution gradually becomes thicker or more viscous as the sugar concentration is increased. In nonchocolate candies such as caramels or peanut brittle, you will also note a change in color. As the candy temperature climbs ever higher, there will be some caramelization of the sugar with the result that the color becomes a darker and darker brown. There is also a chemical change that takes place in sugar as it is cooked in candy mixtures, particularly in those containing an acid such as cream of tartar. Candy contains sucrose, which is the chemical name for the granulated sugar used in food preparation. This sugar undergoes a slow chemical change as the candy cooks, and some of it will be broken into two other sugars, glucose and fructose. Actually, a molecule of sucrose consists of a molecule of glucose and one of fructose chemically linked by the elimination of a molecule of water. It is thus not surprising that a breakdown of sucrose gives equal quantities of these two component sugars. This breakdown of sucrose into glucose and fructose is called inversion. The mixture of glucose and fructose resulting from inversion is often referred to as invert sugar.

When sucrose is broken down during cooking, the resulting product (called _____ _____) contains equal quantities of _____ and _____.

Invert sugar, glucose, fructose.

416

During sugar cookery, only a part of the sucrose is broken down. This means that the finished mixture contains sucrose, fructose, and glucose. The amount of inversion that takes place is influenced by the rate of heating and the addition of acid. A slow rate of heating permits more inversion to take place than will occur when candy is heated at a fast rate. Slow heating means a long cooking period to reach the final temperature and there is more time for inversion to proceed. Acid in the form of cream of tartar speeds the breakdown of sucrose. The amount of invert sugar formed is of importance because candies containing a mixture of sugars, rather than simply the sucrose called for in the uncooked recipe, are less firm and are less likely to have a coarse texture.

*Inversion of sucrose is aided by the addition of an _____ such as cream of tartar and by a **fast/slow** rate of heating.*

Acid, slow.

*

Some recipes include a mixture of sugars and, consequently, are easier to make into a smooth candy than are basic fondant recipes. Corn syrup is commonly used in candies to help make them smoother. It provides a mixture of two sugars, maltose and glucose, along with other larger carbohydrate molecules. Sometimes recipes use honey, which is a source of fructose and glucose. These sweeteners are often used in combination with sucrose and thus offer a mixture of sugars in the finished product despite the rate of cooking.

*One important reason for including corn syrup as a part of a candy recipe is to provide a **single/mixture** of sugars to aid in producing a smooth-textured candy.*

Mixture.

*

CANDIES

There are two basic types of candies. Crystalline candies have an organized crystalline structure. These candies characteristically are easily bitten. They feel soft and creamy in the mouth. Familiar examples of crystalline candies include fudge, panocha, mints, fondant, and the cream centers in chocolate creams.

_____ candies have an organized crystalline network; examples of this type of candy include _____, _____, _____, _____, and _____.

Crystalline, fudge, panocha, mints, fondant, creams.

*

Amorphous candies do not have an organized crystalline structure. These candies are cooked to a higher final temperature than are crystalline candies. They also contain interfering substances that make crystallization difficult. Caramels, taffy, and brittles of various types are all examples of amorphous candies. These candies vary from caramels, with their very chewy consistency, to taffy and brittles that are extremely hard to the point of being brittle.

*Amorphous candies are cooked to a **lower/higher** final temperature than crystalline candies and their consistency is **softer/harder.**

Higher, harder.

*

Preparing Crystalline Candies

For the best success in making crystalline or amorphous candies, it is necessary to have an accurate candy thermometer so that you can carefully determine the final temperature to which the candy mixture is cooked. It is wise to check your candy thermometer by placing it in actively boiling water and noting the temperature. If your thermometer does not show that water boils at 212°F, it will be necessary to make a correction in the final temperature indicated in the candy recipes. For example, if your recipe says that the candy should be boiled to a final temperature of 238°F but your thermometer registers 210°F when water boils, then the candy should only be boiled until your thermometer reads 236°F. Such calculations are essential if your candy is to be the correct firmness.

If you were making a candy that should be boiled to 240°F and your thermometer showed that boiling water was at 214°F, what temperature should your thermometer be indicating when you remove the candy from the range?

242°F. Your thermometer registers 2°F higher than it should at sea level, so the recipe must be corrected by adding 2°F to the final temperature.

*

Some persons may wish to make candy or icings without a thermometer. They rely on the appearance of the cooked syrup when a small amount is dropped into cold water. This method is not very accurate, but it is better than simply guessing by the length of time the mixture has boiled. The following descriptions and their corresponding temperature range will help you if it is ever necessary for you to try to make candy without a thermometer.

Appearance in Cold Water	Temperature Range
Soft ball, flattens on removal	234–238°F
Medium ball, holds shape fairly well	239–244°F
Firm ball, holds shape	245–248°F
Hard ball, yields to pressure	249–254°F
Very hard ball, resists pressure	255–265°F
Crack, rigid	266–288°F
Hard crack, syrup spins long brittle thread	289–300°F

For the best control in candy production, you should use a **thermometer/cold water test.** *A thermometer is* **more/less** *accurate than is the test in which a drop of candy is placed in cold water.*

Thermometer, more.

*

The ingredients for crystalline candies should be placed in a heavy pan that will heat uniformly. The candy is then heated at a moderate rate of speed controlled to cook the candy in a reasonable length of time without letting it get so hot that the ingredients become scorched. To help distribute the ingredients uniformly and to reduce the likelihood of scorching, the candy is stirred slowly while being heated. Stirring helps to ensure that all the sugar goes into solution during the cooking period. When making crystalline candies, stirring should be discontinued one or two degrees below the final temperature.

True or false. Candy should be stirred while boiling so that it will be uniformly thick and free of lumps throughout.

False. Stirring is done to avoid scorching, to aid in dissolving sugar, and to be certain that all ingredients are uniformly mixed.

*

Crystalline candies contain a high concentration of sugar and must be cooled very carefully if they are to be smooth. Care must be taken to avoid disturbing the candy in any way while it is cooling. As soon as the candy is removed from the heat, it should be poured into a platter for cooling or should simply be left undisturbed in the original pan. Butter and vanilla are added but are not stirred in.

419

The butter melts and spreads across part of the surface to form a bit of protection against undesired crystallization during cooling. These two ingredients are blended in adequately during the beating that is done before the candy structure is finally set. It only disturbs the candy unnecessarily to stir these ingredients in before the candy has had an opportunity to cool. Absolutely nothing must happen to disturb the candy while it is cooling if you wish to make perfectly smooth crystalline candies.

When the vanilla and butter are added at the end of the cooking period, they **should/should not** *be stirred into the candy.*

Should not.

*

When candy is removed from the heat, the solution is said to be saturated. It is incapable of dissolving any more sugar at that temperature. However, as the candy solution cools under ideal circumstances, the solution becomes supersaturated. This means that some sugar is actually in solution that really should not be held in solution. Supersaturation is achieved by heating a mixture to a higher temperature and then carefully cooling the solution so that the crystals that are in excess do not precipitate. The cooler the candy solution becomes, the greater is the degree of supersaturation and the more unstable is the solution. As a candy mixture cools below 150°F, it becomes extremely unstable and many things may cause crystallization to begin. When crystallization begins, it continues fairly rapidly until all the excess sugar is precipitated. You then have a saturated solution and precipitated sugar crystals. As the solution continues to cool to room temperature, the excess sugar continues to precipitate on the existing crystals.

When candy is being removed from the heat, it is a **saturated/supersaturated** *solution; as it cools under carefully controlled conditions, the solution becomes* **saturated/supersaturated.**

Saturated, supersaturated.

*

Many things can cause a supersaturated solution to begin to crystallize. Anything that might fall on the surface of a supersaturated solution will seed the solution and cause crystals to start to form. It is important that no lint or dust fall into the solution nor any undissolved sugar crystals remain to seed and start crystallization. Absolutely nothing must disturb the unstable supersaturated solution until it has cooled to approximately 110°F. Then beating should be started to initiate crystal formation, and it should be continued until the crystallization is virtually complete. At that point, the candy will soften very slightly. Immediately spread it as quickly as possible to the desired thickness so that it will be ready to be cut into squares. If you should happen to miss this point for spreading the crystalline candy mass, it will suddenly become a solid yet crumbly product. This candy can be kneaded with the fingers to give a malleable mass that can be worked until all loose portions are blended in and then the whole candy can be manipulated into the desired thickness for cutting.

Crystalline candies should be cooled **with/without any** *disturbance until they reach a temperature of* **110°F/212°F**, *at which time they should be* **beaten/poured** *into the serving dish.*

Without any, 110°F, beaten.

*

The procedure outlined above assumes that no undesirable circumstances caused crystals to start forming before the candy cooled to 110°F. If somebody stirred the cooling candy or touched the surface in any way, crystallization will have begun when the candy was disturbed. It is then necessary to begin beating the candy and to continue beating it until it softens just before it sets in final form. If you do not begin to beat a candy in which crystallization is taking place, you will find that the candy has a very gritty coarse feel on the tongue. This sandy feel is due to large sugar crystals that can be detected readily in the mouth. The purposes of beating are to break up sugar crystal aggregates and to prevent large clumps of crystals from joining together. The tongue cannot detect small sugar crystals, but it can easily signal the presence of large aggregates of sugar crystals.

The purpose of beating is to encourage formation of many **small/large** *crystals of sugar in a crystalline candy.*

Small.

*

It is important to continue beating without stopping from the time crystallization begins. If you beat for awhile and then stop to rest, you allow large aggregates to form and the product will not be good. Many people have the idea that beating is done to make the candy firm. This is not

421

true. The purpose of beating is to control crystal growth and produce a candy with a velvety smooth texture. The firmness of a candy is determined by the final temperature of the candy and by the amount of inversion that took place during the cooking period. A candy that is too soft probably has not been boiled to a high enough temperature and, consequently, has an inadequate concentration of sugar. It is possible, however, that too soft a candy might be caused by an unusually slow rate of heating which allowed too much conversion of sucrose to glucose and fructose.

True or false.   It is reasonable to assume that a crystalline candy that is too soft should have been beaten longer to make it firmer.

False. The candy would still be too soft even if it were beaten longer. The remedy for this problem is to reboil the candy until it reaches the correct final temperature.

*

The slight softening of a crystalline candy just before it finally hardens is caused by the release of heat as the sugar crystallizes in the mixture. The formation of sugar crystals is an exothermic change; that is, heat is released when sugar crystallizes. This slight temperature increase explains the almost imperceptible softening that takes place at the moment when the candy crystallizes. You will notice also that at this same time the candy loses some of its sheen and has a duller appearance. This is also a signal that crystallization is taking place.

When candy is just about to set, you will notice that it will **soften/harden**   *very slightly for a brief moment and the surface will lose some of its*   **sheen/dullness.**

Soften, sheen.

*

Perhaps you are wondering why it is wise to cool crystalline candies carefully to 110°F without disturbing them. If you can cool the candy this much, you will have a very highly supersaturated candy and many, many small crystals will form all at once when you begin beating this relatively cool product. With this sudden development of many crystals, there is little opportunity for growth of crystal aggregates and the candy will have a very fine texture. Of course, you would not want to test the temperature of the cooling candy by inserting a thermometer into it, because this would start crystallization. There are two ways of determining if the candy has cooled sufficiently. You may wish to leave the thermometer in the candy rather than removing it when the candy is removed from the heat. This is perfectly satisfactory providing you do not wiggle the thermometer at any time while the candy is cooling. Of course, the candy must be deep enough to cover the bulb of the thermometer completely if the reading is to have any meaning. The second way of determining temperature is to touch the bottom of the dish. Candy that has reached approximately 110°F will feel slightly warm to the hand. You can check this by carefully moving the dish so that you can feel underneath it. Be sure that the dish is not jolted or tipped because crystallization will begin even though the candy will perhaps not be cooled to the desired temperature. Do not touch the surface of the candy with your finger to see if the candy is cool enough. This would naturally start crystallization.

The exact temperature of candy can be checked by using a thermometer, but you can determine the approximate temperature of cooling candy adequately by touching the **candy/bottom of the plate or pan** *with your hand.*

Bottom of the plate or pan.

*

To obtain crystalline candies that are the desired firmness, it is necessary to reach the exact sugar concentration that will give you a firm yet easily cut product. The humidity or moisture in the air is one factor that does not show up on the thermometer when you are attempting to boil the candy to just the right final sugar concentration. On a very humid day, you will need to boil the candy 1–2°F hotter than you would if it were a very dry day. In some climates it is not unusual to have days with a relative humidity of 90 per cent or more. On rainy or humid days, the final temperature should be about 2°F higher than indicated in the recipe because the candy will attract a considerable amount of moisture from the air, with the result that the actual concentration of sugar is reduced in the candy. If you do not boil the candy to a higher final temperature, the candy will be too soft because of this moisture absorbed from the air. This phenomenon has been noticed by many persons who have tried to make candy on a rainy day only to have it be too soft. Failure to allow for this hygroscopic property of sugar has caused many persons to subscribe to the idea that candy cannot be made on a rainy day.

True or false. Candy cannot be made satisfactorily on a rainy day.

False. However, it is necessary to boil the candy to a higher final temperature (about 2°F higher) to achieve the desired firmness.

423

Crystalline candies are evaluated on flavor, firmness, and texture. A good crystalline candy has an excellent flavor with no suggestion of scorching. It is firm enough to hold its shape easily, yet it is not hard to bite. Crystalline candies should be so smooth textured that they feel almost velvety in the mouth.

True or false. Some crystalline candies just naturally have a grainy texture.

False. Grainy crystalline candies simply have not been crystallized properly but have been allowed to form crystal aggregates slowly.

*

How many times have you wondered how commercial candy manufacturers are able to dip those very soft chocolate cream centers into their chocolate baths? Those soft centers would simply be too hard to manage if they were that soft when they were dipped. Instead, these centers are made to be reasonably firm for dipping. An enzyme, invertase, is mixed with the candy before dipping and then the candy is allowed to age in its chocolate case. During the ripening period, the enzyme causes inversion of some of the sucrose with the result that the candy now contains a mixture of sugars which causes a softening of the center. So you can see that chocolate creams are not really made by magic at all. They are simply a practical application of chemical technology.

Soft centers in chocolate creams are made by adding _____ to the candy to be dipped in chocolate.

Invertase.

*

Making Amorphous Candies

Amorphous candies are considered by some to be easier to make than crystalline candies, because you do not have the problem of controlling crystal growth to achieve a smooth product. The real problem in making amorphous candies is controlling the heat uniformly throughout the mixture so that none of the candy is scorched. The temperature range for amorphous candies is much wider than for crystalline candies. Most crystalline candies are cooked to an end temperature between 234°F and 242°F. The various temperatures for amorphous candies range from as low as 245°F for some caramels to 300°F for brittles. At these very high temperatures, it is easy to scorch the candy. Scorching is a particular problem if a pan develops hot spots. By selecting a heavy-weight pan, such as a thick aluminum pan, you increase your chance of producing a candy free from any scorched flavor. It is also important to stir throughout the cooking period to help keep the candy from sticking and scorching in any part of the pan.

When making amorphous candy, it is important to stir the boiling mixture to avoid **lumps/burning or scorching** *in any part of the pan.*

Burning or scorching.

*

To achieve the desired chewiness of a caramel, the hardness of taffy, or the brittleness of a nut brittle, it is important to use a thermometer. The same adjustments need to be made in the final temperature that are made when making crystalline candies. You will need to allow for altitude and errors in your thermometer by adjusting the final temperature to which you cook the candy. If your thermometer is inaccurate or if you are making candy at a high elevation, the boiling temperature of water will not be 212°F. You will need to subtract from or add to the temperature given in the recipe. At 5000 feet elevation, the boiling temperature of water is 10°F lower than at sea level, so water will boil at only 202°F. This means that you would have to subtract 10°F from the temperature indicated in the recipe.

Suppose you were making candy in Death Valley at an elevation of −280 feet (280 feet below sea level). This elevation would cause water to boil approximately ½°F **higher/lower** *than the 212°F at which water boils at sea level. You should* **add/subtract** *this ½°F to adjust the candy recipe.*

Higher, add.

*

The flavor and the firmness of amorphous candies are the main criteria to judge. All amorphous candies should have pleasing full flavors with no suggestion of scorching. They should also be the expected firmness. Caramels should be chewy yet easily bitten. Taffy should be difficult to bite yet not brittle. Amorphous candies boiled to higher temperatures will be brittle and fairly easy to bite because of this brittle character.

What would you say is the most difficult problem in making amorphous candies?

The most difficult problem is to avoid scorching the candy when it is being heated to such high temperatures.

FROSTINGS

Some frostings are actually crystalline candies that are spread on a cake before crystallization is quite complete. Others are defined as glazes; still others are egg white and sugar syrup foams. The problems involved in making crystalline frostings are identical to those encountered in making crystalline candies. If you wish to make this type of icing, review the section on making crystalline candies.

Why do we not find frostings that are actually amorphous candies?

Amorphous candies are too hard to cut and, consequently, are not suitable toppings for a cake.

*

Glazes are thin icings that are drizzled on the top of a cake and allowed to run down the sides. This application should be done artistically to enhance the appearance as well as the flavor of a cake. If glazes are being prepared for rolls and for some cakes, they are simply a mixture of confectioners sugar, milk, and a flavor concentrate or extract. Such glazes become firm and tend to be hard. Other glazes may have butter added to the preceding ingredients. The butter keeps the glaze softer and retards the drying process. Glazes are fast to make and can be quickly applied to either a hot or cold cake. However, glazes containing butter should only be poured on cool cakes because the butter melts and runs on a hot cake.

The basic ingredients in a glaze are ____ ____, ____, and ____.

Confectioners sugar, milk (or other liquid), flavoring.

*

Some icings are firmer than a glaze yet softer than a crystalline frosting. These are perhaps the most common frostings in home use because they do not require any cooking and there is no problem with a grainy texture. These icings are the popular confectioners sugar icings. They are actually just a glaze that is thickened to a spreading rather than pouring consistency by the addition of powdered sugar. It is important to use butter or other fat in these icings. Otherwise, they will be too dry and brittle to cut well after they have stood a short time. The chief advantage of these confectioners sugar icings is their easy preparation; the chief disadvantage is the raw starch flavor in evidence as a result of the uncooked cornstarch in the confectioners sugar.

Confectioners sugar icings **do/do not** have as good a flavor as crystalline icings, but they are **easier/harder** to make.

*

Egg whites can be beaten to the foamy stage and then a cooked sugar syrup is slowly poured into the egg white foam while it is being beaten vigorously. This type of icing often is called seven-minute frosting. To achieve success with this frosting, it is necessary to cook the syrup to the right stage and then pour it very slowly into the egg whites. It is easiest to make this product if you have an electric mixer and an outlet for it close enough to the range so that you can beat the whites with the mixer while the icing is being held over boiling water. Without these aids, it is difficult to make this icing. A modification of this icing can be made by using corn syrup in place of the cooked sugar syrup. The corn syrup is beaten into the whites without heating over boiling water. This latter icing is easier to make, but the seven-minute frosting holds up better. Neither type of icing lasts as well as a crystalline or a confectioners sugar icing, but the fluffy light appearance makes these icings favorites for cakes that will be eaten quickly.

A cooked frosting made by beating a cooked sugar syrup into egg whites held over boiling water is often called a _____ frosting.

Seven-minute.

*

To be attractive, iced cakes must not only have a well-made icing, but the icing must be properly applied to give a pleasing appearance. If crumbs from the cake get stuck in the icing, the cake has a careless, sloppy appearance. This can be avoided by carefully dusting the loose crumbs from a cake and then applying a thin layer of icing. This first layer of icing is often referred to as the crumb coating. After this layer has hardened, the crumbs are held firmly to the cake and the final layer of frosting can be applied easily without fear of dislodging crumbs. The final layer of icing should be applied with an artistic touch to give a pleasing rhythm to the icing.

For best results, apply _____ layer(s) of frosting to a cake.

Two.

427

1. *True or false.* It is better to use sugar produced from beets than from cane.

False. Both crops are sources of the sugar known as sucrose. It is not possible to distinguish between these two sources of sugar after the sugar has been refined.

2. Molasses is a by-product in the manufacture of *cane/beet* sugar.

Cane.

3. *True or false.* Molasses is a good source of iron.

True.

4. *True or false.* Sugar is hygroscopic.

True.

5. When we say that sugar is hygroscopic, we are saying that sugar *attracts/repels* water.

Attracts.

6. When sucrose is acted upon by heat and an acid, *invert/convert* sugar is formed.

Invert.

7. Invert sugar is made up of *equal/unequal* parts of _____ and _____.

Equal, glucose, fructose.

8. The enzyme that causes the inversion of sucrose is called _____.

Invertase.

9. When a sugar syrup is boiled, the sugar concentration *increases/decreases.*

Increases.

10. As the concentration of sugar increases, the boiling temperature of the solution *increases/decreases.*

Increases.

11. An increase in the concentration of sugar in a solution causes the vapor pressure to *increase/decrease.*

Decrease.

12. A saturated sugar solution *can/cannot* dissolve any additional sugar at that temperature.

Cannot.

13. A supersaturated solution is prepared by *adding sugar to a saturated solution/carefully cooling a saturated solution.*

Carefully cooling a saturated solution.

14. If a crystalline candy begins to crystallize at 150°F, you should begin to beat *immediately/when the candy cools to 110°F.*

Immediately.

15. Ideally, a crystalline candy will be cooled undisturbed to a temperature of _____ °F, at which time beating should start.

110.

16. *True or false.* A crystalline candy that is too soft has not been beaten long enough.

False.

17. When a crystalline candy that is being beaten suddenly becomes firm and crumbly, you should _____ it.

Knead.

18. If you add some sugar crystals or dust particles to a cooling candy syrup, you are _____ the solution.

Seeding.

19. Beating is done to make crystalline candies *harder/more velvety on the tongue.*

More velvety on the tongue.

20. The addition of cream of tartar *favors/retards* inversion of sucrose and thus *aids/hinders* production of a smooth crystalline candy.

Favors, aids.

21. Cream of tartar is an *acid/alkali.*

Acid.

22. Sugar solutions boiled to a high final temperature will be *firmer/softer* than comparable mixtures boiled to a lower final temperature.

Firmer.

23. Candy may be classified as _____ or _____.

Crystalline, amorphous.

24. Classify the following candies as crystalline or amorphous: fudge, peanut brittle, chocolate creams, taffy, caramels, and panocha.

Crystalline candies—fudge, chocolate creams, and panocha. Amorphous—peanut brittle, taffy, and caramels.

15

Meal Management

Effective meal management involves both skill in preparation of food and the ability to use your time, energy, and knowledge well. Your first step will be to plan a meal. Then do the shopping for it, the preparation, and also the service. After you have tried doing an entire meal by yourself from start to finish, you will realize just how many different factors need to be considered. Meal management may seem an imposing challenge at first, but with some guidance in meeting the basic problems involved, you will soon be able to do these tasks effectively.

Meal management is the term used to cover the various steps required: _____, _____, _____, *and* _____.

Planning, shopping, preparation, serving.

*

It is important to learn all aspects of meal management, for any deficiency interferes with maximum enjoyment of the meal. If you have planned well, you will have a meal that is nourishing and pleasing to your diners. Poor planning may result in combinations that lack appeal, or the meal may be low in necessary nutrients. Good shopping is necessary to provide all the ingredients needed in a meal without additional trips to the store. These ingredients need to be well selected to obtain the desired quality and type at a reasonable cost suited to the family's budget. Good preparation is essential to the general appeal of a meal. No matter how well you plan and shop, you can reduce the effectiveness of these first two steps if you do not prepare the food well. And finally, any meal is more enjoyable when served attractively.

*If any aspect of meal management is poorly done, the meal will/**will not** be less enjoyable.*

Will.

*

PLANNING

With the ever-rising cost of food, planning meals becomes an increasingly important part of meal management. It is necessary to meet the nutritional needs of all family members while staying within the family's food budget. This is a monumental assignment if you are working on a limited amount of money. When planning meals on a budget, it is practical to select items for the menu by consulting the specials displayed in grocery store advertisements. Many grocery stores have weekend specials (usually Thursday through Sunday) which they prominently advertise in the newspapers. If you can arrange your schedule so that you can plan your week's menus with the new ad in hand, it will be far easier for you to plan economically.

One good means of planning economically is to use _____.

Newspaper ads.

*

The market price of a food item may vary widely, depending upon the season and the availability. This is particularly true of fresh produce. At times, one head of lettuce may show a variation of as much as 30 cents. Egg prices are also variable. Meat prices vary widely from one part of the country to another and from one kind of meat to another. Usually, beef and lamb prices are proportionately higher than pork, ham, and poultry prices. Salmon, shellfish, and a few other fish are far more expensive than ocean perch, for example. Frozen foods are less variable in price because of their longer storage life and constant availability. However, there is variation related to the original cost of the item to be frozen. Frozen cherries are one illustration of this situation. When the summer growing season has been good and the crop is large, the price of the frozen fruit during the next winter is much lower than it is following the harvesting of a poor crop. A similar example is provided by frozen orange juice concentrates.

The price of foods in the market is greatly influenced by _____ and _____.

Season, availability.

431

When planning menus, keep in mind that the price is not the only reason for relying heavily on fresh foods that are in season. The quality of fresh produce is usually far better when it is in season than when it is out of season. Tomatoes, for instance, are much less expensive in the summer than they are in the winter, and their quality in the summer is usually far superior. Lettuce also shows this inverse relationship between quality and price. When lettuce is the most expensive, it is the poorest in quality.

Fresh produce bought when it is in season will usually be **less/more** *expensive and* **lower/higher** *in quality than when it is purchased out of season.*

Less, higher.

*

You usually will use your time best if you plan menus for one-week blocks. You may wish to do this on a sheet of paper divided into seven days, each containing three columns. This will leave a block for each meal of the week. Start by blocking out any meals when none of the family will be home to eat. It is practical to plan all breakfasts next, since there is frequently less variation involved in this meal than in the other meals. Now, you have an encouraging start on the week's menu plans and it is easier to proceed with the more complex meals remaining. Many people prefer to plan the lunches next and do the dinners last.

Do you think it is necessary to use the above plan, starting with breakfasts and then lunches and dinners?

Other plans may work just as well or better for you, but it is usually most efficient to work out a system and then continue to use it. This saves time and helps to eliminate oversights.

*

The complete menus for each day should include the recommended number of servings of each of the four food groups in the Basic Four Food Plan. These items serve as the skeleton plan for the day and are supplemented by additional servings of foods from these categories as well as by other foods needed to meet each individual's nutritional requirements. Keep in mind that protein is best utilized when offered at intervals throughout the day rather than at only one meal. You will also want to keep in mind nutrients that have been shown to be low in the diets of many Americans. Ascorbic acid and other vitamins found in good quantities in fruits and vegetables are often inadequate in the diet. The minerals most likely to be low are iron and calcium. Fats, which should contribute approximately 40 per cent of the Calories, are usually present in too great an amount. Calories are also available in excess in many diets.

For best utilization by the body, proteins should be eaten once/several times during the day.

Several.

＊

With these general recommendations in mind, you are now ready to proceed with your menu planning. Breakfast is an excellent time to serve the daily citrus fruit. Oranges may be served as juice or in a variety of other ways; grapefruit might be served chilled or broiled. On occasion you might wish to vary the breakfast menu by substituting strawberries as your source of ascorbic acid. A large glass of tomato juice is another possibility for variety. Of course, you could use the citrus fruit as a salad or dessert item at another meal. It is not mandatory that ascorbic acid be supplied for the day at breakfast.

Usually _____ _____ are used as the source of ascorbic acid, but suitable substitutes include _____ or _____.

Citrus fruits, strawberries, tomatoes.

＊

Some source of protein should be included at breakfast. The protein food may be simply a glass of milk. You may like to serve eggs, bacon, sausage, or other meat for breakfast. These are all excellent sources of complete protein and provide a particularly appropriate start toward meeting the day's nutritional needs for people whose lunch may be a bit uncertain. Cereal and/or bread in various forms are popular breakfast items that contribute to the day's intake of the B vitamins. Coffee is a popular breakfast beverage despite the fact that it offers little nutritive value, particularly when consumed black. Studies have shown that a breakfast consisting solely of black coffee causes people to be able to turn out less work in a morning than they can if they have nothing at all to eat or drink. As you would expect, a good breakfast provides the start for a more productive morning than is possible with either no breakfast or simply coffee. A desirable breakfast will contribute between one-fourth and one-third of the day's nutritional needs.

Do you think that a good breakfast will provide one-fourth to one-third of each nutrient needed by the body?

No, it will not. You will probably have almost all of the ascorbic acid requirement for the day included in this one meal. The vitamin A requirement will most likely be met by foods later in the day rather than at breakfast. The calcium intake will depend upon whether or not milk is part of the breakfast menu.

It is wise to serve milk to all family members at breakfast. Cocoa or hot chocolate is a popular way of serving milk, particularly on cold winter mornings. Adults do need to have at least two glasses of milk a day, and breakfast is often a suitable meal for one of these glasses. This practice sets a good example for children in the family. If milk is preferred at the two other meals of the day rather than at breakfast, be sure to plan some other protein source for breakfast. From the preceding suggestions, you should now be able to plan several suitable breakfast menus. A good breakfast menu perhaps will include the following: orange juice, scrambled eggs, muffins, and milk. For sleepy adults you might wish to add coffee.

Now, it is your turn to plan some breakfast menus. Plan a breakfast menu for yourself. Do you think you would eat and enjoy your menu?

Did you include a citrus fruit, milk, and the other foods you need for your morning's activities? If not, try again.

*

If you are planning menus for a family with children eating lunch at school, it is wise to plan the other meals of the day around the menu prepared at school. In many cities, this information is printed in the newspaper, or monthly menu plans are sent home with the children. You can more adequately meet the nutritional needs of school children if the other meals of the day are varied rather than unintentional repeats of lunch items. Lunch is an important meal regardless of where it is eaten. If lunch is to be eaten at home, plan to serve a meal large enough to give a lift, but not so large that it is burdensome to the afternoon activities. Lunch should be a nourishing, planned meal and not just the pickup affair that it often is when husbands are not home for this meal. From a nutritional viewpoint, lunch is certainly as important as dinner and should not be slighted.

Do you think your lunches are usually nutritionally adequate?

If your answer is no, how do you think you might improve your lunches? Although you are perhaps sitting a great deal while studying and in classes, you still need adequate quantities of proteins, vitamins, and minerals.

*

Now, let us see just what a lunch might include for you. If you are not very active, it is necessary to plan lunches that are low in Calories and high in nutrients. A main dish high in protein is an excellent starting point in your plans. Remember that cheese of all types, eggs, and legumes are suitable foods that lend their flavors well to main dish items.

Of course, meats of all types, including poultry and fish, are also excellent choices for lunch. Salads containing a wide variety of foods are good main dishes for lunch. Such a salad bowl might contain hard-cooked eggs, cheese and ham in julienne strips, tomato wedges, and artichoke hearts on a bed of mixed greens. This is just one illustration of a salad that forms the basis of a meal. You can doubtless think of many other tempting salads. Notice that in the salad just described, there is a good source of protein in the eggs, cheese, and ham; the various vegetables provide important vitamins including some ascorbic acid from the tomato and vitamin A from the darker mixed greens. If this salad were served with a popover and a glass of milk, you would have a satisfying lunch, yet one not too high in Calories for relatively inactive people. A dessert could easily be added to this menu for those needing more Calories.

Does the above lunch menu include something from each of the four food groups?

Yes. It has milk from the milk and dairy products category, meat, cheese, and egg from the meat group, mixed greens, tomato wedges, and artichoke hearts from the fruit and vegetables, and a popover from the bread and cereal group.

*

One fallacy that many people believe is that hot soup of any type, when served along with fruit and milk, is a satisfactory lunch. Such a menu is sufficient to fill inactive individuals at the time of the meal, but it may lack long-term satiety value later in the afternoon. The adequacy of this lunch is determined by the soup served. A hearty split pea soup with ham would provide much needed Calories along with protein; a chicken noodle soup, on the other hand, is frequently very long on noodles and may be barely garnished with a small bit of chicken. Soups high in rice, noodles, and other pasta products are quickly used by the body because they are rich in carbohydrate but low in protein and fat. These soups are suitable as a part of a lunch but should be accompanied by a good source of protein. Some homemade soups are good sources of protein because the recipes include a considerable amount of meat; commercial soup products have sufficient meat for flavoring purposes but do not meet the body's need for protein.

If you wished to serve a low-protein soup for lunch, what are some suggestions for foods that might be served to complement it?

High-protein salads such as tomato stuffed with cottage cheese or a raw spinach salad heavily topped with bacon bits and chopped egg are appropriate. An open-face broiled cheese sandwich is another possibility that provides a pleasing textural contrast in addition to the desired protein. Of course, any meat or fish salad or serving of meat as the main course can also meet this need.

When planning lunch and dinner menus, keep in mind the time you can comfortably spare for their preparation. This will frequently need to influence your choice of foods. We might all eat quite differently if time were not the deficient quantity in today's formula for living. If time is short for lunch, it is wise to equate the preparation time with the consumption time. It is of little value to have planned and prepared a beautiful lunch and then not have adequate time to enjoy eating it. Perhaps the time factor is a greater problem at lunch than at dinner. If so, plan around it rather than ignoring it. Many convenience food items are presently available at the market. Some of these are palatable, while others might well be returned to the devlopment laboratory for further refinements. If time is your problem, acquaint yourself with the timesavers that are available and draw upon those that are pleasing to you when you are planning menus with time problems. When you have a day with more time available, you may enjoy the luxury of a more elegant lunch featuring items you enjoy making when there is time. Always keep your own satisfactions in meal preparation and service in mind when planning menus so that you have provided for your own pleasures as well as those of the rest of the family. Good food in a home is the result of the efforts of a person who enjoys food preparation and who has built pleasure into this ever-recurring responsibility.

True or false. A good cook today never will draw upon commercially prepared food items.

False. Most people today will have some time pressures which require that some shortcuts be taken in the kitchen. At such times, draw upon the better commercial products to ease your load.

*

Dinner is considered by many to be the most satisfying meal of the day to plan. Since the entire family is usually home for this meal, this is frequently the one meal when everyone is available for a more formal gathering. It is common practice to begin dinner menu plans by first deciding upon the cut of meat and its preparation. The other foods can then be planned to enhance the meat. When selecting the additional food items, check back through the other two menus for the day to see what deficiencies may still exist in meeting the recommendations for good nutrition as outlined by the Basic Four Food Plan. You will frequently find that two servings of fruits or vegetables remain to be planned in the dinner menu to just meet the basic minimal recommendations. On alternate days you may wish to include the dark-green leafy vegetable or a yellow vegetable in the dinner menu. This need not be as repetitious as it sounds. One day you might use baked sweet potatoes as the source of vitamin A. Another day the vitamin A source might be buttered spinach, while still another time you might serve carrots fixed in any of a large variety of ways. Sweet potatoes or yams are also useful as means of varying the high carbohydrate item in a dinner menu as well as being an outstanding source of vitamin A.

Sweet potatoes, carrots, spinach, and squash are some familiar examples of vegetables that are good sources of the precursor of _____.

Vitamin A.

*

The fruits and vegetables at a meal are important for other reasons besides their nutritive value. These foods can be fixed in a variety of ways to provide desirable modifications in the texture and flavor of these types of food. Their usual intense colors are important color accents in a meal. Without fruits and vegetables, meals would be drab affairs indeed. Keep in mind at all times the color, flavor, and texture variations that can be introduced to a meal through clever selection of vegetables and vegetable recipes. Always work to visualize the impact of their colors on the rest of the meal. Avoid duplication of colors in vegetables. For instance, broccoli and string beans served at the same meal are distinctly monotonous because their colors are so similar. On the other hand, there is such a range of shades of green among the salad greens available in many places that you may be able to generate a subtly fascinating monochromatic salad that attracts attention. Such selections need to be made with an artist's eye. Plan them carefully for maximum inpact.

Aside from their important nutritional contributions, fruits and vegetables are used in a meal to add interesting variations in _____, _____, and _____.

*Color, flavor,
texture.*

*

Pay attention to the flavor impact of a food as well as its color. You will find such repetitious flavors as those of cauliflower and broccoli quite uncomplementary to each other, despite the fact that they have quite different colors.

437

Picture flavor as well as color in your planning. The vegetables must be selected not only to enhance each other, but also to bring out desirable flavor characteristics of other foods in the same course. For instance, a highly seasoned meat such as sauerbraten has a distinctive flavor of its own and is well-suited to service with rather mild-flavored vegetables. However, a bland fish may achieve real distinction when accompanied by the bright flavor of vegetables served with some imagination.

Vegetable flavors need to **complement/compete with** *each other and the flavors of other foods in the meal.*   *Complement.*

*

Since vegetables are characteristically high in cellulose and are suitably served either raw or cooked, it is possible to achieve just about any texture you need in a meal by simply selecting the correct preparation for the right vegetable. Water chestnuts are an outstanding example of the constant crispness that a few vegetables can retain throughout preparation of a recipe. Although celery is less resistant to changes in texture during cooking than is the water chestnut, a considerable crispness can still be given to oven main dishes by including uncooked celery as one ingredient. If the baking period is brief, the celery will retain a distinctly crisp character at a far less expensive price than the water chestnut commands in the market. At other times, you may wish to have a soft tender vegetable to offset the crispness of fried chicken or French-fried shrimp. This is easily done by selecting such items as mashed squash or broiled tomatoes. When considering textures, keep in mind the textural characteristics of a well-prepared vegetable rather than an overcooked one. Most vegetables, when properly prepared, will have just the slightest trace of crispness remaining in them. No mushiness is in evidence. Remember also that deep-fat fried vegetables such as French-fried onion rings are a well-accepted means of introducing an element of crispness to a meal intermingled with the softness of the fried onion itself. Raw vegetables are also popular either as a bright garnish on the plate or as a salad item. Men as well as women in the United States are usually avid salad eaters and may even be more receptive to raw vegetables in a salad than to cooked vegetables alongside the meat.

If a vegetable is served as a salad rather than as a main course item, the nutritive value from the vegetable **is/ is not** *as great as it would be if the vegetable were eaten along with the meat.*

Is.

*

For additional satisfaction at dinner, many persons may wish to have a slightly larger serving than the specified three ounces of meat. This is a good time to use hot breads of any type, ranging from fancy yeast rolls to simple quick breads. You may also wish to include a dessert. These items are added to provide the needed satiety value and caloric intake for the various family members. Before planning a dessert for dinner, you should certainly consider the total impact of the meal. If you have elected to include meat, a couple of vegetables, a salad, milk, and rolls or other bread product, some people will not wish to have anything more to eat. Very active individuals will want a filling dessert. Sedentary people will perhaps not wish a dessert or may want simply a sherbet or very small serving of some type of sweet. These plans for dinner need to be geared to the needs of the family members. Keep in mind that some people will want a dessert even when they do not need it. You may be doing heavier people a favor if you neglect to plan dessert much of the time. If nothing is served, there is far less temptation to have dessert. In this case, lack of planning is a wise solution to the dessert menu. Since one widespread nutrition problem in the United States is overweight (among children as well as adults), you would be wise to consider omitting dessert regularly.

True or false.   For today's population it frequently is unnecessary to serve a dessert to help meet their day's total need for Calories.

True.

*

When planning menus for a family, keep in mind some of the individual members' food preferences as well as the nutritional needs of all. It is very flattering to have some special food favorite prepared to please a member of the family. At different meals, plan the favorite food of all family personnel. This helps everyone to learn to appreciate the individuality of food preferences. However, do not be limited by the list of food preferences each family member may indicate to you. One pleasure of food preparation is

439

trying new recipes, and family members can expand their own food preferences when you serve many different foods. Add interest to meal preparation by experimenting with new recipes and gradually add your good experimental items to your regular repertoire. A wide variety of menu items provides a more interesting mealtime and also contributes a better range of nutrients.

True or false. When planning family menus, it is a good idea to use only food items and recipes that have been well accepted by the family previously.

False. This practice leads to an ever-narrowing menu in the home, is too restrictive for good nutrition in many cases, and also is too limiting in the opportunities afforded for new challenges in the kitchen and at the dining table.

*

MARKETING

Marketing is an important part of meal management because of its influence on total cost of feeding the family and its role in determining maximum food quality at a meal. Although it has been pointed out frequently that, compared to 20 years ago, a proportionately smaller part of a family's spendable income is used for food, there is still a large total sum required to feed a family each month. Careful marketing can lower this expense a bit, thus leaving more money available for other items you may wish to buy. Do not construe this remark to mean that you should always seek out the least expensive items in the market. There is a difference between wise and miserly shopping. It is true that some of these decisions are more important on a low income than on a high one, but in any income bracket wise shopping is important.

False. Families with low incomes face a greater challenge in managing the food budget, but all persons will receive the greatest value for their money if they shop wisely.

True or false. Careful shopping is only of importance for people with limited incomes.

*

Shoppers are faced with two important tasks. They need to select foods that will provide good nutrition for their families within the budget they have available for food purchases. Money management in the market is a big task today because of the multitudes of items available and because of the numerous variations in package sizes. One has to be an agile mathematician or carry around a pocket-sized computer to work out the price per ounce of the various-sized packages available for just one item. You can spend a discouragingly long time in the market just estimating the package that is the most economical size to buy. In some cases, you also have to estimate the shelf life of the product to determine whether the giant jumbo sizes will last satisfactorily until you use all of the package. Obviously, it is poor economy to buy a huge package of something for a smaller price per ounce and then have to throw out part of the package. Such considerations are factors when buying potatoes, cereal products, and coffee, to cite a few examples.

To determine the best package size for you to buy, calculate the _____ per ounce and estimate the _____ to be certain the product will still be edible at the bottom of the package.   *Price, shelf life.*

*

Other important considerations when shopping are how much and what kind of storage space you have available. A particularly obvious illustration of the importance of storage is the purchase of three gallons of ice cream that you found on sale when you only have frozen storage for two gallons. The troubles here are readily apparent. Apartment or home kitchens with very limited cupboard space are very frustrating to work in if you have bought larger supplies of canned good and staples than you can store comfortably. Of course, refrigerator storage for fresh produce and meats is essential. It is not difficult to learn to shop with your storage limitations in mind. Just beware when you are shopping to avoid storage problems when you unpack at home.

True or false. It may prove to be false economy if you purchase more food than you can store properly.   *True.*

*

To make your dollars have more buying power at the market, select quality of items according to their eventual use in a recipe. Perhaps the best example of this is tomatoes. For use in a salad or for broiling, you would certainly wish to use fresh tomatoes of excellent quality. However, in many casseroles a canned tomato product is often very appropriate and may be a good moneysaver. Among the canned tomato products, you have several choices available to you. You can select peeled whole tomatoes at a premium price if a well-known, high-quality brand or you may select

441

peeled whole tomatoes of a somewhat lower grade. The lower grade usually will include smaller tomatoes, often with a lighter color and variation in size. There may also be a higher ratio of juice to tomato in a less expensive whole tomato product. If you were planning to serve whole tomatoes as a vegetable in a meal, you might select the top-quality can. For a casserole, the less expensive product might be perfectly suitable. You will also notice that even within the same brand name, there is a lower price tag attached to stewed tomatoes, tomato sauce, or tomato puree in comparison with peeled whole tomatoes. If you do not need large pieces of tomato in a recipe, you can save a few pennies per can by selecting the stewed tomatoes rather than the whole ones.

At times, you may suitably save money by buying a **higher/lower** *quality product or by buying the same grade of food but packaged in* **larger/smaller** *pieces of fruit or vegetable.*

Lower, smaller.

*

An educated shopper can help her family have better quality at the table for the same money because she knows how to make good selections within a given group of items. This knowledge is particularly valuable at the meat counter. If you can judge meat quality well, you can select the best package of meat in the display of each meat item you purchase and pay the same price per pound as is charged for packages containing meat not quite so desirable. One thing to look for is even thickness of the cut. Chops and steaks need to be cut straight across so that all portions of the cuts are the same thickness. If part of the cut is very thick and another area is very thin, it is impossible to cook the meat properly. Part of the cut probably will be overdone, while other areas will be undercooked. Notice also the ratio of meat to bone, and select cuts that have a large amount of the desired muscle. When selecting bacon or other pork products, look for packages that have a large quantity of lean and as little fat as possible.

Within the display of a particular cut of meat, it **is/is not** *possible to obtain a better buy if you know something about meat selection.*

Is.

Knowledge is also useful when selecting fresh produce. If you are buying lettuce by the head, a firm head will give you more lettuce for the money than will a loose head. Another illustration is the purchase of sweet corn by the ear. A well-filled ear of corn that still has its fresh plumpness may sell for the same price as a wormy ear or an ear that is skimpily filled and beginning to become somewhat dehydrated. Softness in grapefruit may indicate the pithiness that results when the fruit is lightly frozen while on the tree. This grapefruit is not nearly as pleasing or economical a purchase as a grapefruit heavy with juice. The appearance of the skin of many fruits and vegetables can be a helpful guide to selection. Skins that are shriveling a bit are an indication of soft produce that has begun to dehydrate and lose its maximum fresh quality. Obviously, you will want to examine fresh produce for blemishes because these imperfect food items will not store well.

Careful selection of fresh produce gives you maximum value for your money and also increases _____ life of the produce.   Storage.

*

Your selection of a market can significantly influence your food budget. Few stores do a charge business today, but those that do will have to charge a little more on their foods to cover the cost of such a service. The size of the store and its membership in a chain versus local ownership also may influence price. Supermarkets can afford to take a smaller markup on individual items than can a small market because they do a much larger dollar volume of business. Some chain markets can maintain lower prices than local businesses because of the large volume of business done in chains as well as the savings that large chains can make by purchasing centrally for all their stores. Sometimes the cost of market promotions is met by slight increases in the price of many food items. Comparison shopping in the newspaper ads may give you an indication of this practice in your area.

*Usually **large/small** and **chain/locally owned** markets are able to sell food at the lowest prices.*   Large, chain.

*

Although it is wise to consult grocery ads when doing your planning, it is frequently impractical and uneconomical to do your shopping in several markets to take advantage of

443

the various specials each advertises. Remember, it is necessary to include transportation costs between markets plus the value of your time while doing the marketing. You may save money by doing all of your shopping in a well-selected market, even though the market does not have the lowest price on each item. Sometimes shoppers become so engrossed with the grocery ads that they forget transportation costs and time need to be considered in total cost of the food.

To determine the cost of food marketing accurately, add the cost of _____ and your _____ to the actual figure shown on the sales slip from the store.

Transportation, time.

*

Usually your time is spent most economically if you make only one trip to the market each week. This saves travel time plus time standing in line at the check-out counter. Both of these items may add up to a significant figure at the end of a week if you dash to the store for any item that you feel you need in the middle of the week. In addition, most grocers are well aware of human weaknesses in the grocery store and display tempting items at points where consumers are sure to see them. Impulse buying at grocery stores can play havoc with a budget, but it can do far less harm in a once-a-week marketing expedition than it can in a once-a-day shopping pattern. If you happen to be the unusual person who can walk through a store and buy only what you had on the list, this last argument is not important. However, how many people do you know with such will-power?

Once-a-week marketing is recommended because it saves time and reduces the chance of **impulse/planned** *buying.*

Impulse.

*

You can save considerable time in the grocery store if you have an accurate list arranged in the order you encounter the items in your store. This list needs to be based on the menus planned for the week and should include all staple items that are not going to last through the week. The first few times, it will take you a fairly long time to make up the list because of need to check carefully the menus and the supplies remaining in the kitchen. You will soon get this developed into an accurate yet rapid routine and shopping

444

trips need not be too lengthy. It is convenient to start a list in the kitchen as you approach the bottom of containers of spices and other staple items. Add to this list as you think of things throughout the week. By doing this first, you will be far less likely to forget small items that are not weekly purchases.

*For the easiest use of a shopping list in the grocery store, arrange the list in **alphabetical order/sequence as you move around the store.***

Sequence as you move around the store.

*

PREPARATION

Specific pointers about the preparation of the many different types of items in a meal are given in the appropriate chapters throughout this book. If you need to review the information about a particular item before preparing it, read that section quickly before beginning your meal preparation. Meals are much easier to prepare if you are well acquainted with the foods. This should not be interpreted to mean that it is a waste of time to try new things; rather, it means that you can approach foods with confidence if you are familiar with the fundamentals involved in their preparation. You can apply the information learned from reading this volume to new recipes and proceed without difficulty. Simply stop to see what principles are involved; then even a new recipe can be prepared readily with no particular difficulty. However, unless you study to see what knowledge is needed in a new recipe, you may find yourself making mistakes in procedure or taking unnecessary time simply because you have not analyzed just what problems are involved in the recipe.

*It is wise to study a recipe briefly before beginning preparation. This is likely to save mistakes in procedure and will usually **cost/save** you time during meal preparation.*

Save.

*

Experienced homemakers usually have developed sufficient skill in meal management to prepare a meal without writing out a formal time plan, but they still have a time schedule outlined in their minds as they prepare a meal. A time plan is essential to the timing of a meal. Few people forget how chaotic their first meal was when they were on their own with nobody to aid them with decisions or last-minute details. Frequently, in class there are several people working together on one meal, and this is a somewhat different experience compared with working all alone. These first attempts at solo preparation of complete meals should be planned thoroughly ahead of time and should include fairly simple fare. You can work up to fancier meals as you gain confidence with simple ones.

*To achieve success with even your very first attempts at preparing meals alone, plan a **complicated/uncompli-cated** menu and then work out a _____ plan for its preparation.*

Uncomplicated, time.

*

There are different ways of approaching a time plan for a meal, but perhaps the most workable way for many people is to start the plan by calculating just how long each menu item must cook. Then, subtract each of these values from a time five minutes prior to when you plan to serve the meal. This extra five minutes at the end allows time for serving the food on plates or into bowls and platters. It also provides a small cushion of time in case something takes a little longer to cook than you thought it should. The values that you have for each menu item tell you when the food itself should start to cook. If you need to bring water to a boil before adding foods such as vegetables, the water has to be put on to heat several minutes before the food itself should be started. Ovens should be preheated so that they are up to temperature when baked items should be started. Fat for deep-fat frying or for shallow-fat frying needs some time for heating to the correct temperature. Of course, the larger the quantity of fat or water you are heating, the more time you will need to allow for this preheating. Time plans also should include any preparation done on the food prior to cooking. For example, it takes several minutes to bread pork chops for frying. You must build into your time plan the time for actually breading the meat and, in addition, the time for beating up the egg and rolling the crumbs for the breading operation. These items all sound trivial, but in meal preparation many trivial items can combine to wreak havoc with your time plan. Try to anticipate all operations when planning your first few meals.

446

A time plan for a meal should indicate the time when:
A. *Foods should start to cook.*
B. *Food should be served.*
C. *You will go to the market.*
D. *You need to prepare ingredients for the various recipes.*
E. *Fats and cooking water as well as ovens should be started for preheating.*

All except C. Marketing is assumed to be a separate responsibility from actual meal preparation. In most menus, there is little available time for making a quick trip to the grocery store. The other items are all important and should be carefully figured until you have gained considerable experience.

*

In your time plan, you must allow time to set the table, pour the beverage, and take care of any small details you may have forgotten to include. For a family meal, only a rather small amount of time, a matter of a few minutes, may be needed to set the table, including planning and preparing a centerpiece. However, for elaborate dinner parties, you may find that it requires a surprising amount of time to arrange the centerpiece and set the individual covers for each guest. Despite the occasion, do allow enough time to set the table attractively and neatly. Regardless of the quality of the food served, it will seem to be distinctly better if it is served on a table that has been well laid. Of course, the opposite is also true. If you have labored over a particularly difficult meal to prepare it to perfection but have neglected to allow time to do justice to the dining table, the meal will not seem quite so special. When working with food, never ignore the psychological subtleties of service and appearance. And speaking of appearance, don't ignore yours at the table. Your time schedule should include adequate time for you to put finishing touches on you as well as your food.

The appearance of the _____ and the _____ are both very important at a meal. Any good time plan allows time to prepare both adequately for the meal.

Table, cook.

447

With some menus, you may be able to do several preparation steps early in the day or even the day before. This can be particularly important if you are planning an elaborate meal. You also may wish to use items from your freezer to ease the meal schedule on particularly busy days. With the ample refrigerator and freezer space available in many homes today, it is frequently practical to do pre-preparation of some items for elaborate meals. Do be sure to include in your plans time to thaw any items you may have frozen earlier. Pre-preparation of some items such as gelatin salads, marinated foods, and frozen desserts is essential if the food is to have the desired characteristics when served at the meal.

*Pre-preparation of items needing a long time to   **bake/set** is important to the quality of the products at the meal.*   *Set.*

*

On your first time plans, you will want to put down all the important details you think you might neglect if they were not on the list. After each meal, it is helpful if you take time to check back through your time plan to see what parts of it worked out well and what changes might have been made to make things go more smoothly for you. Incorporate these suggestions into your next time plan. You will quickly learn to time your meal toward a specific serving time and avoid the panic that may occur with inappropriate or inadequate time planning.

True or false.   It is wise to review each time plan soon after the meal to see what can be learned that will help you prepare the next meal.   *True.*

*

Even with a well-conceived time plan, there may be times when you simply cannot serve at the hour expected. Perhaps someone is late coming in for a meal or some other unpredictable circumstance arises. With modern equipment, this is no longer a great strain on the cook. If the meal is ready, but the diners are not ready for it, simply dish it up and place it in the oven set at the low keep-warm temperatures between 140°F and 150°F. You can hold many foods at these low temperatures for a couple of hours if necessary. Of course, the most fragile foods such as souffles do not endure this treatment without considerable loss of quality. These items ideally will be prepared to be

448

served at a specific time when you anticipate that it will be convenient for all.

Keep-warm temperatures in today's ovens are in the temperature range of 90-100°F/140-150°F.

140–150.

*

Meal management also may involve the scheduling of more than one baked item into an oven. This presents little problem if you have a double oven, but persons with single ovens may need to coordinate their preparations well to get all items baked in time for the meal. You may have to bake breads and quick breads earlier and then warm them briefly in a foil wrap so they may be served hot at the meal. Many desserts may be baked ahead. Large roasts can be scheduled so that they are on the counter for the half hour just prior to serving. This makes the roast easier to carve and has the added advantage of freeing the oven to warm last-minute items. Of course, it is often possible to bake more than one item in the oven at the same time. Oven meals of meat loaf, squash, and baked potatoes are one illustration of the convenience of baking several items at one temperature. Although the vegetables can be baked at a higher temperature, they are still very palatable when baked for a longer time at the lower temperature needed for the meat. Items that do not need to be browned can be placed on the lower rack, but any foods requiring browning have the best appearance if they are placed on the top rack in the center of the oven.

Do you think it would be practical to bake muffins and a souffle in the same oven?

This would be an unfortunate combination for two reasons. Muffins bake at a temperature 100°F higher than souffles so the two temperatures are obviously incompatible. And it would be necessary to remove the muffins from the oven before the souffle had finished baking; this would almost certainly cause the souffle to fall, unless done with extreme care.

*

Throughout preparation, attention needs to be given toward just how a food will be served so that foods to be served cold will be adequately chilled and hot foods will be done just right. Greens to be used as salad liners are easier to assemble just before serving if you have washed and divided them into the desired pieces, wrapped them in a damp towel, and refrigerated them for last-minute use. Plan also for the garnishes you wish to use on the platter and serving bowls or on the individual plates. Although these touches take only a minute to add if you have provided for their preparation earlier in the plans, they mean a great deal to the final appearance of a meal.

449

Would you say that the preparation of a meal with its many details and concerns for appearance, flavor, and texture is an overwhelming, almost impossible task?

At first it may seem like it, but you will quickly learn some details so that they are almost habit and require little thought on your part.

*

Table settings are actually a part of the preparation for a meal. Some require considerable planning and time to arrange, whereas less formal meals are quickly set. Despite the formality of the occasion, you will achieve the greatest success with table settings by working to avoid a cluttered table. Each person's silverware should be neatly arranged at each cover. Even when a large amount of silverware is being used for a formal dinner, you can readily avoid a jumble of silverware by proper arrangement of the various pieces according to the sequence in which they are to be used. You may even wish to place the dessert silverware when that course is served. Silverware should be arranged at each cover with the handles all lined up evenly approximately one inch from the edge of the table. Although various arrangements of silver are accepted, the one most commonly used is the placement of the silverware to be used first on the outer portion of the cover farthest from the plate, followed in sequence by the other pieces in the order of their use. Spoons usually are placed in this sequence farthest to the right of the plate; immediately to the right of the plate the knives are arranged. On the left of the plate, the forks are arranged. The fork to be used first is farthest to the left, the next in sequence is the second from the left, and so forth. This arrangement is good because it gives diners a good indication of what utensil the hostess expects to have them use. However, it is still best manners to observe the hostess and follow her lead at all times.

If you were served a fruit cup as an appetizer, where would you expect to find that spoon placed at your cover? Where would you find this spoon if the fruit cup were served as a dessert?

The spoon for the fruit cup served as an appetizer usually would be the spoon farthest to the right of the silverware at your cover. When the fruit cup is served as a dessert, the spoon for this item would be the spoon nearest your plate.

*

Sometimes at very elaborate dinners, the meal is served in several courses. This, of course, requires a considerable

450

quantity of silverware for each person and the table will have too cluttered an appearance if all the silverware is placed on both sides of the plate. In such cases, it is perfectly appropriate either to place the dessert silverware above the plate with the handles parallel to the edge of the table or to place the dessert silverware when that course is served. It is generally recommended that there be no more than three pieces of silverware on either side of the plate at any time. The place settings have a more pleasing appearance if the quantity of silverware on either side of the plate is balanced.

Would it be considered correct to set a place with three forks on the left of the plate and only a knife on the right?

No. This would give an unbalanced appearance to the cover.

*

The choice of a tablecloth or placemats is an individual decision; either may be correct for practically any occasion. The important thing is to gear the fabric used to the type of meal served. A snowy white damask cloth obviously is very inappropriate for a patio meal as are Italian cutwork placemats. For such an occasion, a more casual table covering in bright colors or coarser weave would be a far better choice. Placemats appear to be gaining constantly in popularity; this favor bestowed on them may be due to the greater ease of laundering. However, many tablecloths are also available in easy-care fabrics. If you wish to use placemats, it is necessary to have a table with a nice surface because much of it will be exposed between the mats.

True or false. For formal occasions it is necessary to use a tablecloth, but placemats may be used for less formal occasions.

False. The formality of the table setting is influenced more by the fabric and design of the linens than by the mere fact of using a tablecloth.

*

Napkins are always placed at each cover. Their most appropriate position is to the left of the forks and set in so that the bottom edge of the napkin is one inch from the edge of the table. When napkins are laundered, they should be folded in half and then in half again, with each fold being pressed in. To set the table, simply fold the napkin, which has been pressed into quarters, so that the napkin is now in eighths. Position it with the last light fold to the left and the open corners of the napkin on the lower right toward the plate. Napkins arranged in this manner look neat and can be readily picked up by the diner and un-folded to the correct half fold. Occasionally, you may find the table rather crowded and have to resort to placing the napkin immediately in front of the diner where the plate will be. This is less desirable than placement to the left of the silverware but is considered correct. Fancy folding of napkins is not generally appropriate.

The napkin should be folded in **halves/eighths** *and placed* **to the left of the forks/in the center of the cover.**

Eighths, to the left of the forks.

*

The water glass should be placed just above the tip of the knife. Arrange any additional glasses to the right of the water glass and just slightly angling toward the edge of the table. Cups and saucers are placed immediately to the right of the spoons. The salad plate can be placed just above the napkin to the left of the forks. If the napkin is large, it may be more convenient to place the salad plate directly above the forks. The bread and butter plate, if used is to the right of the salad plate. This places the bread and butter plate either immediately above the forks or above the dinner plate, depending upon the position of the salad plate. If a bread and butter plate is used, it is common practice to place the butter spreader on it with the handle parallel to the edge of the table.

The water glass is placed **above/to the right of** *the tip of the knife; other glasses are placed to the* **right/left** *of the water glass.*

Above, right.

*

Centerpieces are an important part of any table setting. They can be used to carry out the theme of a particular meal or occasion, or they may be used simply to beautify the table. When planning a centerpiece, always keep in mind the location of the centerpiece. If it is to be in the center of the table with diners on all sides, it is important to keep the centerpiece low so that all diners can see each other easily. However, if you are planning a centerpiece for a buffet table or for a table with diners only on two or three sides and the centerpiece well toward the back on the fourth side, a higher arrangement may be very effectively used without obstructing anyone's view. Centerpieces add the most to the appearance of the table when they are scaled to the size of the table and to the space available for them. A very long table requires an elongated, rather large centerpiece; a small table is best dressed with a small arrangement. Be sure there is ample space on the table for your arrangement. There is little to be gained by planning such a large centerpiece that it is crowded into the food.

When planning a centerpiece, keep in mind the:
A. *Size of the table.*
B. *Arrangement of guests at the table.*
C. *Height of the food items to be served.*
D. *Space actually available for the centerpiece.*
E. *Occasion.*

All except C.

*

SERVING

There are many different ways in which you might choose to serve the meal you have prepared. Service may vary from a very formal meal with servants waiting on the table to a highly informal picnic served buffet style. You will want to work out the type of service that best fits your particular situation and the number of persons to be served. The important thing to remember at any meal is that the host and hostess should direct their attention toward maintaining an interesting conversation and helping all persons to feel at ease. Hospitality should be the keynote; service should be inconspicuously, but correctly, done.

True or false.   Attention of the hostess at a meal obviously should be directed toward serving her guests.

False. Guests should be the prime concern of the hostess and she will want to be sure that all their needs are met, but the gracious hostess will be meeting these needs while outwardly maintaining a stimulating conversation. Nothing wears out a guest more quickly than an overanxious, hovering hostess!

*

For large groups of people, you may find it convenient to set a buffet table and let each person help himself from large platters of food. This type of service is particularly pleasant for everyone when the buffet table is supplemented with small tables set with each person's cover, including silverware, napkin, and beverage. The buffet table itself can be arranged so that persons pass along only one side or down both sides. If down one side, you may wish to use a dramatic centerpiece positioned toward the back of the table. For service on both sides, the centerpiece needs to be in the center of the table. Arrange the food on a buffet table for easy service. Sometimes it is convenient to have the host carve the meat at one end of the table and serve it directly onto the plate, which he hands to the guest. The guest then proceeds down the table, pausing to help himself to any of the foods he wishes. At other times, the casserole or main meat course may be arranged so that the guest can take a plate and serve himself all along the table.

*In today's homes where it is usually the responsibility of the hostess to serve dinner, it **may/may not** be practical to plan to use buffet service for large groups.*

May.

*

Family service is a popular style of service in this country. You will notice in different families that there are many variations of family service. The one outlined here is a workable arrangement that allows a minimum of disruption of the conversation and enables you to serve the food while it is still hot. If there is to be an appetizer course, it can be on the table before people are seated. When everyone has finished this course, someone may clear the table by removing the first-course dishes to a service table or cart, if available, or by carrying them to the kitchen directly. Then the dinner plates are placed in front of the host and the platters or bowls containing the main course are arranged where the host can reach them easily. The first plate is served and passed to the hostess, after which the other plates are prepared and passed to the specified person. To clear the table after the main course, first remove the serving dishes, salt and pepper, and any other items such as trivets. Next, the dinner plates are all removed and transported to the kitchen. The dessert is then served. Following dinner, the table is cleared after everyone has left the table. As you can see, this type of service is tailored to the family or to a family and a few guests. Family members may be enlisted to help with this type of service. The accent in this type of service is on informality and graciousness.

The type of service in which the first course is already on the table when the diners sit down, the second course is served by the host, and the dessert is served from the kitchen after the table has been cleared is known as ____ service.

Family.

More formal types of service can only be done effectively with trained help. Russian service, English, and compromise service all rely heavily on competent help because the hostess remains at the table throughout the meal. In Russian service, considered to be the most formal, the food is served by the butler. The first course, served after everyone has been seated, is placed directly on the service plate. There may be more than one course before the entree is served; in this case, each course is served on the service plate. Then, the service plate is removed and a warm dinner plate is substituted. The entree is then served onto the warm dinner plates by the butler. This course is customarily followed with dessert; demitasse is served in the living room.

Service for Russian or any other type of service is customarily done from the left for everything except the beverage, which is traditionally done from the right because of the position of the glasses and cups. Left-hand service means that plates are placed or removed from the left of the diner and that any food served to an individual is served from the left. Some persons prefer to use right-hand service. This is considered to be acceptable as long as it is done consistently from the right.

*Service and removal of plates are traditionally done from the **right/left** and beverages are refilled from the **right/left**.*

Left, right.

*

English service is considered to be somewhat less formal than the impersonal Russian service, but it still is necessary to have at least one person serve the meal. The food is served at the table by either the host or hostess in English service, but a waiter is needed to carry the filled plate to the correct individual. The waiter is also responsible for removing the plates and bringing in the plates and food for the next course. In typical English service, the hostess serves the appetizer, salad, dessert, and beverage; the host serves the main course consisting of meat and vegetables.

Indicate whether it is usually the host or the hostess who serves the following courses in English service: appetizer, salad, entree (meat and vegetables), dessert, and beverage.

Hostess: appetizer, salad, dessert, and beverage. Host: entree.

455

Compromise service is a relatively formal service utilizing some aspects of both Russian service and English service. It is at the discretion of the hostess just how she wishes to carry out compromise service. She should have it clearly understood by the servant (s) which courses are to be served at the table and which are to be served from the kitchen. There is no set pattern regarding the courses that are served at the table. However, it is often desirable to serve salad or dessert and beverage at the table with the other courses being served from the kitchen.

Rank the following forms of service from most formal to least formal: compromise, English, buffet, Russian, and family.

Russian, compromise, English, family, and buffet.

*

REVIEW

1. Meal management is an inclusive term that covers _____, _____, _____, and _____ the food.

   Planning, marketing, preparing, serving.

2. If you are planning menus according to the Basic Four Food Plan, provide the specified number of servings in each food group in one   *meal/day.*

   Day.

3. Name the four groups in the Basic Four Food Plan.

   Milk and dairy products, meat and meat substitutes, fruits and vegetables, and breads and cereals.

4. *True or false.* The largest size available in a commodity is always the best buy.

   False. Usually the largest size is the least expensive per ounce, but if you cannot use it up before it spoils, this is poor economy.

5. For efficient use of time, it is wise to shop once a   *day/week.*

   Week.

6. When shopping, it is easiest and most efficient to use a list arranged according to   *the sequence in the store/each day's meals.*

   The sequence in the store.

7. *True or false.* The wise shopper ordinarily will do her shopping in one store where she knows there are good products at reasonable prices.

True. Few people today have the time to shop in several stores to save a penny or two on the items being featured that week. Besides, it costs money to drive the distance between stores.

8. To prepare a meal, it is practical to develop a *specific/ general* time plan when you are first studying meal management.

Specific.

9. A good time plan allows time for:
   A. Cooking all items.
   B. Preparing all items for cooking.
   C. Setting the table.
   D. Serving the food.
   E. Freshening the cook a little.

All of these.

10. The handles of all silverware should be *one inch/two inches* from the edge of the table.

One inch.

11. *True or false.* For a formal dinner, it is essential to use an elaborate white cloth.

False. Formal placemats are also appropriate if you have a pretty table on which to use them.

12. For a very formal dinner, you might wish to use *Russian/ family* service if you had a servant.

Russian.

13. The form of service most commonly used in the United States is _____ service.

Family.

14. Family service *does/does not* require a servant.

Does not.

15. *True or false.* For family service, the appetizer is usually on the table when people are seated.

True.

16. To remove the dishes after the main course of a meal served family style, first remove the *serving dishes/dinner plates.*

Serving dishes.

17. When dinner plates are served or bread is passed, service ordinarily should be from the *right/left.*

Left.

18. Beverages are always served from the *right/left.*

Right.

19. If plates are served from the left, they should be removed from the *right/left.*

Left.

A

Glossary

The following list gives commonly accepted definitions of some cooking verbs.

Bake—To heat without a cover in an oven or similar equipment; synonymous with "roast" if meat is cooked uncovered.

Barbecue—To bake in a covered pit or roast or broil slowly on a spit, using glowing heat of some type to cook an item; foods prepared in this way are frequently basted with a spicy sauce or served with such a sauce.

Bard—To lay strips of salt pork, bacon, or fatty tissue on top of or in gashes in fish, poultry, or meat to prevent it from drying out while roasting.

Baste—To moisten foods during cooking with melted fat, meat drippings, juices, sauces, or liquids to add flavor and to retain moisture.

Batter—To immerse in a soft batter before frying or deep-fat frying; usually stated as "batter-dipping."

Beat—To use a vigorous brisk motion, lifting up and over with a spoon or other utensil, or to achieve a similar result using an egg beater, electric mixer, or other tool.

Blanch—To immerse briefly in boiling water; purpose may be to partially cook, to aid in the removal of skins, or to inactivate enzymes.

Blend—To combine two or more ingredients thoroughly.

Boil—To cook immersed in water or liquid that is boiling; you could say that boiling is making water or liquid "laugh out loud" while simmering is making water "only smile." In other words, bubbles rise and break the surface in boiling; in simmering, bubbles come toward the surface but do not break through.

Braise—To cook foods in a small quantity of liquid; often meat is browned before the liquid is added. Some foods may be called braised when they are lightly fried or sauteed in butter or fat such as "braised vegetables." In braising, the food is cooked with a cover to trap the moisture. "Fricasseeing," "pot roasting," "casseroling," and "Swissing" are other terms occasionally used to denote this process.

Bread—To coat with crumbs of bread or other finely divided food; usually a breaded item is first dipped into seasoned flour, then into a liquid such as lightly beaten eggs or milk, and finally coated with crumbs before being fried.

Broil—To cook by radiant (glowing) heat.

Brush—To brush liquid butter or other liquid on food with a pastry brush.

Butter—To cover or brush with butter.

Candy—To cook in a sugar syrup (or in a mixture of sugar and butter or margarine) until plump.

Caramelize—To heat sugar or foods that contain sugar until they are golden brown and have a caramel flavor.

Casserole—To cook in a casserole dish either covered or uncovered; frequently stated of entree items baked in a sauce.

Chill—To cool to below 45°F but not freeze; may be done by placing in cold water or by storing in the refrigerator.

Chop—To cut into pieces with a knife, cutter, chopper, or other sharp tool; the size should be specified since foods may be coarsely or finely chopped. If the latter, the word "mince" is preferable.

Clarify—To remove cloudy materials by decanting, filtering, or using foods that pick up the flocculent material and leave a clear liquid.

Coagulate—To heat protein foods until the protein becomes firm; often this process is accompanied by a color change.

Coat—To cover with a liquid, crumbs, or other substance; the coating process may be done by shaking the food in a bag with the coating material, pouring or sifting the coating material over the food, or dipping the food into the coating.

Coddle—To cook by pouring boiling water over the food and covering while the food is allowed to stand until the proper doneness is achieved; the food most commonly cooked in this manner is eggs.

Cool—To reduce the temperature; the temperature is not lowered as much as it is in chilling.

Cream—To work a food with a spoon, utensil, or machine until it is soft and creamy such as creaming butter, sugar, and eggs until they are light and fluffy; to cream also means to cook in or serve with a white cream sauce.

Crimp—To gash the surface of fish or meat.

Crisp—To soak or moisten and then chill foods to make them crisp; also to fry or cook so that the outer coating or surface is crisp.

Cube—To cut into pieces, usually from one-fourth to one-half inch on a side; if foods such as meat are to be cut into larger pieces, this should be stated in the recipe.

Cut-in—To cut butter, margarine, or shortening into a dry mixture until the fat is in small pieces; knives, a pastry blender, mixer, or other utensil can be used.

Deep-fry—To cook immersed in hot fat or oil.

Devil—To cook or bake in tangy sauce; deviled eggs or some other deviled foods are foods that are made tangy.

Dice—To cut into small cubes, usually about one-fourth inch in size.

Dissolve—To add a dry substance to a liquid and mix or stir until it is completely in solution and no longer can be seen.

Dock—To prick, pierce, or cut to assist in proper baking; a single pie crust is pricked to prevent steam bubbles forming under it during baking; rye bread or hard breads may be slashed (docked) to prevent them from bursting during baking.

Dot—To scatter small bits, such as nuts, fruits, butter, or margarine, over the surface of food.

Dredge—To coat with flour.

Dress—To garnish; to cover.

Dust—To lightly sprinkle until thinly coated.

Eggs, beaten—To create a foam by whipping eggs. The following terms are often used to describe various stages to which white, yolks, and whole eggs may be beaten.

Whites—beaten stiff: Beaten until they stand in peaks that just bend over when when the beater or whip is lifted from the bowl; surface is still moist and glossy.

Whites—very stiff: Beaten until the surface is dry and the peaks stand up straight.

Yolks—well beaten: Beaten until thick enough to pile and are lemon yellow.

Whole eggs—slightly beaten: Beaten until the yolks and whites are just blended.

Whole eggs—well beaten: Eggs are light and frothy and color is light.

Emulsify—To make two immiscible (nonmixing) liquids such as water and oil into a uniform mixture by dispersing one of the liquids as very fine drops in the other liquid.

Fillet—To cut into strips; to remove bone from raw meat or fish. (also spelled filet).

Flour—To dredge or coat with flour; the flour is usually seasoned.

Fold—To use a delicate over-and-under motion that carries and buries light delicate substances such as beaten egg whites, whipped cream, or whipped gelatin into a product to give lightness. This term also can mean to place in layers by lapping or laying one part over another such as is done with bread.

Freeze—To subject to such a low temperature that ice crystals form throughout the food.

Fricassee—To cook by braising, such as is done with fowl, rabbit, or veal; frequently the item is not browned before liquid is added.

Fry—To cook in shallow fat; also called sautéing, pan-frying, and skillet- or griddle-frying.

Gel—To thicken into a solid mass such as occurs in gelatin or pectin mixtures.

Glaze—To cover with a shiny coating such as glazing with sugar or covering a pie crust or bread with an egg and milk mixture and allowing it to develop a rich, golden-brown, shiny crust; covering meat or other foods with a rich sauce that gives a pleasing sheen to the food.

461

Grate—To rub on a grater to produce fine, medium, or coarse particles; to rasp.

Gratinee—To cover with crumbs, cheese, or other food and delicately brown under an open flame or glowing heat; often called *"au gratin."*

Grease—To coat a pan or food with oil or fat.

Griddle—To sauté on a griddle.

Grill—Used to mean to broil but no longer is used to indicate this; it now means to cook on a griddle.

Grind—To divide food by cutting or crushing with a grinder or chopper.

Homogenize—To divide into small particles so that the particles stay in suspension; usually done to keep cream from rising and forming a separate layer in milk; also frequently done in the making of mayonnaise to develop very fine oil droplets, resulting in a more stable emulsion.

Ice—To chill; sometimes, to freeze. Frequently, to ice may mean chill by covering with chopped ice.

Jugged—To braise in a covered casserole such as "jugged hare."

Julienne—To cut into strips one-fourth to one-eighth inch thick.

Knead—To work with a vigorous, pressing motion in which the pressure folds and stretches the food at the same time; commonly done to bread dough to develop better texture.

Lard—To pull strips of fat through a food (usually very lean meat).

Marinate—To soak food in a marinade (acidic, seasoned liquid) to add flavor or to tenderize; *sauerbraten,* a German pot roast, is made by marinating a roast in such a mixture for several days before braising; sometimes salad foods are marinated a short time to give them extra flavor.

Mask—To cover or coat completely with a sauce, mayonnaise or other substance.

Measure—To divide into a desired quantity.

Melt—To liquefy.

Mince—To cut, chop, or divide into very small pieces.

Mix—To incorporate foods into an even mixture.

Mold—To shape by putting foods into a utensil that gives form.

Mound—To heap so that the food keeps some shape.

Pan broil—To cook uncovered in a frying pan or griddle without fat; any fat coming from the food is removed as it collects.

Pan fry—To sauté or cook in shallow fat.

Parboil—To boil or blanch until partially cooked.

Parch—To brown or dry out completely with dry heat, usually applied to grains, but sometimes nuts.

Pare—To remove the outside skin or covering with a knife or peeler.

Pasteurize—To destroy pathogenic bacteria or other microorganisms with moist heat.

Peel—To pare; usually we say a potato is pared and an orange is peeled.

Pipe—To put a border around food.

Pit—To remove seeds or pits.

Plank—To broil meat, fish, or other food on a wooden plank with a border of Duchess potatoes around the outside and vegetables around the broiled food.

Poach—To cook immersed in hot liquid, usually under boiling.

Poeler—To sauté in butter or fat until browned and then to cover and braise.

Pot roast—To braise; usually said of larger cuts of meat that are browned before liquid is added.

Preheat—To bring up to a desired temperature before baking or cooking.

Preserve—To retard spoilage by the use of sugar, salt, or other preservatives. Preserves are small whole fruit or uniform pieces of larger fruit cooked in a sugar syrup and stored in the thick syrup or jellied juice.

Punch—To fold over or press down a bread dough during the rising period.

Puree—To press through a fine sieve or food mill.

Rasp—To grate; to divide into very fine particles.

Reduce—To remove liquid by evaporation; the quantity of reduction should be stated.

Render—To separate fat from connective tissue, using low heat.

Rice—To put cooked potatoes through a ricer, which divides the food into small rice-like particles.

Rissole—To oven-brown.

Roast—To bake in dry heat in an uncovered pan with the meat on a rack above the collecting drippings.

Roll-in—To layer or dot with butter or other fat and then fold dough over, and to continue the process until many layers of dough and butter are obtained; done for Danish pastry, puff pastry, and some other baked goods.

Saute—To fry in shallow fat.

Scald—To heat until all bacteria and enzymes are inactivated; to dip into very hot liquid (see blanch). Also, to heat a liquid, usually milk, until bubbles appear around the edges—not boiling.

Scale—To weigh or measure.

Scallop—To bake food in a sauce or other liquid.

Scorch—To burn lightly; a slight off-flavor may result.

Score—To cut or mark; an omelet may be dusted with powdered sugar and then scored by caramelizing some of the sugar with a hot instrument; sometimes used to describe the marking of ham, but crimping is the term commonly applied to this process with most meats and fish.

Scrape—To remove thin layers from food with a sharp or blunt instrument; similar to pare and peel.

Sear—To brown the surface of food with dry, intense heat.

Season—To flavor.

Shake—To toss in a container.

Shape—To mold; to give form.

Shirr—To bake, as eggs.

Shred—To cut or tear into fine pieces; sometimes to cut into small julienne strips.

Sift—To put through a flour sifter or fine sieve; often dry ingredients are sifted once and then measured and sifted three times to thoroughly blend them before they are added to a batter or dough that will not be mixed vigorously.

Simmer—To cook in liquid that is just below boiling.

Skewer—To hold in place by means of metal or wooden skewers; to string onto a wooden or metal rod such as in making shishkabobs.

Skim—To remove the surface layer.

Slice—To cut into layers; to carve.

Sliver—To cut or split into long thin pieces.

Smother—To cook covered with some food; liver is often served covered (smothered) with onions.

Snip—To cut into small pieces with a scissors.

Soak—To allow to stand immersed in a liquid.

Steam—To cook above boiling water in a closed container.

Steep—To allow a food to stand in a liquid to remove or to take on flavor, color, or other qualities.

Sterilize—To completely destroy microorganisms.

Stew—To cook in a small quantity of liquid; more liquid is used than is added for braising, but the quantity is less than is used for simmering.

Stir—To mix food with a circular motion.

Swish and swirl—To remove drippings by a process in which the drippings are scraped from the pan while being worked into the liquid; usually done over heat.

Swiss—To pound meat to aid in tenderizing it and then to braise; usually a tomato sauce is the liquid added.

Toast—To brown with heat.

Toss—To flip lightly to mix ingredients or to coat them with a dressing.

Try-out—To render; to remove fat by low heat.

Warm—To heat to a temperature of 105° to 115°F.

Whip—To beat rapidly to bring air into a product, thus giving a light texture and increased volume; to make foamy.

B

Nutritive Values of the Edible Part of Food

The source of the data in this appendix table is *Nutritive Value of Foods,* Home and Garden Bulletin 72, revised, U. S. Department of Agriculture, Washington, D. C. Data for some cooked and prepared foods are taken from Church and Church, *Food Values of Portions Commonly Used—Bowes and Church,* 9th ed., Lippincott, Philadelphia.

The abbreviation for trace (tr) is used to indicate fatty acid and vitamin values that would round to zero with the number of decimal places carried in these tables. For other components that would round to zero, a zero is used. Dashes show that no basis could be found for imputing a value although there was some reason to believe that a measurable amount of the constituent might be present. Other abbreviations used in this table are:

av—average	oz—ounce
c—cup	%—per cent
diam—diameter	pc—piece
hp—heaping	qt—quart
jc—juice	sc—section
lb—pound	serv—serving
lg—large	sl—slice
lv—leaves	sm—small
med—medium	sq—square
tbsp—tablespoon	

Food	Weight, gm	Approximate Measure	Food Energy, Cal.	Pro-tein, gm	Fat (total lipid), gm
Almonds, shelled	142	1 c	850	26	77
Apple, raw	150	1 med	70	tr	tr
Apple brown betty	230	1 c	345	4	8
Apple butter	20	1 tbsp	37	tr	tr
Apple juice, bottled or canned	249	1 c	120	tr	tr
Applesauce, sweetened	254	1 c	230	1	tr
Apricots:					
raw	114	3 apricots	55	1	tr
sirup pack	259	1 c	220	2	tr
dried, uncooked	150	1 c	390	8	1
dried, cooked	285	1 c	240	5	1
Asparagus:					
fresh, cooked	175	1 c	35	4	tr
canned, green	96	6 spears	20	2	tr
Bacon:					
broiled or fried	16	2 sl	100	5	8
Canadian, cooked	21	1 sl	65	6	4
Banana, raw	150	1 med	85	1	tr
Beans:					
baked, with tomato sauce, with pork	261	1 c	320	16	7
baked, with tomato sauce, without pork	261	1 c	310	16	1
green snap, fresh cooked	125	1 c	30	2	tr
green snap, canned	239	1 c	45	2	tr
Lima, fresh, cooked	160	1 c	180	12	1
red kidney, canned	256	1 c	230	15	1
wax, canned	125	1 c	27	2	tr
Beef, cooked:					
cuts, braised, simmered, pot-roasted	72	2.5 oz, lean	140	22	5
cuts, braised, simmered, pot-roasted	85	3 oz, lean and fat	245	23	16
hamburger, ground lean	85	3 oz	185	23	10
hamburger, regular	85	3 oz	245	21	17
rib roast	51	1.8 oz, lean	125	14	7
rib roast	85	3 oz, lean and fat	375	17	34
round	78	2.7 oz, lean	125	24	3
round	85	3 oz, lean and fat	165	25	7
steak, sirloin	56	2 oz, lean	115	18	4
steak, sirloin	85	3 oz, lean and fat	330	20	27
Beef, canned:					
corned beef	85	3 oz	185	22	10
corned beef hash	85	3 oz	155	7	10

466

Fatty Acids Saturated (total), gm	Unsaturated Oleic, gm	Unsaturated Linoleic, gm	Carbohydrate, gm	Calcium, mg	Iron, mg	Vitamin A Value, IU	Thiamine, mg	Riboflavin, mg	Niacin, mg	Ascorbic Acid, mg
6	52	15	28	332	6.7	0	0.34	1.31	5.0	tr
—	—	—	18	8	0.4	50	0.04	0.02	0.1	3
4	3	tr	68	41	1.4	230	0.13	0.10	0.9	3
—	—	—	9	3	0.1	0	tr	tr	tr	tr
—	—	—	30	15	1.5	—	0.01	0.04	0.2	2
—	—	—	60	10	1.3	100	0.05	0.03	0.1	3
—	—	—	14	18	0.5	2890	0.03	0.04	0.7	10
—	—	—	57	28	0.8	4510	0.05	0.06	0.9	10
—	—	—	100	100	8.2	16350	0.02	0.23	4.9	19
—	—	—	62	63	5.1	8550	0.01	0.13	2.8	8
—	—	—	6	37	1.0	1580	0.27	0.32	2.4	46
—	—	—	3	18	1.8	770	0.06	0.10	0.8	14
3	4	1	1	2	0.5	0	0.08	0.05	0.8	—
—	—	—	3	4	—	0	0.18	0.03	1.1	0
—	—	—	23	8	0.7	190	0.05	0.06	0.7	10
3	3	1	50	141	4.7	340	0.20	0.08	1.5	5
—	—	—	60	177	5.2	160	0.18	0.09	1.5	5
—	—	—	7	62	0.8	680	0.08	0.11	0.6	16
—	—	—	10	81	2.9	690	0.08	0.10	0.7	9
—	—	—	32	75	4.0	450	0.29	0.16	2.0	28
—	—	—	42	74	4.6	tr	0.13	0.10	1.5	—
—	—	—	6	45	2.1	150	0.05	0.06	0.5	6
2	2	tr	0	10	2.7	10	0.04	0.16	3.3	—
8	7	tr	0	10	2.9	30	0.04	0.18	3.5	—
5	4	tr	0	10	3.0	20	0.08	0.20	5.1	—
8	8	tr	0	9	2.7	30	0.07	0.18	4.6	—
3	3	tr	0	6	1.8	10	0.04	0.11	2.6	—
16	15	1	0	8	2.2	70	0.05	0.13	3.1	—
1	1	tr	0	10	3.0	tr	0.06	0.18	4.3	—
3	3	tr	0	11	3.2	10	0.06	0.19	4.5	—
2	2	tr	0	7	2.2	10	0.05	0.14	3.6	—
13	12	1	0	9	2.5	50	0.05	0.16	4.0	—
5	4	tr	0	17	3.7	20	0.01	0.20	2.9	—
5	4	tr	9	11	1.7	—	0.01	0.08	1.8	—

Food	Weight, gm	Approximate Measure	Food Energy, Cal.	Protein, gm	Fat (total lipid), gm
Beef, dried or chipped	57	2 oz	115	19	4
Beef and vegetable stew	235	1 c	210	15	10
Beef potpie	227	1 pie, 4¼″ diam	560	23	33
Beer, av 3.6% alcohol	240	1 c	100	1	0
Beets, cooked, diced	165	1 c	50	2	tr
Beet greens, cooked	100	½ c	27	2	tr
Beverages, carbonated:					
cola type	240	1 c	95	0	0
gingerale	230	1 c	70	0	0
Biscuit, enriched flour	38	1, 2½″ diam	140	3	6
Blackberries, raw	144	1 c	85	2	1
Blueberries, raw	140	1 c	85	1	1
Bluefish, baked or broiled	85	3 oz	135	22	4
Bouillon cubes	4	1 cube	5	1	tr
Brains, all kinds, raw	85	3 oz	106	9	7
Bran, raisin	28	⅔ c	99	2	tr
Bran flakes, 40%	28	1 oz	85	3	1
Brazilnuts, shelled	140	1 c	915	20	94
Bread:					
Boston, enriched	48	1 sl	100	3	1
cracked-wheat	23	1 sl	60	2	1
French or Vienna, enriched	454	1 lb	1315	41	14
Italian, enriched	454	1 lb	1250	41	4
raisin, enriched	23	1 sl	60	2	1
rye, American	23	1 sl	55	2	tr
rye, pumpernickle	454	1 lb	1115	41	5
white, enriched	23	1 sl	60	2	1
white, unenriched	23	1 sl	60	2	1
whole wheat	23	1 sl	55	2	1
Breadcrumbs, dry	88	1 c	345	11	4
Broccoli, cooked	150	1 c	40	5	tr
Brussels sprouts, cooked	130	1 c	45	5	1
Buckwheat flour, light	98	1 c	342	6	1
Butter:					
stick, ⅛	14	1 tbsp	100	tr	11
pat or square	7	1 pat	50	tr	6
Buttermilk, cultured, skim	246	1 c	90	9	tr
Cabbage:					
raw	100	1 c	25	1	tr
cooked	170	1 c	35	2	tr
Chinese, raw	100	1 c	15	1	tr

[1] Calcium may not be usable because of presence of oxalic acid.

[2] Year-round average.

| Fatty Acids | | | | | | | | | | |
| Saturated (total), gm | Unsaturated | | Carbohydrate, gm | Calcium, mg | Iron, mg | Vitamin A Value, IU | Thiamine, mg | Riboflavin, mg | Niacin, mg | Ascorbic Acid, mg |
	Oleic, gm	Linoleic, gm								
2	2	tr	0	11	2.9	—	0.04	0.18	2.2	—
5	4	tr	15	28	2.8	2310	0.13	0.17	4.4	15
9	20	2	43	32	4.1	1860	0.25	0.27	4.5	7
—	—	—	9	12	tr	—	0.01	0.07	1.6	—
—	—	—	12	23	0.8	40	0.04	0.07	0.5	11
—	—	—	6	118[1]	3.2	6700	0.08	0.18	0.4	34
—	—	—	24	—	—	0	0	0	0	0
—	—	—	18	—	—	0	0	0	0	0
2	3	1	17	46	0.6	tr	0.08	0.08	0.7	tr
—	—	—	19	46	1.3	290	0.05	0.06	0.5	30
—	—	—	21	21	1.4	140	0.04	0.08	0.6	20
—	—	—	0	25	0.6	40	0.09	0.08	1.6	—
—	—	—	tr	—	—	—	—	—	—	—
—	—	—	1	14	3.1	0	0.20	0.22	3.7	15
—	—	—	22	0	1.0	0	0.10	—	1.1	0
—	—	—	23	20	1.2	0	0.11	0.05	1.7	0
19	45	24	15	260	4.8	tr	1.34	0.17	2.2	—
—	—	—	22	43	0.9	0	0.05	0.03	0.6	0
—	—	—	12	20	0.3	tr	0.03	0.02	0.3	tr
3	8	2	251	195	10.0	tr	1.26	0.98	11.3	tr
tr	1	2	256	77	10.0	0	1.31	0.93	11.7	0
—	—	—	12	16	0.3	tr	0.01	0.02	0.2	tr
—	—	—	12	17	0.4	0	0.04	0.02	0.3	0
—	—	—	241	381	10.9	0	1.05	0.63	5.4	0
tr	tr	tr	12	16	0.6	tr	0.06	0.04	0.5	tr
tr	tr	tr	12	16	0.2	tr	0.02	0.02	0.3	tr
tr	tr	tr	11	23	0.5	tr	0.06	0.03	0.7	tr
1	2	1	65	107	3.2	tr	0.19	0.26	3.1	tr
—	—	—	7	132	1.2	3750	0.14	0.29	1.2	135
—	—	—	8	42	1.4	680	0.10	0.18	1.1	113
—	—	—	78	11	1.0	0	0.08	0.04	0.4	0
6	4	tr	tr	3	0	460[2]	—	—	—	0
3	2	tr	tr	1	0	230[2]	—	—	—	0
—	—	—	13	298	0.1	10	0.09	0.44	0.2	2
—	—	—	5	49	0.4	130	0.05	0.05	0.3	47
—	—	—	7	75	0.5	220	0.07	0.07	0.5	56
—	—	—	3	43	0.6	150	0.05	0.04	0.6	25

Food	Weight, gm	Approximate Measure	Food Energy, Cal.	Pro-tein, gm	Fat (total lipid), gm
Cakes:					
angelfood	40	2″ sc, ¹⁄₁₂ of 8″ diam	110	3	tr
chocolate, chocolate icing	120	2″ sc, ¹⁄₁₆ of 10″ diam	445	5	20
cupcake, with chocolate					
icing	50	1, 2¾″ diam	185	2	7
cupcake, without icing	40	1, 2¾″ diam	145	2	6
fruitcake, dark	30	1 pc, 2″ x 2″ x ½″	115	1	5
plain, with chocolate icing	100	2″ sc, ¹⁄₁₆ of 10″ diam	370	4	14
plain, without icing	55	1 pc, 3″ x 2″ x 1½″	200	2	8
pound	30	1 sl, 2¾″ x 3″ x ⅝″	140	2	9
sponge	40	2″ sc, ¹⁄₁₂ of 8″ diam	120	3	2
Candy:					
butterscotch	5	1 pc	21	0	tr
caramels	28	1 oz	115	1	3
chocolate almond bar	32	1 bar	176	3	12
chocolate, milk	28	1 oz	150	2	9
chocolate cream	13	1 pc	51	1	2
fondant	11	1 av	4	—	—
fudge, plain	28	1 oz	115	1	3
hard	28	1 oz	110	0	tr
marshmallow	28	1 oz	90	1	tr
peanut brittle	25	1 pc	110	2	4
Cantaloupe, raw	385	½ of 5″ melon	60	1	tr
Carrots:					
raw	110	1 c, grated	45	1	tr
cooked	145	1 c, diced	45	1	tr
Cashew nuts	135	1 c	760	23	62
Cauliflower:					
raw	100	1 c	25	2	tr
cooked	120	1 c	25	3	tr
Celery:					
raw	100	1 c, diced	15	1	tr
cooked	65	½ c, diced	12	1	tr
Cheese:					
blue or Roquefort type	28	1 oz	105	6	9
Camembert	28	1 oz	84	5	7
Cheddar or American	17	1″ cube	70	4	5
Cheddar or American	112	1 c, grated	445	28	36
Cheddar, process	28	1 oz	105	7	9

[3] If the fat used in the recipe is butter or fortified margarine, the vitamin A value for chocolate cake with fudge icing will be 490 IU; 100 IU for fruit cake; 300 IU for plain cake without icing; 220 IU per cupcake; 400 IU for plain cake with icing; 220 IU per cupcake with icing; and 300 IU for pound cake.

Fatty Acids										
Satur-ated (total), gm	Unsaturated		Carbo-hydrate, gm	Cal-cium, mg	Iron, mg	Vitamin A Value, IU	Thia-mine, mg	Ribo-flavin, mg	Niacin, mg	Ascorbic Acid, mg
	Oleic, gm	Lino-leic, gm								
—	—	—	24	4	0.1	0	tr	0.06	0.1	0
8	10	1	67	84	1.2	190[3]	0.03	0.12	0.3	tr
2	4	tr	30	32	0.3	90[3]	0.01	0.04	0.1	tr
1	3	tr	22	26	0.2	70[3]	0.01	0.03	0.1	tr
1	3	1	18	22	0.8	40[3]	0.04	0.04	0.2	tr
5	7	1	59	63	0.6	180[3]	0.02	0.09	0.2	tr
2	5	1	31	35	0.2	90[3]	0.01	0.05	0.1	tr
2	5	1	14	6	0.2	80[3]	0.01	0.03	0.1	0
1	1	tr	22	12	0.5	180[3]	0.02	0.06	0.1	tr
—	—	—	4	1	0.1	0	0	tr	tr	0
2	1	tr	22	42	0.4	tr	0.01	0.05	tr	tr
—	—	—	16	68	0.9	40	0.03	0.16	0.3	tr
5	3	tr	16	65	0.3	80	0.02	0.09	0.1	tr
—	—	—	9	—	—	—	—	—	—	—
—	—	—	10	—	—	—	—	—	—	—
2	1	tr	21	22	0.3	tr	0.01	0.03	0.1	tr
—	—	—	28	6	0.5	0	0	0	0	0
—	—	—	23	5	0.5	0	0	tr	tr	0
—	—	—	18	10	0.5	7	0.02	0.12	1.2	0
—	—	—	14	27	0.8	6540[4]	0.08	0.06	1.2	63
—	—	—	11	41	0.8	12100	0.06	0.06	0.7	9
—	—	—	10	48	0.9	15220	0.08	0.07	0.7	9
10	43	4	40	51	5.1	140	0.58	0.33	2.4	—
—	—	—	5	22	1.1	90	0.11	0.10	0.6	69
—	—	—	5	25	0.8	70	0.11	0.10	0.7	66
—	—	—	4	39	0.3	240	0.03	0.03	0.3	9
—	—	—	2	33	0.3	0	0.03	0.02	0.2	3
5	3	tr	1	89	0.1	350	0.01	0.17	0.1	0
—	—	—	1	29	0.1	286	0.01	0.21	0.3	0
3	2	tr	tr	128	0.2	220	tr	0.08	tr	0
20	12	1	2	840	1.1	1470	0.03	0.51	0.1	0
5	3	tr	1	219	0.3	350	tr	0.12	tr	0

[4] Value based on varieties with orange-colored flesh, for green-fleshed varieties value is about 540 IU per ½ melon.

Food	Weight, gm	Approximate Measure	Food Energy, Cal.	Pro-tein, gm	Fat (total lipid), gm
foods, Cheddar	28	1 oz	90	6	7
cottage, creamed	225	1 c	240	31	9
cream	15	1 tbsp	55	1	6
Limburger	28	1 oz	97	6	8
Parmesan	28	1 oz	110	10	7
Swiss	28	1 oz	105	8	8
Cherries:					
raw, sweet, with stems[5]	130	1 c	80	2	tr
canned, red, sour, pitted,					
heavy sirup	260	1 c	230	2	1
Chicken:					
broiled	85	3 oz, flesh only	115	20	3
canned, boneless	85	3 oz	170	18	10
creamed	118	½ c, sm serv	208	18	12
fryer, breast, fried	94	½ breast, with bone	155	25	5
fryer, leg, fried	59	with bone	90	12	4
potpie	227	1 pie, 4¼″ diam	535	23	31
roasted	80	2 sl, 3″ x 3″ x ¼″	158	23	7
Chili con carne (no beans)	255	1 c	510	26	38
Chili sauce	17	1 tbsp	20	tr	tr
Chocolate:					
bitter or baking	28	1 oz	145	3	15
sweet	28	1 oz	150	1	10
Chocolate-flavored milk drink	250	1 c	190	8	6
Chocolate sirup	20	1 tbsp	50	tr	tr
Clams:					
raw	85	3 oz	65	11	1
canned, solids and liquid	85	3 oz	45	7	1
Cocoa beverage with milk	242	1 c	235	9	11
Coconut:					
dried, sweetened	62	1 c, shredded	340	2	24
fresh	97	1 c, shredded	335	3	34
Coleslaw	120	1 c	120	1	9
Cookies:					
plain and assorted	25	1 cooky, 3″ diam	120	1	5
wafers	10	2 wafers, 2⅛″ diam	49	1	2
Corn:					
fresh, cooked	140	1 ear, 5″ long	70	3	1
canned	256	1 c	170	5	2
Corn flakes	28	1 oz	110	2	tr

[5] Measure and weight apply to entire vegetable or fruit including parts not usually eaten.

Fatty Acids										
Satur- ated (total), gm	Unsaturated		Carbo- hydrate, gm	Cal- cium, mg	Iron, mg	Vitamin A Value, IU	Thia- mine, mg	Ribo- flavin, mg	Niacin, mg	Ascorbic Acid, mg
	Oleic, gm	Lino- leic, gm								
4	2	tr	2	162	0.2	280	0.01	0.16	tr	0
5	3	tr	7	212	0.7	380	0.07	0.56	0.2	0
3	2	tr	tr	9	tr	230	tr	0.04	tr	0
—	—	—	1	165	0.2	358	0.02	0.14	0.1	0
—	—	—	1	325	0.1	297	tr	0.20	0.1	0
4	3	tr	1	262	0.3	320	tr	0.11	tr	0
—	—	—	20	26	0.5	130	0.06	0.07	0.5	12
—	—	—	59	36	0.8	1680	0.07	0.06	0.4	13
1	1	1	0	8	1.4	80	0.05	0.16	7.4	—
3	4	2	0	18	1.3	200	0.03	0.11	3.7	3
—	—	—	7	83	1.1	328	0.04	0.18	3.8	tr
1	2	1	1	9	1.3	70	0.04	0.17	11.2	—
1	2	1	tr	6	0.9	50	0.03	0.15	2.7	—
10	15	3	42	68	3.0	3020	0.25	0.26	4.1	5
—	—	—	0	16	1.7	0	0.06	0.14	7.2	0
18	17	1	15	97	3.6	380	0.05	0.31	5.6	—
—	—	—	4	3	0.1	240	0.02	0.01	0.3	3
8	6	tr	8	22	1.9	20	0.01	0.07	0.4	0
6	4	tr	16	27	0.4	tr	0.01	0.04	0.1	tr
3	2	tr	27	270	0.4	210	0.09	0.41	0.2	2
tr	tr	tr	13	3	0.3	—	tr	0.01	0.1	0
—	—	—	2	59	5.2	90	0.08	0.15	1.1	8
—	—	—	2	47	3.5	—	0.01	0.09	0.9	—
6	4	tr	26	286	0.9	390	0.09	0.45	0.4	2
21	2	tr	33	10	1.2	0	0.02	0.02	0.2	0
29	2	tr	9	13	1.6	0	0.05	0.02	0.5	3
2	2	5	9	52	0.5	180	0.06	0.06	0.3	35
—	—	—	18	9	0.2	20	0.01	0.01	0.1	tr
—	—	—	7	—	—	—	—	—	—	—
—	—	—	16	2	0.5	310[6]	0.09	0.08	1.0	7
—	—	—	40	10	1.0	690[6]	0.07	0.12	2.3	13
—	—	—	24	5	0.4	0	0.12	0.02	0.6	0

[6] Based on yellow varieties; white varieties contain only a trace of cryptoxanthin and carotenes, the pigments in corn that have biological activity.

Food	Weight, gm	Approximate Measure	Food Energy, Cal.	Protein, gm	Fat (total lipid), gm
Corn grits:					
enriched, cooked	242	1 c	120	3	tr
unenriched, cooked	242	1 c	120	3	tr
Corn muffin, enriched	48	1 med, 2¾″ diam	150	3	5
Cornmeal, white or yellow, dry:					
enriched	145	1 c	525	11	2
unenriched	118	1 c	420	11	5
Crabmeat, canned	85	3 oz	85	15	2
Crackers:					
Graham	14	2 med	55	1	1
saltines	8	2 crackers	35	1	1
soda, plain	11	2 crackers	50	1	1
Cranberry sauce, sweetened	277	1 c	405	tr	1
Cream:					
half-and-half	15	1 tbsp	20	tr	2
heavy or whipping	15	1 tbsp	55	tr	6
light or coffee	15	1 tbsp	30	tr	3
Cucumber, raw	50	6 sl	5	tr	tr
Custard, baked	248	1 c	285	13	14
Dandelion greens, cooked	180	1 c	60	4	1
Dates, fresh and dried	178	1 c	490	4	1
Doughnut, cake type	32	1 doughnut	125	1	6
Eggs:					
raw, whole	50	1 med	80	6	6
boiled	100	2 med	160	13	12
scrambled	64	1 med	110	7	8
Farina, enriched, cooked	238	1 c	100	3	tr
Fats, cooking, vegetable	12.5	1 tbsp	110	0	12
Figs, dried	21	1 fig	60	1	tr
Fig bars	16	1 sm	55	1	1
Fishsticks, breaded, cooked	227	10 sticks	400	38	20
Fruit cocktail, canned	256	1 c	195	1	1
Gelatin, dry, plain	10	1 tbsp	35	9	tr
Gelatin dessert:					
plain	239	1 c	140	4	0
with fruit	241	1 c	160	3	tr
Gingerbread	55	1 pc, 2″ x 2″ x 2″	175	2	6

[7] Vitamin A value based on yellow product; white product contains only a trace.

[8] Iron, thiamine, riboflavin, and niacin are based on the minimal level of enrichment specified in standards of identity promulgated under the Federal Food, Drug, and Cosmetic Act.

[9] Based on recipe using white cornmeal; if yellow cornmeal is used, the vitamin A value is 140 IU per muffin.

| Fatty Acids | | | | | | | | | | |
Satur- ated (total), gm	Unsaturated Oleic, gm	Lino- leic, gm	Carbo- hydrate, gm	Cal- cium, mg	Iron, mg	Vitamin A Value, IU	Thia- mine, mg	Ribo- flavin, mg	Niacin, mg	Ascorbic Acid, mg
—	—	—	27	2	0.7[8]	150[7]	0.10[8]	0.07[8]	1.0[8]	0
—	—	—	27	2	0.2	150[7]	0.05	0.02	0.5	0
2	2	tr	23	50	0.8	80[9]	0.09	0.11	0.8	tr
tr	1	1	114	9	4.2[8]	640[7]	0.64[8]	0.38[8]	5.1[8]	0
1	2	2	87	24	2.8	600[7]	0.45	0.13	2.4	0
—	—	—	1	38	0.7	—	0.07	0.07	1.6	—
—	—	—	10	6	0.2	0	0.01	0.03	0.2	0
—	—	—	6	2	0.1	0	tr	tr	0.1	0
tr	1	tr	8	2	0.2	0	tr	tr	0.1	0
—	—	—	104	17	0.6	40	0.03	0.03	0.1	5
1	1	tr	1	16	tr	70	tr	0.02	tr	tr
3	2	tr	tr	11	tr	230	tr	0.02	tr	tr
2	1	tr	1	15	tr	130	tr	0.02	tr	tr
—	—	—	2	8	0.2	tr	0.02	0.02	0.1	6
6	5	1	28	278	1.0	870	0.10	0.47	0.2	1
—	—	—	12	252	3.2	21060	0.24	0.29	—	32
—	—	—	130	105	5.3	90	0.16	0.17	3.9	0
1	4	tr	16	13	0.4[10]	30	0.05[10]	0.05[10]	0.4[10]	tr
2	3	tr	tr	27	1.1	590	0.05	0.15	tr	0
4	5	1	1	54	2.3	1180	0.09	0.28	0.1	0
3	3	tr	1	51	1.1	690	0.05	0.18	tr	0
—	—	—	21	10	0.7[11]	0	0.11[11]	0.07[11]	1.0[11]	0
3	8	1	0	0	0	—	0	0	0	0
—	—	—	15	26	0.6	20	0.02	0.02	0.1	0
—	—	—	12	12	0.2	20	0.01	0.01	0.1	tr
5	4	10	15	25	0.9	—	0.09	0.16	3.6	—
—	—	—	50	23	1.0	360	0.04	0.03	1.1	5
—	—	—	—	—	—	—	—	—	—	—
—	—	—	34	—	—	—	—	—	—	—
—	—	—	40	—	—	—	—	—	—	—
1	4	tr	29	37	1.3	50	0.06	0.06	0.5	0

[10] Based on product made with enriched flour. With unenriched flour, approximate values per doughnut are: iron, 0.2 mg; thiamine, 0.01 mg; riboflavin, 0.03 mg; niacin, 0.2 mg.
[11] Iron, thiamine, riboflavin, and niacin are based on the minimum levels of enrichment specified in standards of identity promulgated under the Federal Food, Drug, and Cosmetic Act.

Food	Weight, gm	Approximate Measure	Food Energy, Cal.	Pro-tein, gm	Fat (total lipid), gm
Grapefruit:					
raw, white	285	½ med, 4¼" diam	55	1	tr
raw, white	194	1 c, sc	75	1	tr
juice, canned	247	1 c, unsweetened	100	1	tr
juice, dehydrated,					
water added	247	1 c	100	1	tr
Grapes:					
Concord, Niagara	153	1 c	65	1	1
Muscat, Thompson, Tokay	160	1 c	95	1	tr
Grape juice, bottled	254	1 c	165	1	tr
Grapenut flakes	28	1 oz	110	3	tr
Gravy, meat, brown	18	1 tbsp	41	tr	4
Haddock, fried	85	3 oz	140	17	5
Heart, beef, lean, braised	85	3 oz	160	27	5
Herring:					
Atlantic, broiled	85	1 med	217	21	14
smoked, kippered	100	½ fish	211	22	13
Honey, strained or extracted	21	1 tbsp	65	tr	0
Honeydew melon	150	1 wedge, 2" x 6½"	48	1	0
Ice cream, plain	71	1 sl, or ⅛ qt brick	145	3	9
Ice milk	187	1 c	285	9	10
Jams and preserves	20	1 tbsp	55	tr	tr
Jellies	20	1 tbsp	55	tr	tr
Kale, cooked	110	1 c	30	4	1
Kohlrabi, cooked	75	½ c	23	2	tr
Lamb:					
chop, cooked	137	1 chop, 4.8 oz	400	25	33
leg, roasted	71	2.5 oz, lean	130	20	5
shoulder, roasted	64	2.3 oz, lean	130	17	6
Lard	14	1 tbsp	125	0	14
Lemon	106	1 med	20	1	tr
Lemon juice, fresh	15	1 tbsp	5	tr	tr
Lettuce:					
head, Iceberg	454	1 head, 4¼" diam	60	4	tr
leaves	50	2 lg	10	1	tr
Lime juice, fresh	246	1 c	65	1	tr
Liver:					
beef, fried	57	2 oz	130	15	6
calf, cooked	72	2 sl, 3" x 2¼" x ⅜"	147	16	7
pork, fried	74	2 sl, 3" x 2¼" x ⅜"	170	18	7
Lobster:					
boiled or broiled	334	1 (¾ lb) + 2 tbsp butter	308	20	25
canned	85	½ c	75	15	1

| Fatty Acids | | | | | | | | | | |
| Satur-ated (total), gm | Unsaturated | | Carbo-hydrate, gm | Cal-cium, mg | Iron, mg | Vitamin A Value, IU | Thia-mine, mg | Ribo-flavin, mg | Niacin, mg | Ascorbic Acid, mg |
	Oleic, gm	Lino-leic, gm								
—	—	—	14	22	0.6	10	0.05	0.02	0.2	52
—	—	—	20	31	0.8	20	0.07	0.03	0.3	72
—	—	—	24	20	1.0	20	0.07	0.04	0.4	84
—	—	—	24	22	0.2	20	0.10	0.05	0.5	92
—	—	—	15	15	0.4	100	0.05	0.03	0.2	3
—	—	—	25	17	0.6	140	0.07	0.04	0.4	6
—	—	—	42	28	0.8	—	0.10	0.05	0.6	tr
—	—	—	23	—	1.2	0	0.13	—	1.6	0
—	—	—	2	—	0.2	0	0.15	0.01	tr	—
1	3	tr	5	34	1.0	—	0.03	0.06	2.7	2
—	—	—	1	5	5.0	20	0.21	1.04	6.5	1
—	—	—	0	—	1.2	130	0.01	0.15	3.3	0
—	—	—	0	66	1.4	0	tr	0.28	2.9	0
—	—	—	17	1	0.1	0	tr	0.01	0.1	tr
—	—	—	13	26	0.6	60	0.08	0.05	0.3	34
5	3	tr	15	87	0.1	370	0.03	0.13	0.1	1
6	3	tr	42	292	0.2	390	0.09	0.41	0.2	2
—	—	—	14	4	0.2	tr	tr	0.01	tr	tr
—	—	—	14	4	0.3	tr	tr	0.01	tr	1
—	—	—	4	147	1.3	8140	—	—	—	68
—	—	—	5	35	0.5	tr	0.03	0.03	0.2	28
18	12	1	0	10	1.5	—	0.14	0.25	5.6	—
3	2	tr	0	9	1.4	—	0.12	0.21	4.4	—
3	2	tr	0	8	1.0	—	0.10	0.18	3.7	—
5	6	1	0	0	0	0	0	0	0	0
—	—	—	6	18	0.4	10	0.03	0.01	0.1	38
—	—	—	1	1	tr	tr	tr	tr	tr	7
—	—	—	13	91	2.3	1500	0.29	0.27	1.3	29
—	—	—	2	34	0.7	950	0.03	0.04	0.2	9
—	—	—	22	22	0.5	30	0.05	0.03	0.3	80
—	—	—	3	6	5.0	30280	0.15	2.37	9.4	15
—	—	—	3	5	9.0	19130	0.13	2.39	11.7	15
—	—	—	8	10	15.6	12070	0.25	2.30	12.4	10
—	—	—	1	80	0.7	920	0.11	0.06	2.3	0
—	—	—	0	55	0.7	—	0.03	0.06	1.9	—

Food	Weight, gm	Approximate Measure	Food Energy, Cal.	Pro-tein, gm	Fat (total lipid), gm
Macaroni:					
enriched, cooked	130	1 c	190	6	1
unenriched, cooked	130	1 c	190	6	1
Macaroni & cheese, baked	220	1 c	470	18	24
Mackerel, canned	85	3 oz	155	18	9
Malted milk beverage	270	1 c	280	13	12
Mangos	100	1 sm	66	1	tr
Margarine:					
stick, ⅛	14	1 tbsp	100	tr	11
pat or sq	7	1 pat	50	tr	6
Metrecal	237	8 oz	225	18	5
Milk:					
whole	244	1 c	160	9	9
nonfat, skim	246	1 c	90	9	tr
dry, nonfat, instant	70	1 c	250	25	tr
condensed	306	1 c	980	25	27
evaporated	252	1 c	345	18	20
Molasses:					
light	20	1 tbsp	50	—	—
blackstrap	20	1 tbsp	45	—	—
Muffins, white, enriched	48	1 med, 2¾″ diam	140	4	5
Mushrooms, canned	244	1 c	40	5	tr
Mustard greens, cooked	140	1 c	35	3	1
Noodles:					
enriched, cooked	160	1 c	200	7	2
unenriched, cooked	160	1 c	200	7	2
Oats, puffed	28	1 oz	115	3	2
Oatmeal, cooked	236	1 c	130	5	2
Oils, salad, corn	14	1 tbsp	125	0	14
Okra, cooked	85	8 pods	25	2	tr
Olives:					
green, pickled	16	4 med	15	tr	2
ripe, pickled	10	3 sm	15	tr	2
Onions:					
raw	110	1 onion, 2½″ diam	40	2	tr
cooked	210	1 c	60	3	tr
young green	50	6 onions	20	1	tr
Orange:					
navel	180	1 med	60	2	tr

[12] Iron, thiamine, riboflavin, and niacin are based on the minimum levels of enrichment specified in standards of identity promulgated under the Federal Food, Drug, and Cosmetic Act.

[13] Based on the average vitamin A content of fortified margarine. Federal specifications for

| Fatty Acids | | | | | | | | | | |
Satur-ated (total), gm	Unsaturated Oleic, gm	Lino-leic, gm	Carbo-hydrate, gm	Cal-cium, mg	Iron, mg	Vitamin A Value, IU	Thia-mine, mg	Ribo-flavin, mg	Niacin, mg	Ascorbic Acid, mg
—	—	—	39	14	1.4[12]	0	0.23[12]	0.14[12]	1.9[12]	0
—	—	—	39	14	0.6	0	0.02	0.02	0.5	0
11	10	1	44	398	2.0	950	0.22	0.44	2.0	tr
—	—	—	0	221	1.9	20	0.02	0.28	7.4	—
—	—	—	32	364	0.8	670	0.17	0.56	0.2	2
—	—	—	17	9	0.2	6350	0.06	0.06	0.9	41
2	6	2	tr	3	0	460[13]	—	—	—	0
1	3	1	tr	1	0	230[13]	—	—	—	0
—	—	—	28	500	3.8	1250	0.50	0.75	3.8	25
5	3	tr	12	288	0.1	350	0.08	0.42	0.1	2
—	—	—	13	298	0.1	10	0.10	0.44	0.2	2
—	—	—	36	905	0.4	20	0.24	1.25	0.6	5
15	9	1	166	802	0.3	1090	0.23	1.17	0.5	3
11	7	1	24	635	0.3	820	0.10	0.84	0.5	3
—	—	—	13	33	0.9	—	0.01	0.01	tr	—
—	—	—	11	137	3.2	—	0.02	0.04	0.4	—
1	3	tr	20	50	0.8	50	0.08	0.11	0.7	tr
—	—	—	6	15	1.2	tr	0.04	0.60	4.8	4
—	—	—	6	193	2.5	8120	0.11	0.19	0.9	68
1	1	tr	37	16	1.4[14]	110	0.23[14]	0.14[14]	1.8[14]	0
1	1	tr	37	16	1.0	110	0.04	0.03	0.7	0
tr	1	1	21	50	1.3	0	0.28	0.05	0.5	0
tr	1	1	23	21	1.4	0	0.19	0.05	0.3	0
1	4	7	0	0	0	—	0	0	0	0
—	—	—	5	78	0.4	420	0.11	0.15	0.8	17
tr	2	tr	tr	8	0.2	40	—	—	—	—
tr	2	tr	tr	9	0.1	10	tr	tr	—	—
—	—	—	10	30	0.6	40	0.04	0.04	0.2	11
—	—	—	14	50	0.8	80	0.06	0.06	0.4	14
—	—	—	5	20	0.3	tr	0.02	0.02	0.2	12
—	—	—	16	49	0.5	240	0.12	0.05	0.5	75

fortified margarine require a minimum of 15000 IU of vitamin A per pound.
[14] Iron, thiamine, riboflavin, and niacin are based on the minimum levels of enrichment specified in standards of identity promulgated under the Federal Food, Drug, and Cosmetic Act.

Food	Weight, gm	Approximate Measure	Food Energy, Cal.	Pro- tein, gm	Fat (total lipid), gm
other varieties	210	1 med	75	1	tr
sections	97	½ c	44	1	tr
juice, fresh	247	1 c	100	1	tr
juice, frozen	248	1 c	110	2	tr
juice, dehydrated,					
water added	248	1 c	115	1	tr
Orange and grapefruit					
juice, frozen	248	1 c	110	1	tr
Ocean perch, breaded, fried	85	3 oz	195	16	11
Oyster meat, raw	240	1 c	160	20	4
Oyster stew	230	1 c with 3–4 oysters	200	11	12
Pancakes:					
white, enriched	27	1 cake, 4″ diam	60	2	2
buckwheat	27	1 cake, 4″ diam	55	2	2
Papayas, raw	182	1 c	70	1	tr
Parsley, raw, chopped	3.5	1 tbsp	1	tr	tr
Parsnips, cooked	155	1 c	100	2	1
Peaches:					
raw	114	1 med	35	1	tr
raw	168	1 c, sliced	65	1	tr
canned, sirup pack	257	1 c	200	1	tr
dried, cooked	270	1 c	220	3	1
frozen	340	12-oz carton	300	1	tr
Peanuts, roasted	9	1 tbsp	55	2	4
Peanut butter	16	1 tbsp	95	4	8
Pears:					
raw	182	1 med	100	1	1
canned, sirup pack	255	1 c	195	1	1
Peas, green:					
fresh, cooked	160	1 c	115	9	1
canned	249	1 c	165	9	1
Pecans, chopped	7.5	1 tbsp	50	1	5
Peppers, green, raw	62	1 med	15	1	tr
Pickles:					
dill	135	1 pickle, 4″ long	15	1	tr
relish	13	1 tbsp	14	tr	tr
sour	30	1 sl, 1½″ diam x 1″	3	tr	tr
sweet	20	1 pickle, 2¾″ long	30	tr	tr
Pies:					
apple	135	⅐ of 9″ pie	345	3	15

[15] Based on yellow-fleshed varieties; for white-fleshed varieties value is about 50 IU per 114-gm peach and 80 IU per cup of sliced peaches.

[16] Average weight in accordance with commercial freezing practices. For products without

| Fatty Acids | | | | | | | | | | |
| Saturated (total), gm | Unsaturated | | Carbohydrate, gm | Calcium, mg | Iron, mg | Vitamin A Value, IU | Thiamine, mg | Riboflavin, mg | Niacin, mg | Ascorbic Acid, mg |
	Oleic, gm	Linoleic, gm								
—	—	—	19	67	0.3	310	0.16	0.06	0.6	70
—	—	—	11	32	0.4	180	0.08	0.03	0.3	48
—	—	—	23	25	0.5	490	0.22	0.06	0.9	127
—	—	—	27	22	0.2	500	0.21	0.03	0.8	112
—	—	—	27	25	0.5	500	0.20	0.06	0.9	108
—	—	—	26	20	0.2	270	0.16	0.02	0.8	102
—	—	—	6	28	1.1	—	0.08	0.09	1.5	—
—	—	—	8	226	13.2	740	0.33	0.43	6.0	—
—	—	—	11	269	3.3	640	0.13	0.41	1.6	—
tr	1	tr	9	27	0.4	30	0.05	0.06	0.3	tr
1	1	tr	6	59	0.4	60	0.03	0.04	0.2	tr
—	—	—	18	36	0.5	3190	0.07	0.08	0.5	102
—	—	—	tr	7	0.2	300	tr	0.01	tr	6
—	—	—	23	70	0.9	50	0.11	0.13	0.2	16
—	—	—	10	9	0.5	1320[15]	0.02	0.05	1.0	7
—	—	—	16	15	0.8	2230[15]	0.03	0.08	1.6	12
—	—	—	52	10	0.8	1100	0.02	0.06	1.4	7
—	—	—	58	41	5.1	3290	0.01	0.15	4.2	6
—	—	—	77	14	1.7	2210	0.03	0.14	2.4	135[16]
1	2	1	2	7	0.2	—	0.03	0.01	1.5	0
2	4	2	3	9	0.3	—	0.02	0.02	2.4	0
—	—	—	25	13	0.5	30	0.04	0.07	0.2	7
—	—	—	50	13	0.5	tr	0.03	0.05	0.3	4
—	—	—	19	37	2.9	860	0.44	0.17	3.7	33
—	—	—	31	50	4.2	1120	0.23	0.13	2.2	22
tr	3	1	1	5	0.2	10	0.06	0.01	0.1	tr
—	—	—	3	6	0.4	260	0.05	0.05	0.3	79
—	—	—	3	35	1.4	140	tr	0.03	tr	8
—	—	—	3	2	0.2	14	0	tr	tr	1
—	—	—	1	8	0.4	93	tr	0.02	tr	2
—	—	—	7	2	0.2	20	tr	tr	tr	1
4	9	1	51	11	0.4	40	0.03	0.02	0.5	1

added ascorbic acid, value is about 37 mg per 12-oz carton and 50 mg per 16-oz carton; for those with added ascorbic acid, 139 mg per 12-oz carton and 186 mg per 16-oz carton.

Food	Weight, gm	Approximate Measure	Food Energy, Cal.	Pro-tein, gm	Fat (total lipid), gm
cherry	135	⅐ of 9″ pie	355	4	15
custard	130	⅐ of 9″ pie	280	8	14
lemon meringue	120	⅐ of 9″ pie	305	4	12
mince	135	⅐ of 9″ pie	365	3	16
pumpkin	130	⅐ of 9″ pie	275	5	15
Piecrust, plain, baked	135	1, 9″ crust	675	8	45
Pimentos, canned	38	1 med	10	tr	tr
Pineapple:					
raw	140	1 c, diced	75	1	tr
canned, sirup pack	260	1 c, crushed	195	1	tr
canned, sirup pack	122	2 sm sl + 2 tbsp jc	90	tr	tr
juice, canned	249	1 c	135	1	tr
Pizza, cheese	75	5½″ sector	185	7	6
Plums:					
raw	60	1 plum, 2″ diam	25	tr	tr
canned, sirup pack	122	3 plums + 2 tbsp jc	100	tr	tr
Popcorn, popped	14	1 c	65	1	3
Pork:					
chop, cooked	98	1 chop, 3.5 oz	260	16	21
ham, cured	85	3 oz	245	18	19
ham, fresh, lean	107	2 sl, 2″ x 1½″ x 1″	254	40	9
boiled ham	57	2 oz	135	11	10
Potatoes:					
baked	99	1 med	90	3	tr
French fried	57	10 pc	155	2	7
hash-browned	100	½ c	241	3	12
mashed	195	1 c with milk	125	4	1
mashed	195	1 c with milk & butter	185	4	8
Potato chips	20	10 chips	115	1	8
Pretzels	5	5 sm sticks	20	tr	tr
Prunes:					
dried, uncooked	32	4 prunes	70	1	tr
dried, cooked, sirup	270	1 c (17–18 prunes)	295	2	1
juice, canned	256	1 c	200	1	tr
Puddings:					
chocolate	144	½ c	219	5	7
lemon snow	130	1 serv	114	3	tr
tapioca	132	½ c	181	5	5
vanilla	248	1 c	275	9	10
Pumpkin, canned	228	1 c	75	2	1
Radishes, raw	40	4 sm	5	tr	tr
Raisins, dried	160	1 c	460	4	tr
Raspberries, red, raw	123	1 c	70	1	1
Rhubarb, cooked, sugar added	272	1 c	385	1	tr

Fatty Acids										
	Unsaturated					Vitamin				
Satur- ated (total), gm	Oleic, gm	Lino- leic, gm	Carbo- hydrate, gm	Cal- cium, mg	Iron, mg	A Value, IU	Thia- mine, mg	Ribo- flavin, mg	Niacin, mg	Ascorbic Acid, mg
4	10	1	52	19	0.4	590	0.03	0.23	0.6	1
5	8	1	30	125	0.8	300	0.07	0.21	0.4	0
4	7	1	45	17	0.6	200	0.04	0.10	0.2	4
4	10	1	56	38	1.4	tr	0.09	0.05	0.5	1
5	7	1	32	66	0.6	3210	0.04	0.15	0.6	tr
10	29	3	59	19	2.3	0	0.27	0.19	2.4	0
—	—	—	2	3	0.6	870	0.01	0.02	0.1	36
—	—	—	19	24	0.7	100	0.12	0.04	0.3	24
—	—	—	50	29	0.8	120	0.20	0.06	0.5	17
—	—	—	24	13	0.4	50	0.09	0.03	0.2	8
—	—	—	34	37	0.7	120	0.12	0.04	0.5	22
2	3	tr	27	107	0.7	290	0.04	0.12	0.7	4
—	—	—	7	7	0.3	140	0.02	0.02	0.3	3
—	—	—	26	11	1.1	1470	0.03	0.02	0.5	2
2	tr	tr	8	1	0.3	—	—	0.01	0.2	0
8	9	2	0	8	2.2	0	0.63	0.18	3.8	—
7	8	2	0	8	2.2	0	0.40	0.16	3.1	—
—	—	—	0	7	2.5	0	0.69	0.33	5.4	0
4	4	1	0	6	1.6	0	0.25	0.09	1.5	—
—	—	—	21	9	0.7	tr	0.10	0.04	1.7	20
2	2	4	20	9	0.7	tr	0.07	0.04	1.8	12
—	—	—	32	18	1.2	30	0.08	0.06	1.7	7
—	—	—	25	47	0.8	50	0.16	0.10	2.0	19
4	3	tr	24	47	0.8	330	0.16	0.10	1.9	18
2	2	4	10	8	0.4	tr	0.04	0.01	1.0	3
—	—	—	4	1	0	0	tr	tr	tr	0
—	—	—	18	14	1.1	440	0.02	0.04	0.4	1
—	—	—	78	60	4.5	1860	0.08	0.18	1.7	2
—	—	—	49	36	10.5	—	0.02	0.03	1.1	4
—	—	—	37	147	0.2	196	0.05	0.22	0.2	0
—	—	—	27	4	0.1	0	tr	0.02	tr	10
—	—	—	28	151	0.6	195	0.05	0.21	0.6	0
5	3	tr	39	290	0.1	390	0.07	0.40	0.1	2
—	—	—	18	57	0.9	14590	0.07	0.12	1.3	12
—	—	—	1	12	0.4	tr	0.01	0.01	0.1	10
—	—	—	124	99	5.6	30	0.18	0.13	0.9	2
—	—	—	17	27	1.1	160	0.04	0.11	1.1	31
—	—	—	98	212	1.6	220	0.06	0.15	0.7	17

Food	Weight, gm	Approximate Measure	Food Energy, Cal.	Pro- tein, gm	Fat (total lipid), gm
Rice:					
parboiled, cooked	176	1 c	185	4	tr
puffed	14	1 c	55	1	tr
white, cooked	168	1 c	185	3	tr
Rice flakes	30	1 c	115	2	tr
Rolls:					
plain, enriched	38	12 per lb	115	3	2
plain, unenriched	38	12 per lb	115	3	2
sweet	43	1 roll	135	4	4
Rutabagas, cooked	100	½ c	38	1	tr
Rye flour, light	80	1 c	285	8	1
Salads:					
apple, celery, walnut	154	3 hp tbsp, 2 lv lettuce	137	2	8
carrot & raisin	134	3 hp tbsp, 2 lv lettuce	153	2	6
fruit, fresh	195	3 hp tbsp, 2 lv lettuce	174	2	11
gelatin with fruit	188	1 sq, 2 lv lettuce	139	2	6
gelatin with vegetable	164	1 sq, 2 lv lettuce	115	2	6
lettuce, tomato,		4 lv lettuce,			
mayonnaise	115	3 sl tomato	80	2	6
potato	123	½ c, French dressing	184	2	11
Salad dressings:					
blue cheese	16	1 tbsp	80	1	8
commercial, plain	15	1 tbsp	65	tr	6
French	15	1 tbsp	60	tr	6
home cooked, boiled	17	1 tbsp	30	1	2
mayonnaise	15	1 tbsp	110	tr	12
Thousand Island	15	1 tbsp	75	tr	8
Salmon, pink, canned	85	3 oz	120	17	5
Sardines, Atlantic	85	3 oz	175	20	9
Sauerkraut, canned	235	1 c	45	2	tr
Sausage:					
bologna	227	8 sl	690	27	62
frankfurter, cooked	51	1 frankfurter	155	6	14
liverwurst	30	1 sl, 3″ diam x ¼″	79	5	6
pork, links or patty, cooked	113	4 oz	540	21	50
Vienna	18	1 av, 2″ x ¾″ diam	39	3	3
Scallops, fried	145	5–6 med pc	427	24	28
Shad, baked	85	3 oz	170	20	10
Sherbet, orange	193	1 c	260	2	2
Shortbread	16	2 pc, 58 per lb	78	1	3

[17] Iron, thiamine, and niacin are based on the minimum levels of enrichment specified in standards of identity promulgated under the Federal Food, Drug, and Cosmetic Act. Ribo- flavin based on unenriched rice. When the minimum level of enrichment for riboflavin speci-

Saturated (total), gm	Oleic, gm	Linoleic, gm	Carbohydrate, gm	Calcium, mg	Iron, mg	Vitamin A Value, IU	Thiamine, mg	Riboflavin, mg	Niacin, mg	Ascorbic Acid, mg
—	—	—	41	33	1.4[17]	0	0.19[17]	0.02[17]	2.0[17]	0
—	—	—	13	3	0.3	0	0.06	0.01	0.6	0
—	—	—	41	17	1.5[17]	0	0.19[17]	0.01[17]	1.6[17]	0
—	—	—	26	9	0.5	0	0.10	0.02	1.6	0
tr	1	tr	20	28	0.7	tr	0.11	0.07	0.8	tr
tr	1	tr	20	28	0.3	tr	0.02	0.03	0.3	tr
1	2	tr	21	37	0.3	30	0.03	0.06	0.4	0
—	—	—	9	55	0.4	330	0.07	0.08	0.9	36
—	—	—	62	18	0.9	0	0.12	0.06	0.5	0
—	—	—	16	32	0.8	355	0.08	0.08	0.4	5
—	—	—	28	48	1.5	4708	0.08	0.08	0.5	6
—	—	—	21	45	0.8	685	0.08	0.09	0.4	32
—	—	—	22	23	0.5	391	0.04	0.05	0.3	16
—	—	—	15	24	0.5	1977	0.04	0.06	0.3	8
—	—	—	7	20	0.8	1115	0.06	0.07	0.5	19
—	—	—	21	21	0.8	243	0.07	0.04	0.8	16
2	2	4	1	13	tr	30	tr	0.02	tr	tr
1	1	3	2	2	tr	30	tr	tr	tr	—
1	1	3	3	2	0.1	—	—	—	—	—
1	1	tr	3	15	0.1	80	0.01	0.03	tr	tr
2	3	6	tr	3	0.1	40	tr	0.01	tr	—
1	2	4	2	2	0.1	50	tr	tr	tr	tr
1	1	tr	0	167[18]	0.7	60	0.03	0.16	6.8	—
—	—	—	0	372	2.5	190	0.02	0.17	4.6	—
—	—	—	9	85	1.2	120	0.07	0.09	0.4	33
—	—	—	2	16	4.1	—	0.36	0.49	6.0	—
—	—	—	1	3	0.8	—	0.08	0.10	1.3	—
—	—	—	1	3	1.6	1725	0.05	0.34	1.4	0
18	21	5	tr	8	2.7	0	0.89	0.39	4.2	—
—	—	—	0	2	0.4	0	0.02	0.02	0.6	0
—	—	—	19	41	3.1	0	0.09	0.17	2.3	0
—	—	—	0	20	0.5	20	0.11	0.22	7.3	—
—	—	—	59	31	tr	110	0.02	0.06	tr	4
—	—	—	11	2	tr	0	0.01	tr	tr	0

fied in the standards of identity becomes effective the value will be 0.12 mg per cup of parboiled rice and of white rice.

[18] Based on total contents of can. If bones are discarded, value will be greatly reduced.

Food	Weight, gm	Approximate Measure	Food Energy, Cal.	Protein, gm	Fat (total lipid), gm
Shrimp, canned	85	3 oz	100	21	1
Sirups, table blends	20	1 tbsp	60	0	0
Soups:					
bean	250	1 c	170	8	6
beef	250	1 c	100	6	4
beef noodle	250	1 c	70	4	3
beef bouillon, broth,					
consomme	240	1 c	30	5	0
chicken	250	1 c	75	4	2
chicken noodle	250	1 c	65	4	2
clam chowder	255	1 c	85	2	3
cream, mushroom	240	1 c	135	2	10
pea, green	245	1 c	130	6	2
tomato	245	1 c	90	2	2
vegetable with beef broth	250	1 c	80	3	2
vegetable-beef	203	1 serv, 3 from can	64	6	2
Soy flour, medium fat	88	1 c	232	37	6
Spaghetti:					
enriched, cooked	140	1 c	155	5	1
unenriched, cooked	140	1 c	155	5	1
in tomato sauce	250	1 c with cheese	260	9	9
Italian style	292	1 serv, with meat sauce	396	13	21
Italian style	302	1 serv, as above with grated cheese	436	15	24
Spinach	180	1 c	40	5	1
Squash:					
summer, cooked	210	1 c	30	2	tr
winter, cooked	205	1 c	130	4	1
Strawberries:					
raw	149	1 c	55	1	1
frozen	284	10-oz carton	310	1	1
Sugar:					
brown	14	1 tbsp	50	0	0
maple	15	1 pc, 1¼" x 1" x ½"	52	—	—
white, granulated	12	1 tbsp	45	0	0
white, powdered	8	1 tbsp	30	0	0
Sweet potatoes:					
baked	110	1 med, 5" x 2"	155	2	1
candied	175	1 sm, 3½" x 2¼"	295	2	6
Tangerine	114	1 med	40	1	tr

[19] Iron, thiamine, riboflavin, and niacin are based on the minimum levels of enrichment specified in standards of identity promulgated under the Federal Food, Drug, and Cosmetic Act.

| Fatty Acids | | | | | | | | | | |
| Satur-ated (total), gm | Unsaturated | | Carbo-hydrate, gm | Cal-cium, mg | Iron, mg | Vitamin A Value, IU | Thia-mine, mg | Ribo-flavin, mg | Niacin, mg | Ascorbic Acid, mg |
	Oleic, gm	Lino-leic, gm								
—	—	—	1	98	2.6	50	0.01	0.03	1.5	—
—	—	—	15	9	0.8	0	0	0	0	0
1	2	2	22	62	2.2	650	0.14	0.07	1.0	2
2	2	tr	11	15	0.5	—	—	—	—	—
1	1	1	7	8	1.0	50	0.05	0.06	1.1	tr
0	0	0	3	tr	0.5	tr	tr	0.02	1.2	—
1	1	tr	10	20	0.5	—	0.02	0.12	1.5	—
tr	1	1	8	10	0.5	50	0.02	0.02	0.8	tr
—	—	—	13	36	1.0	920	0.03	0.03	1.0	—
1	3	5	10	41	0.5	70	0.02	0.12	0.7	tr
1	1	tr	23	44	1.0	340	0.05	0.05	1.0	7
tr	1	1	16	15	0.7	1000	0.06	0.05	1.1	12
—	—	—	14	20	0.8	3250	0.05	0.02	1.2	—
—	—	—	6	5	0.5	2340	0.03	0.04	0.8	—
—	—	—	33	215	11.4	100	0.72	0.30	2.3	0
—	—	—	32	11	1.3[19]	0	0.19[19]	0.11[19]	1.5[19]	0
—	—	—	32	11	0.6	0	0.02	0.02	0.4	0
2	5	1	37	80	2.2	1080	0.24	0.18	2.4	14
—	—	—	39	27	2.1	901	0.12	0.12	3.0	24
—	—	—	40	99	2.2	1041	0.12	0.16	3.0	24
—	—	—	6	167	4.0	14580	0.13	0.25	1.0	50
—	—	—	7	52	0.8	820	0.10	0.16	1.6	21
—	—	—	32	57	1.6	8610	0.10	0.27	1.4	27
—	—	—	13	31	1.5	90	0.04	0.10	0.9	88
—	—	—	79	40	2.0	90	0.06	0.17	1.5	150
—	—	—	13	12	0.5	0	tr	tr	tr	0
—	—	—	14	27	0.5	—	—	—	—	—
—	—	—	12	0	tr	0	0	0	0	0
—	—	—	8	0	tr	0	0	0	0	0
—	—	—	36	44	1.0	8910	0.10	0.07	0.7	24
2	3	1	60	65	1.6	11030	0.10	0.08	0.8	17
—	—	—	10	34	0.3	350	0.05	0.02	0.1	26

Food	Weight, gm	Approximate Measure	Food Energy, Cal.	Protein, gm	Fat (total lipid), gm
Tomatoes:					
raw	150	1 med	35	2	tr
canned	242	1 c	50	2	tr
Tomato juice, canned	242	1 c	45	2	tr
Tomato catsup	17	1 tbsp	15	tr	tr
Tongue, beef, simmered	85	3 oz	210	18	14
Tuna, canned, drained	85	3 oz	170	24	7
Turkey, roasted	100	3 sl, 3″ x 2½″ x ¼″	200	31	8
Turnips, cooked, diced	155	1 c	35	1	tr
Turnip greens, cooked	145	1 c	30	3	tr
Veal:					
chop, loin, cooked	122	1 med	514	28	44
cutlet, broiled	85	3 oz	185	23	9
roast	85	3 oz	230	23	14
Vinegar	15	1 tbsp	2	0	—
Waffles, baked	75	1 waffle, ½″ x 4½″ x 5½″	210	7	7
Walnuts, English	8	1 tbsp, chopped	50	1	5
Watermelon:					
raw	100	½ c cubes	28	1	tr
raw	925	1 wedge, 4″ x 8″	115	2	1
Wheat:					
puffed	28	1 oz	105	4	tr
shredded	28	1 oz	100	3	1
Wheat flakes	28	1 oz	100	3	tr
Wheat flours:					
all-purpose or family, enriched	110	1 c, sifted	400	12	1
all-purpose or family, unenriched	110	1 c, sifted	400	12	1
cake or pastry flour	110	1 c, sifted	365	8	1
self-rising, enriched	110	1 c	385	10	1
whole wheat	120	1 c	400	16	2
Wheat germ	68	1 c	245	18	7
White sauce, medium	265	1 c	430	10	33
Yeast, brewer's, dry	8	1 tbsp	25	3	tr
Yoghurt	246	1 c	120	8	4

| Fatty Acids | | | | | | | | | | |
| Saturated (total), gm | Unsaturated | | Carbo-hydrate, gm | Cal-cium, mg | Iron, mg | Vitamin A Value, IU | Thia-mine, mg | Ribo-flavin, mg | Niacin, mg | Ascorbic Acid, mg |
	Oleic, gm	Lino-leic, gm								
—	—	—	7	20	0.8	1350	0.10	0.06	1.0	34[20]
—	—	—	10	15	1.2	2180	0.13	0.07	1.7	40
—	—	—	10	17	2.2	1940	0.13	0.07	1.8	39
—	—	—	4	4	0.1	240	0.02	0.01	0.3	3
—	—	—	tr	6	1.9	—	0.04	0.25	3.0	—
—	—	—	0	7	1.6	70	0.04	0.10	10.1	—
—	—	—	0	30	5.1	tr	0.08	0.17	9.8	0
—	—	—	8	54	0.6	tr	0.06	0.08	0.5	33
—	—	—	5	267	1.6	9140	0.21	0.36	0.8	100
—	—	—	0	7	3.5	0	0.17	0.26	5.8	0
5	4	tr	—	9	2.7	—	0.06	0.21	4.6	—
7	6	tr	0	10	2.9	—	0.11	0.26	6.6	—
—	—	—	1	1	0.1	—	—	—	—	—
2	4	1	28	85	1.3	250	0.13	0.19	1.0	tr
tr	1	3	1	8	0.2	tr	0.03	0.01	0.1	tr
—	—	—	7	7	0.2	590	0.05	0.05	0.2	6
—	—	—	27	30	2.1	2510	0.13	0.13	0.7	30
—	—	—	22	8	1.2	0	0.15	0.07	2.2	0
—	—	—	23	12	1.0	0	0.06	0.03	1.3	0
—	—	—	23	12	1.2	0	0.18	0.04	1.4	0
tr	tr	tr	84	18	3.2[21]	0	0.48[21]	0.29[21]	3.8[21]	0
tr	tr	tr	84	18	0.9	0	0.07	0.05	1.0	0
tr	tr	tr	79	17	0.5	0	0.03	0.03	0.7	0
tr	tr	tr	82	292	3.2[21]	0	0.49[21]	0.29[21]	3.9[21]	0
tr	1	1	85	49	4.0	0	0.66	0.14	5.2	0
1	2	4	32	49	6.4	0	1.36	0.46	2.9	0
18	11	1	23	305	0.5	1220	0.12	0.44	0.6	tr
—	—	—	3	17	1.4	tr	1.25	0.34	3.0	tr
2	1	tr	13	295	0.1	170	0.09	0.43	0.2	2

[20] Year-round average. Samples marketed from November through May average around 15 mg per 150-gm tomato; from June through October, around 39 mg.

[21] Iron, thiamine, riboflavin, and niacin are based on the minimum level of enrichment specified in the standards of identity promulgated under the Federal Food, Drug, and Cosmetic Act.

*

Index